Processing of Metal and Ceramic Powders

Processing of Metal and Ceramic Powders

Proceedings of a symposium sponsored by the Powder Metallurgy Committee of The Metallurgical Society of AIME and the Basic Science Division of the American Ceramic Society held at the Fall Meeting of The Metallurgical Society of AIME, Louisville, Kentucky, October 12-14, 1981.

Edited by
R. M. German
Materials Engineering Department
Rensselaer Polytechnic Institute
Troy, New York
and
K. W. Lay
Corporate Research and Development
General Electric Company
Schenectady, New York

A Publication of The Metallurgical Society of AIME

A Publication of The Metallurgical Society of AIME

420 Commonwealth Drive
Warrendale, Pa. 15086
(412) 776-9000

Printed in the United States of America.
Library of Congress Card Catalogue Number 82-61009
ISBN Number 0-89520-396-0

Foreword

In October of 1981, the Powder Metallurgy Committee of The Metallurgical Society of AIME and the Basic Science Division of the American Ceramic Society jointly sponsored five sessions on "Powder Processing". This was a part of the first joint meeting of the two societies, held in Louisville, Kentucky. This book represents the proceedings of the symposium and demonstrates the common concerns of these two groups.

The authors provide an excellent demonstration of the common interests of the powder metallurgist and the ceramist in particulate materials. In contrast, the meeting also emphasized the differences in the two types of materials. For example, studies on the basic mechanisms of compaction, sintering, and grain growth appear to be generally material independent. However, differences are apparent in the powder fabrication routes available for the metals and ceramics (e.g., fusion versus chemical techniques). Also, generally the particle size of interest to the ceramist is much finer. On the other hand, plasticity in consolidation is available for the metallurgist. Net-shape forming is a common interest in both types of materials, although the ceramist gives greater emphsisis to densification while dimensional control is more important to the powder metallurgist. We sense a common interest in many of the newer materials which fail to fall into the classical ionic versus metallic divisions. Examples are research and development efforts on the glassy metals, cermets, and non-oxide ceramics. Consequently, we feel this form of interaction will be needed on a periodic basis.

The book is organized into three groupings of papers -- powder production, consolidation, and properties. We believe the book's organization shows the commonality of the interests of the two groups while the details evidence the obvious contrast. We wish to thank our session chairmen for their efforts-- J.D. Hodge, T.H. Courtney, M.D. Sacks, J.W. Halloran, and N.S. Nayar.

R.M. German
Materials Engineering Department
Rensselaer Polytechnic Institute
Troy, New York 12181

K.W. Lay
Corporate Research and Development
General Electric Company
Schenectady, New York 12301

May 1982

Table of Contents

PROPERTIES

Powder Fabrication

PREPARATION AND CHARACTERIZATION OF HIGH PURITY MgO POWDER COMPACTS

C.A. Handwerker, M.M. O'Connor, R.M. Cannon, Jr., and R.L. Coble

Ceramics Division,
Department of Materials Science And Engineering
Massachusetts Institute of Technology
Cambridge, MA 02139 USA

Procedures were developed for forming high-purity powder compacts by decomposition of carbonates and subsequent settling and evaporation of alcohol dispersions. Procedures were selected to minimize contamination by examining the effects of each variable in clean-room processing on the physical and chemical characteristics of the powders. A reagent grade carbonate powder and, subsequently, a high-purity powder and powder compact were studied. Systematic changes in the average crystallite size, the particle/agglomerate morphology and size distribution, the specific surface area, the phase composition, the adsorption/desorption behavior of the MgO surfaces and the purity were observed with changes in the processing parameters.

Introduction

High purity should be the primary criterion for materials selection for many scientific investigations. Too often scientific studies, such as studies of grain growth, sintering, or infrared absorbance, have been performed with either poorly characterized materials, especially with respect to purity, or with high-purity starting materials which have been contaminated during processing. The experimental results from a single study of material of unknown purity are frequently compared with data from other studies of materials of likewise unknown, but probably different, purity. Such comparisons frequently provide an estimate of the effects of contaminants and processing variables on the measured parameters rather than a measure of some basic material property.

There is considerable literature on the effects of high concentrations of dopants on kinetic processes in metal and ceramic systems. There is no information on the effects of low levels of total impurities--cations, anions, and organics in the bulk and on the surface--on the physical and chemical properties of ceramic materials. This has been due, in part, to a lack of samples of high initial purity with which to experiment.

The goal of this study was to prepare such a high-purity compact for subsequent studies of grain growth and other properties. A high-purity magnesium carbonate precursor was prepared and the starting high purity was maintained during the production of the oxide compact. The cationic, anionic, and organic contamination was minimized by design of special procedures to be carried out under clean conditions. Each processing step was examined for its effect on the purity and the physical and chemical properties of the powder.

Experiments

High-purity magnesium carbonate complexes were prepared from reagent grade basic magnesium carbonate under class 100 clean-room conditions. The procedures for the purification and precipitation of this high-purity precursor are listed in Fig. 1. The basic procedures involve dissolution of the reagent grade carbonate, filtration to remove insoluble matter, chelating with diethylammonium diethyldithiocarbamate (DDDC) which removes transition metal impurities, precipitation, and washing to remove surface salts. The details and additional procedures have been discussed previously (1).

Due to fabrication costs of the high-purity material, the reagent grade carbonate from which it was prepared was used as a prototype for the high-purity powder. The effects of each processing step on these physical and chemical characteristics were first determined for the lower purity material. The complete range of experiments was not performed with the high-purity powders, except when the powder characteristics were sensitive to slight changes in the processing procedures.

The processing procedures and the evaluation techniques subsequently used are listed in Fig. 2. The carbonate was transformed to the oxide at various temperatures; the oxide was deagglomerated in various solvents by ultrasonic disruption using a high-power ultrasonic probe immersed in the suspension; compacts were formed by settling and evaporation; and the residual organics and surface species were removed by heating the compacts.

The anionic and organic contaminants due to a processing step were assessed by measurements of weight loss and thermal changes upon heating, by changes in the specific surface area and average crystallite size, by detection of surface species by Fourier transform infrared spectroscopy, and by phase analysis by x-ray diffraction. Changes in the powder morphology and particle size distribution were measured by direct microscopic observation using the SEM and by particle settling experiments.

$MgCO_3 \bullet$ n-hydrate + H_2O $\xrightarrow{P_{CO_2} = 1.7\ MPa}$ $Mg(HCO_3)_2$ aq. solution →

Preliminary Extraction →
DDDC in H_2O
Centrifuge
Filtration (0.25 m)

Primary Liquid-Liquid Extraction →
DDDC in $CHCl_3$ (~ 8 cycles)

Filtration (0.25μm) $\xrightarrow{80°\ C}$ $MgCO_3 \bullet$ n-hydrate →

Filtration (0.25μm)
H_2O washing
Centrifuge
→ High Purity $MgCO_3 \bullet$ n-hydrate

Figure 1 - Preparation of High Purity Basic Magnesium Carbonate Under Class 100 Clean Room Conditions

PROCESSING STEPS	CHARACTERIZATION TECHNIQUES
2 Types of $MgCO_3 \bullet$ n-hydrate Reagent Grade and High Purity	X-Ray Diffraction
↓ MgO Powder	X-Ray Diffraction, DTA, TGA, X-Ray Line Broadening, Observation by SEM, BET Surface Area, IR Spectroscopy Sedigraph
↓ Ultrasonic at Power Level <u>P</u> For Time <u>T</u> in Alcohol <u>A</u>	Sedigraph
↓ Settle and Decant	Sedigraph Sedimentation Volume
↓ Dry in Air	Observation by SEM IR Spectroscopy
↓ Heat to Desorb/ Decompose/Sinter	TGA, DTA, BET Surface Area Observation by SEM

Figure 2 - Clean Room Processing Steps and Characterization Techniques Used in This Study

The purity with respect to cations was measured for the reagent grade and high-purity carbonates, and for the high-purity powder after transformation to the oxide and after compact formation.

A series of clean-room processing procedures for high-purity MgO was chosen which was determined to minimize total contamination and produce an uncracked, high-density green body.

Results

Precursors: Purity, Phase Composition, and Morphology. The reagent grade magnesium carbonate n-hydrate powder contains approximately 40 wt % MgO. The high-purity magnesium carbonate material prepared under clean conditions from a magnesium bicarbonate solution contains approximately 20 wt % MgO. Physically adsorbed water in the high-purity material accounts for the difference in oxide content from the reagent grade material. The chemical composition in p.p.m. (cation basis) of both materials handled in a Class 100 clean room have been measured by plasma coupled emission spectroscopy and are listed in Table I.*

Table I

Composition Analysis by Plasma Coupled Emission Spectroscopy Reported in p.p.m. Cation Basis (μg/g of Mg)

	Reagent Grade $MgCO_3 \cdot$ N-Hydrate	High Purity $MgCO_3 \cdot$ N-Hydrate	MgO	MgO Compact
Ba	1	0.6	2	3
Ca	33	20	109	123
Na	430	15	60	54
Sr	1	1	3	4
Mn	3	<0.6	<0.6	<0.6

Below Detectability Limit for Al, As, B, Be, Cd, Co, Cr, Cu, Fe, Ga, La, Li, Mo, Ni, P, Pb, Pd, Sb, Se, Si, Sn, Ti, V, Zn, Zr, K

For these two materials handled under clean conditions, the measurable purity differences are in Ca, Na, and Mn. The concentrations of two other alkaline earths, Ba and Sr, are similar. The incorporation of Ca can be minimized by precipitating magnesium carbonate in the presence of ethylene diaminetetraacetic acid (2) and the Na can be removed by additional washing. The purity levels measured for the remaining metallic elements are below the detection limits typically (1-8 p.p.m.) for both reagent grade and high-purity powders. Although these materials are nominally of the same purity by emission spectroscopy, visual inspection of the color of the chelated complexes formed during liquid-liquid extraction indicates removal of significant amounts of transition metal ions from the bicarbonate solutions[(3)]. The total concentration of these ions remaining in the aqueous solution after extraction is below one part per million.

The effect of clean-room handling on the purity of the reagent grade carbonate was assessed by preparing samples for chemical analysis in a

*Union Carbide Corporation

standard ceramics laboratory using laboratory glassware cleaned in Alconox ® and rinsed in distilled water and by performing chemical analysis in non-clean room conditions. With short exposure to airborne contaminants characteristic of standard laboratory conditions, the total levels of Fe, Si, Al, and Pb were found to rise to 100-1000 p.p.m. as measured by emission spectroscopy.

The reagent grade powder, shown in Fig. 3(a), is composed of spherical agglomerates of plates and of unagglomerated plate fragments. The average agglomerate size by SEM examination is approximately 30 μm, the plate width distribution is wide, while the distribution of plate thicknesses appears to be quite narrow. The high-purity precursor, shown in Fig. 3(c), has particles of two distinct morphologies: single crystal needles and spherulites. The average agglomerate size of the high-purity spherulites and the average needle length are approximately 100 μm by SEM inspection.

The phase analyses for the precursors are listed in Table II. Both precursor powders are hydromagnesite, approximate composition $4MgCO_3 \cdot Mg(OH)_2 \cdot 4H_2O$. Several low intensity peaks were not identified with any magnesium phase listed in the ASTM Powder Diffraction Index. From the precipitation literature, the two high-purity powder morphologies correspond to two distinct phases--the spherulites to hydromagnesite and the needles to nesquehonite, $MgCO_3 \cdot 3H_2O$ (4). However, the x-ray diffraction data do not indicate the presence of nesquehonite.

Surface area, x-ray line broadening, and sedigraph measurements were difficult to interpret due to the residual water in the carbonate powders.

Transformation to MgO. To transform the carbonate complex to an oxide, batches of approximately fifty grams of the carbonate powders were placed in Al_2O_3 crucibles, heated in air at approximately 10°C/minute to 500°, 800°, and 1000°C, held at temperature for two hours, and cooled at 10°C/minute. The techniques used to evaluate the high purity and reagent grade oxide powders produced by these three transformation cycles are listed in Fig. 2.

The phases detected in the transformed powders are listed in Table II. The two powders transformed at 500°C contain different phases: $MgCO_3$ and MgO for the reagent grade and $MgCO_3$, $Mg(OH)_2$, and MgO for the high-purity powder. After heating to 800°C and 1000°C, only MgO is present.

The magnesium carbonate powders in approximately 2 mg batches were transformed in the TGA and DTA according to the heating schedules listed above. Representative TGA and DTA data are shown in Fig. 4 for the reagent grade material heated to 1000°C at 10°C/minute. The broad endotherms at 130°, 300°, and 428°C indicate loss of CO_2 and H_2O to form various amorphous hydrated magnesium carbonates and, at 428°C to also form magnesium oxide. The exotherm at 510°C corresponds to the crystallization of amorphous magnesite, $MgCO_3$, followed by desorption of CO_2 and H_2O at $T > 510°C$ (4). Upon heating the reagent grade material to 500°C, 800°C, and 1000°C, the weight losses are 57%, 58%, and 60%, respectively. After annealing at temperature for 2 hours, the additional weight losses are 2.5%, 2.2%, and 1%.

The weight loss in the transformation of the compound $4MgCO_3 \cdot Mg(OH)_2 \cdot 4H_2O$ to MgO is 57% of the initial weight. The actual H_2O and CO_2 content varies as a function of the temperature and partial pressure conditions of the precipitation. Molecular water adsorbed on the surface can also contribute to weight losses greater than 57%. Up to 400°C the locations of the endothermic peak minima and the shapes of the endotherms, as well as the weight losses, are sensitive to the original hydromagnesite precipitation conditions (T, pH, excess CO_2, and impurities), and are, therefore, different for the reagent and the high purity powders. At $T > 400°C$, the features of the TGA and DTA data are similar for the two powders and are similar to reported values for other

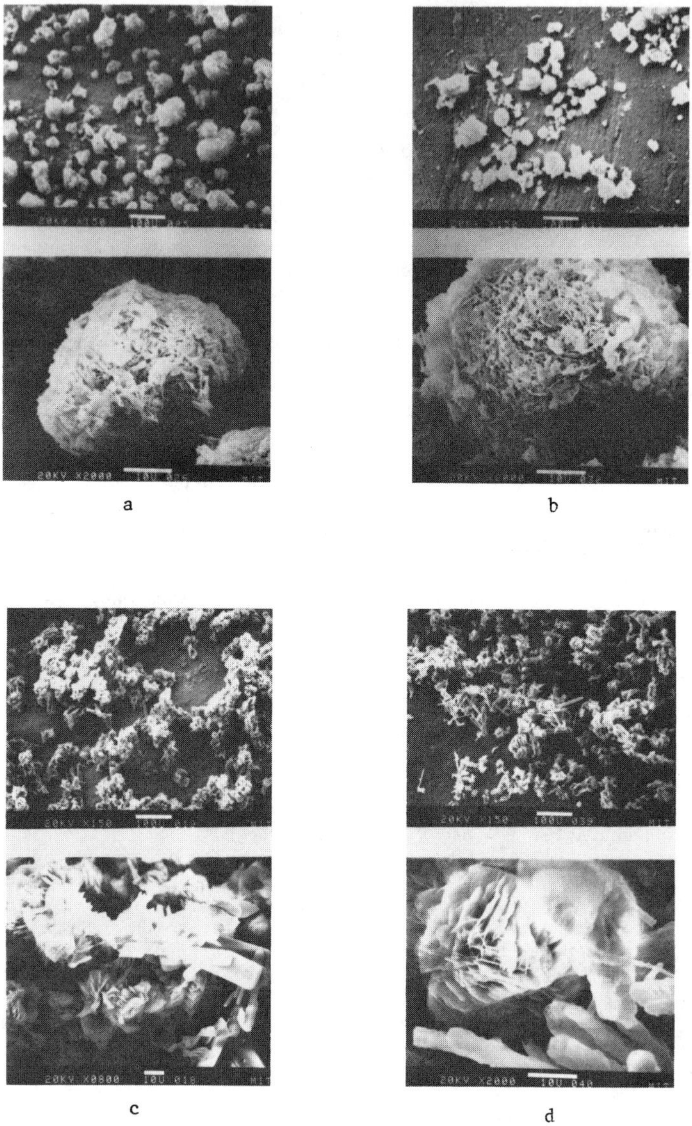

Figure 3 - Micrographs of magnesium carbonae and oxide powders by SEM. Reagent grade carbonate: (a) as received; and (b) after transformation to the oxide at 1000°C. High-purity carbonate: (c) as precipitated; and (d) after transformation to the oxide at 800°C.

Table II

Phase Analyses, Specific Surface Area, and Crystallite Size as a Function of Heat-Treatment

MgO (Reagent Grade)	Phases Identified	BET Surface Area (m^2/g) (Calc. Eq. Sph. Diameter (nm)	Particle Size by X-ray Line Broadening (nm)
As-Received	$4MgCO_3 \cdot Mg(OH)_2 \cdot 4H_2O$	17 (101)	--
500°C	$MgO + MgCO_3$	90 (19)	18
800°C	MgO	26 (65)	28
1000°C	MgO	21 (81)	53
MgO (High Purity)			
As-Precipitated	$4MgCO_3 \cdot Mg(OH)_2 \cdot 4H_2O$	74 (23)	--
500°C	$MgO + MgCO_3 + Mg(OH)_2$	158 (11)	--
800°C	MgO	38 (46)	31
1000°C	MgO	18 (95)	39

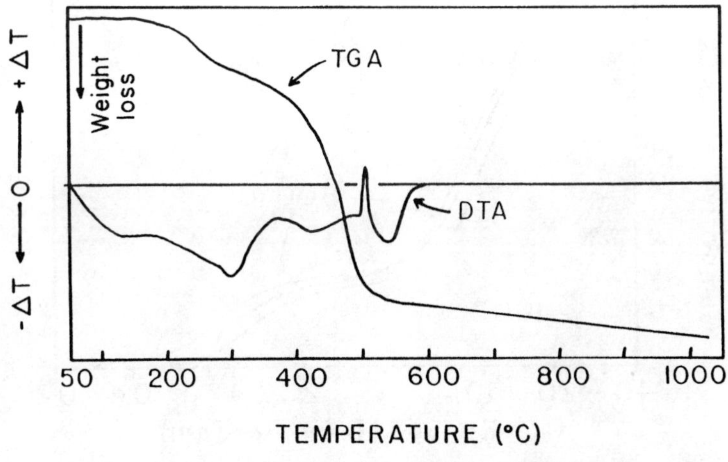

Figure 4 - Representative TGA and DTA reagent grade magnesium carbonate: 50° to 1000°C at 10°C/minute in air.

hydromagnesite powders. From the DTA and x-ray diffraction study of Dell and Weller (4), it was found that the formation of $MgCO_3$ and the crystallization exotherm at 510°C can be avoided by slower heating rates and by more efficient removal of the CO_2 and H_2O desorbed at T > 400°C to avoid the hydrothermal conditions under which $MgCO_3$ is stable.

Thus, considering the poor gas-flow condition with a large batch size (~50 grams) and the amount of CO_2 and H_2O to be removed from the powders (80 wt % for the high purity and 60% for the reagent grade), it is not surprising that the carbonate and hydroxide phases identified by x-ray diffraction were formed during the heating to 500°C.

Although weight loss continues to higher temperatures (T > 1000°C), sintering occurs concurrently, leading to significant agglomeration. Therefore, the maximum transformation temperature was kept at 1000°C.

The transformation to MgO did not significantly change the gross morphology from that of the precursors for all transformation temperatures. Representative microstructures are shown in Fig. 3 for the reagent grade and high-purity powders. The reagent grade powder, Fig. 3(b), appears unchanged by the transformation at these magnifications. The needles of the high-purity oxide, Fig. 3(d), break up upon transformation, whereas the spherulites show no change.

The agglomerate size distributions of the as-transformed powders were measured by settling in isopropanol. The distributions are different for the two powders, but each is not very sensitive to the transformation temperature. Representative distributions for the reagent grade powder are shown in Fig. 5. The distribution for the reagent grade powder is bimodal, corresponding to agglomerates and unagglomerated plates. Agglomerated high-purity powder is composed of particles greater then 40 μm in equivalent spherical diameter. Thus, the distribution is not measured by settling. The particle size distribution values on the graph are calculated assuming a model for the settling of dense spheres (ρ_{MgO} = 3.58 gm/cm^3) and are reported as the cumulative mass percent of particles with an equivalent spherical diameter greater than a given size. Since the powder morphology is dissimilar to the assumed geometry, i.e., the

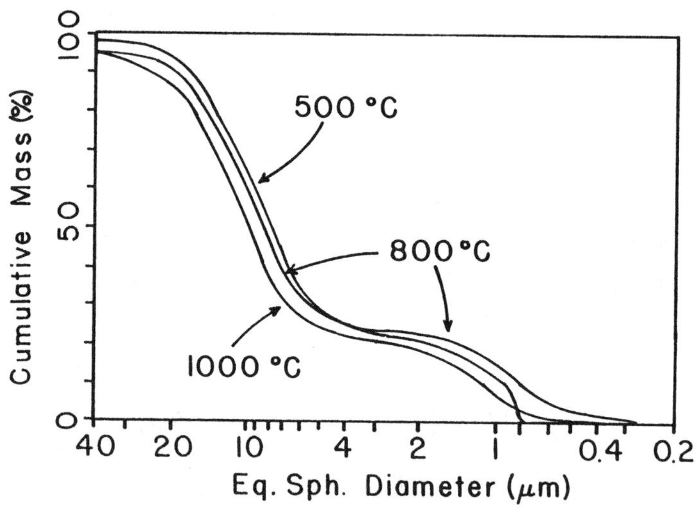

Figure 5 - Particle size distributions of reagent grade powder transformed at 500°, 800°, and 1000°C.

agglomerates are porous and the dense isolated particles are either plates or needles, these calculated values are used to evaluate the changes in powder characteristics with processing rather than to measure true particle size distributions.

The BET surface areas and the average crystallite sizes are listed in Table II. The surface area reaches a maximum at 500°C, where MgO is first observed. The decrease in surface area corresponds to densification within crystallite arrays, i.e., within plates, and not to interagglomerate sintering. This interpretation is suggested by the observation by SEM that there is no significant interagglomerate neck formation. In addition, the particle size distributions measured by settling were nearly identical for each powder for the three heating schedules. The average crystallite sizes calculated from the specific surface areas are larger than the calculated values from x-ray line broadening.

Compact Formation. The hydromagnesite transformed at low temperatures retains water and carbon dioxide as evident from phase analysis and DTA and weight loss measurements at higher temperatures. The high purity and reagent grade powders transformed at 1000°C were, therefore, the highest purity powders with respect to H_2O and CO_2 and were used for the compact formation experiments.

As indicated in Fig. 2, the as-transformed powders must be deagglomerated before compact formation, followed by settling and evaporating to form a green body and heated to desorb and decompose the adsorbed species before high temperature heat-treatment. All experiments were first performed with only the reagent grade powder transformed at 1000°C.

After the effects of processing conditions on compact formation were determined for the lower purity material, the high-purity powder was processed using the conditions determined to be optimal for the range of variables studied.

The reagent powders were placed in high-purity liquids: benzaldehyde, water, and alcohols (methanol, ethanol, and isopropanol) and deagglomerated by ultrasonic agitation using an immersed probe. The pH of the alcohol and water were varied by additions of HCl and NH_4OH. Sedimentation volume was measured for each liquid and are listed in Table III. The particle densities varied from 0.35 to 0.06 gms/cm^3. Upon drying, the densities were significantly higher.

The reagent grade magnesium oxide was found to be a catalyst for benzaldehyde decomposition and formed a gel when combined in all ratios with benzaldehyde. The MgO powder in water formed an unstable suspension and gave high settled volumes. Since the three pure alcohols formed the most stable suspension with the smallest sedimentation volumes, they were studied further. The three alcohols are of comparable purity with respect to dissolved cations and water.

The effects of the ultrasonic power, treatment time, and suspension temperature were examined for the reagent grade powder methanol. The conditions beyond which either the particle size in distribution did not change or the number of fine particles remained constant or decreased were 4 minutes at 60 watts at 50°C.

The particle size distributions of the high-purity and reagent grade MgO (1000°C) were measured after the same ultrasonic treatment in the three alcohols. The distributions for the deagglomerated high-purity MgO (1000°C) and reagent grade MgO (1000°C) in isopropanol are shown in Fig. 6. The bimodal size distribution evident in the as-transformed reagent grade powder (Fig. 5) remains in the deagglomerated material in isopropanol. The distribution for the high-purity material is narrower than for the reagent grade with no particles < 0.2 μm and the average is much larger.

Table III

Sedimentation Volume of Reagent Grade MgO-transformed at 100°C and deagglomerated. Powder weight = 0.5 gm.

Suspension Liquid	Volume of Powder (cm^3)	Density (g/cm^3)
Methanol	1.3	0.38
Methanol + HCl (pH = 4.3)	4.9	0.10
Methanol + NH4OH (pH = 9.2)	4.8	0.10
Ethanol	2.2	0.22
Ethanol + HCl (pH = 6.2)	2.6	0.19
Ethanol + NH_4OH (pH = 8.9)	4.5	0.11
Isopropanol	2.1	0.24
Isopropanol + HCl (pH = 4.3)	2.1	0.24
Isopropanol + NH_4OH (pH = 9.2)	7.7	0.06

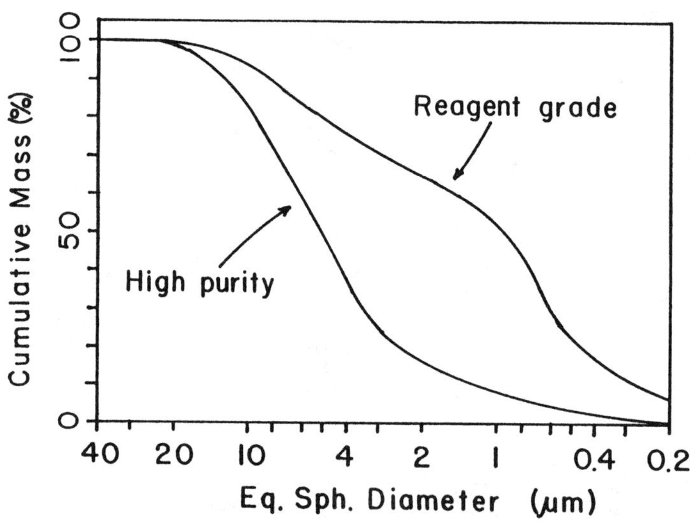

Figure 6 - Particle size distributions of reagent grade and high purity MgO transformed at 1000°C and deagglomerated by ultrasonic disruption.

The particle size distributions of the reagent grade material were dependent upon the alcohol used in the deagglomeration treatment. The results for the reagent grade powder deagglomerated in methanol and isopropanol are shown in Fig. 7. The ethanol and isopropanol suspensions have similar particle size distributions. The particle size for which the cumulative mass percent equals 50% is 1 μm for both suspensions; however, the methanol suspension has a narrower size distribution and no particles < 0.2 μm. The isopropanol suspension has 5% < 0.2 μm. Mechanical deagglomeration of the spherical agglomerates and platey particles shown in Fig. 3(b) should produce a decrease in the large agglomerate sizes and an increase in small particles due to break-up of plates. This behavior is observed in isopropanol and in ethanol but not in methanol. It was found that by cooling the methanol-MgO suspension to 0°C during ultrasonic disruption, some sub-micron particles are produced.

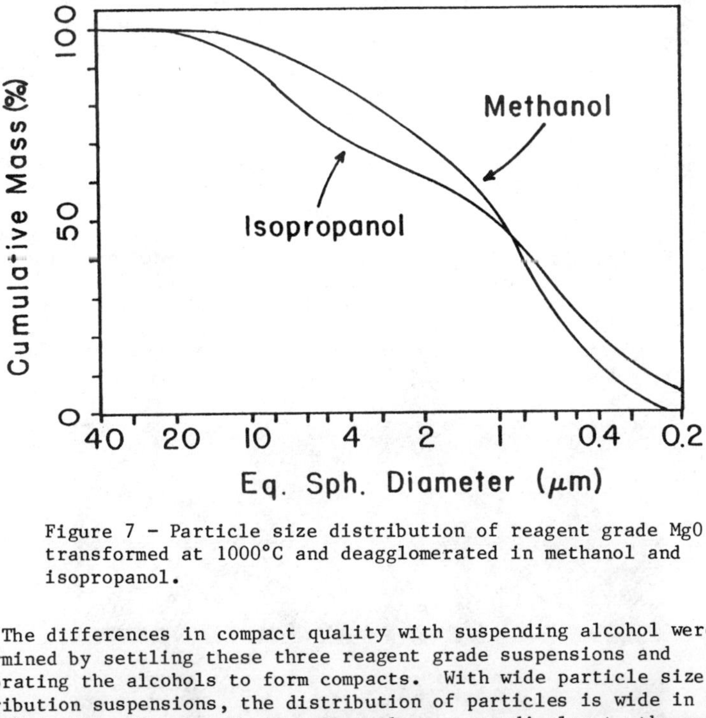

Figure 7 - Particle size distribution of reagent grade MgO transformed at 1000°C and deagglomerated in methanol and isopropanol.

The differences in compact quality with suspending alcohol were determined by settling these three reagent grade suspensions and evaporating the alcohols to form compacts. With wide particle size distribution suspensions, the distribution of particles is wide in the total body, but the distribution in a plane perpendicular to the settling direction is narrow and changes continuously from the bottom of the compact to the top. Settling for various times and decanting the remaining suspensions gave narrower size distributions; centrifuging these suspensions with narrow particle size distributions and small average particle size has been used to form compacts. The characteristics of the narrowly sized compacts are not discussed here. The three reagent grade compacts produced from the entire original particle size distribution are shown in Fig. 8. The compact from methanol was severely delaminated, warped, and hard; the compact from ethanol exhibited moderate delamination and cracking; the compact from isopropanol was monolithic with an approximate green density of 55%. The compacts from ethanol and isopropanol suspension were friable. Micrographs of the methanol-dried

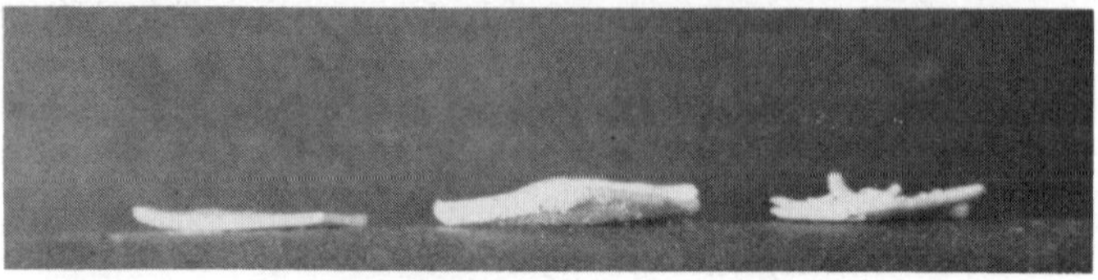

Figure 8 - Reagent grade MgO compacts deagglomerated and settled in (a) isopropanol; (b) ethanol; and (c) methanol.

compact taken at four positions along a fracture surface parallel to the settling direction are shown in Fig. 9. Substantial neck formation is observed through the sample thickness, indicating that significant dissolution/reprecipitation occurs as the methanol evaporates. Slight neck formation was observed for the compact settled in ethanol; no neck formation was observed for the compact settled in isopropanol.

Representative infrared spectra for the three alcohol-treated powders and the untreated powder are shown in Fig. 10. The same amount of powder and mulling media--a fluorinated hydrocarbon or a mineral oil--were used in each infrared measurement. The powders and the dried compacts had been exposed to normal humidity air for several weeks prior to these measurements. The intensity of the broad band adsorbance in the 3700–3100 cm^{-1} range (indicative of surface hydroxyls) varies in the order: untreated > methanol > ethanol = isopropanol. In particular, the 3700 cm^{-1} peak associated with a strongly bonded surface hydroxyl group decreases (5) in the same order for the alcohol-treated powders with no well-defined peak at 3700 cm^{-1} for the untreated powder. Molecular water and $CO_3^{=}$ were also detected in all four spectra. Although other magnesium-organic compounds have been observed to form from MgO in contact with alcohols (6-10), no other magnesium phase or complexes at the surfaces could be unequivocably identified from these spectra.

a b c d

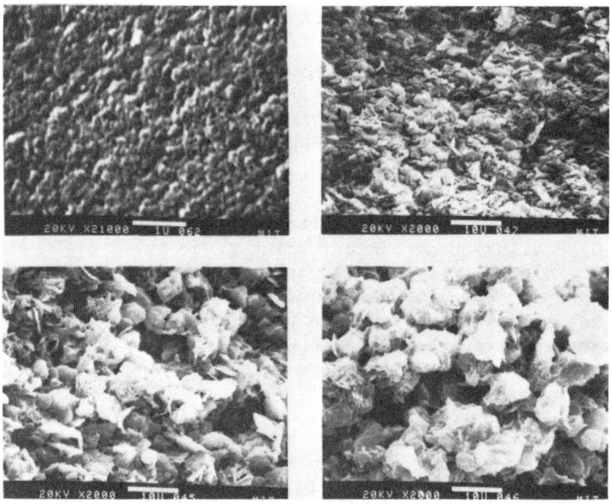

Figure 9 - Micrographs of fracture surface of reagent grade MgO compact settled in methanol: (a) adjacent to top surface; (b) at 1/3 of compact height; (c) at 2/3 of compact height; and (d) adjacent to bottom surface.

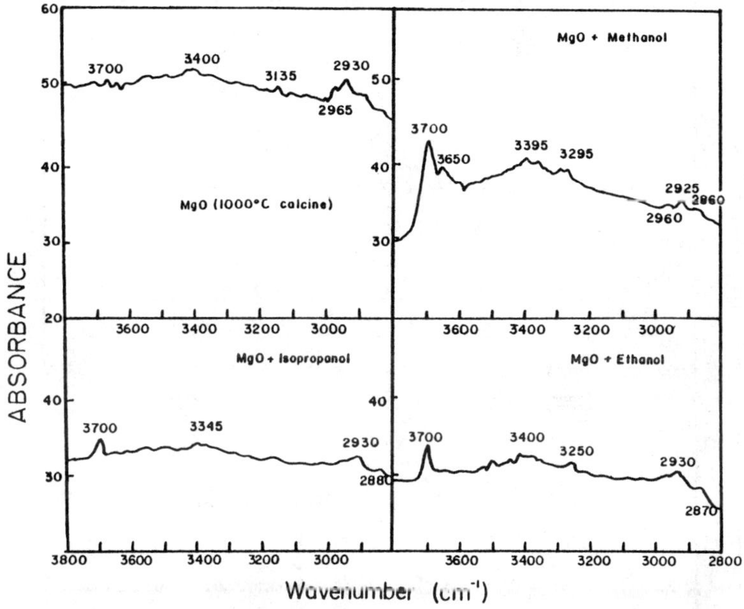

Figure 10 - Infared spectra ;of four MgO powders from 3800 to 2800 cm^{-1}: (a) agglomerated MgO; and deaggomerated MgO: (b) in methanol; (c) in ethanol; and (d) in isopropanol.

Table IV

TGA and BET Results for Reagent Grade MgO (1000°C Transformation), Agglomerated, and Unagglomerated in Alcohols

	TGA Weight Loss 50° → 1000°C	BET Surface Area (m^2/g) After Heating to 600°C in Air
Methanol	17%	80
Ethanol	6%	20
2-Propanol	4%	20
MgO in Laboratory Air Without Alcohol Deagglomeration	3%	20
MgO in Dry, CO_2-Free Air Without Alcohol Deagglomeration	0%	20

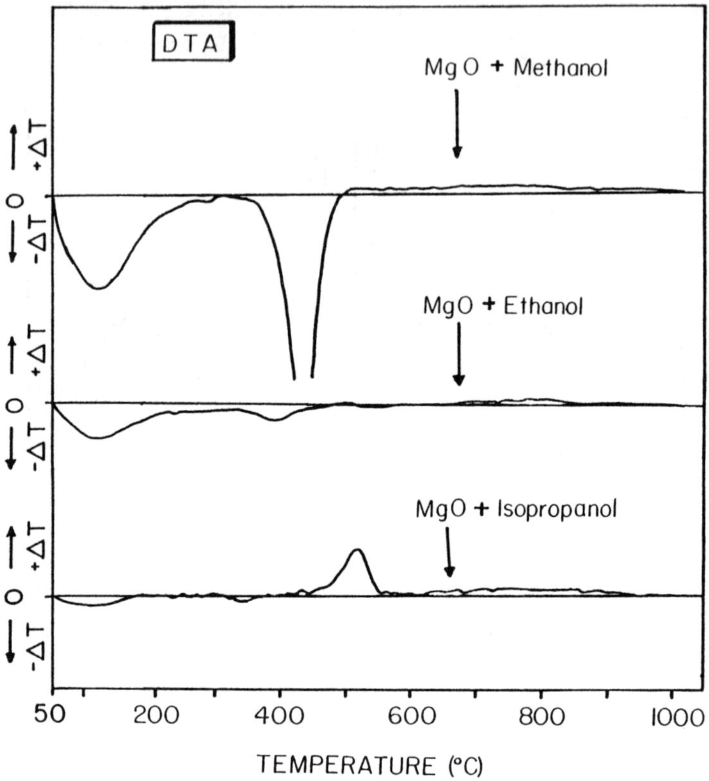

Figure 11 - DTA plots for reagent grade MgO deagglomerated in methanol, in ethanol, and in isopropanol heated from 50°C to 1000°C at 10°C/minute in air.

Phase analysis of the three alcohol treated powders by x-ray diffraction indicates the presence of MgO and $Mg(OH)_2$ for the methanol-dried powder and only MgO for the ethanol- and isopropanol-dried powders. In the methanol-dried powder, there were additional peaks not associated with MgO or $Mg(OH)_2$. For this reason the actual phase composition of the methanol-dried powders is uncertain. Due to the detection limits of x-ray diffraction, other phases may also be present for the powders settled in ethanol and methanol (9). The degree of reaction with the alcohols was assessed further by TGA and DTA in air from 50° to 1000°C at 10°/min and by measurements of the surface area of the powders after heating to 600°C in air. The TGA and surface area results are listed in Table IV. The powder dried from methanol exhibited substantial weight loss and a four-fold increase in surface area accompanying the transformation back to MgO. The weight loss for the powder settled in ethanol was greater than for the powder settled in isopropanol, or the powder exposed to air, but the specific surface areas were identical.

The DTA results are shown in Fig. 11. The methanol-dried powder shows two large endotherms whose minima are at 130° and 430°C; the powder dried from ethanol shows two shallow endotherms at 155° and 385°; and the powder dried from isopropanol shows an exotherm at 520°C. Formation of surface alkoxides has been reported for MgO in contact with methanol and ethanol under a variety of conditions (10). The endotherm at 130°C for the methanol-dried powder could correspond to the pyrolysis of magnesium methoxide, in agreement with the experimentally determined decomposition temperature of magnesium methoxide of approximately 120°C. The endotherm at 430°C could correspond to the decomposition of $Mg(OH)_2$ formed during the deagglomeration, settling, and decomposition of alcoholate.

The pyrolysis of magnesium ethoxide has been observed in the 270°-280°C range (11) which is significantly higher than the endotherm observed at 155°C for the ethanol-dried MgO powder. Although no phase other than MgO was identified by x-ray diffraction, the weight loss upon heating indicates significant adsorption of or reaction with ethanol. The endotherm at 385°C could again correspond to the decomposition of $Mg(OH)_2$.

The exotherm in the TGA for the isopropanol-dried powder indicates combustion of the surface species rather than decomposition (endothermic) followed by vapor phase combustion (exothermic). The total weight loss of this sample indicates significantly less adsorption or reaction with the isopropanol than for the methanol-dried and the ethanol-dried powder, as listed in Table IV. Preliminary experiments indicate that the high-purity MgO is insensitive to the alcohol use for compact formation.

High-Purity Compact Formation. High-purity MgO powder compacts were formed using the parameters determined to give a high green density, uncracked compacts with low levels of OH^- and C contamination. The precursor was transformed at 1000°C in an MgO (98%) crucible, and the oxide was deagglomerated in isopropanol, dried to form a compact, and heated to 600°C to remove the surface species. Chemical analysis of the as-transformed oxide and the as-dried compacts by emission spectroscopy are listed in Table 1. Some degradation has occurred in the sample purity (Ca,Na) possibly due to contact with the crucible at 1000°C. The clean-room processing of the high-purity precursor and oxide has resulted in the formation of a compact with no measureable degradation in purity except for Ca and Na, from the high purity precursor.

Discussion

The procedures developed for preparing high-purity MgO compacts are described in the section on compact formation. The background cation concentration was increased by the processing: Ca changed from 20 to 130 p.p.m. and Na from 15 to 60 p.p.m. This increase is probably due to contamination from the crucible during transformation to MgO. Use of high-purity MgO crucibles for transformation should reduce this contamination source. Anion impurities, with the exception of hydroxyls, were minimized by the choice of an MgO precursor which contains only CO_2 and H_2O as foreign species. Hydroxyl and carbon contamination are indigenous to the precursor and are also introduced into the MgO powder during the deagglomeration process. The degree to which these species are removed during transformation to MgO or reheating of the compact is not known. From DTA and TGA, the MgO powders must be heated to approximately 1000°C for the transformation and 600°C for the re-heating to remove most of these species. Residual OH^- and C after these heat-treatments have not been measured directly.

The effects of the deagglomeration/suspension medium on reactivity, partice size distribution, and compact formability are substantial, and cannot be attributed to differences in the liquid surface tensions. The surface tensions for the three alcohols are similar: pure methanol, 22.61 dynes/cm, 20°C, air; pure ethanol, 22.75 dynes/cm, 20°C, vapor; and pure isopropanol, 21.7 dynes/cm, 20°C, vapor (12).

With deagglomeration in methanol, the reagent grade MgO fines dissolved, leading to reprecipitation, forming particle-particle contacts during alcohol evaporation. The morphology of the powder compact, as well as the presence of $Mg(OH)_2$, and the large weight loss and increase in surface area upon reheating, indicate significant reactivity.

The powder compact formed from the ethanol also showed neck formation, but the weight loss was less than methanol-deagglomerated material and the surface area did not increase upon reheating.

The compact from the isopropanol deagglomerated suspension showed no neck formation. The surface area did not increase upon reheating. The weight loss on heating was much less than for the methanol-dried compact, and less than for the ethanol-dried compact but still greater than for the agglomerated MgO.

The infrared spectra showed more broad band hydroxyl absorption for the agglomerated MgO, but for the methanol-deagglomerated MgO, sharper peaks with lower total IR absorption was found. The ethanol and isopropanol-dried compacts showed the sharp absorption with a lower total hydroxyl level than the methanol-dried powders.

The differences in the apparent reactivity of the three alcohols are due to the differences in the solubility of the alkoxfde in the corresponding alcohol (13), in the free energies of formation of the alkoxides, or in the kinetics of surface alkoxide formation in the alcohol. The room temperature solubility of $Mg(OCH_3)_2$ in CH_3OH is much greater than for $Mg(OC_2H_5)_2$ in C_2H_5OH; the solubility of $Mg(OC_3H_7)_2$ in C_3H_7OH has not been reported. Upon evaporation of the alcohol, the alkoxide precipitates on the remaining MgO, leading to structures as shown in Fig. 9 for the methanol-dried MgO. The presence of these species inhibits rehydration of the surface to the level of the agglomerated powder, as evident from the infrared spectra. A positive effect of isopropanol on oxide compact formation and final sintered density had previously been noted by Rossi and Fulrath (14), although no mechanism was isolated.

The difference in the reactivity of the high-purity and the reagent grade MgO powders is not understood. Further experiments are being performed on these systems to answer these questions on reactivity.

Summary

Clean-room processing techniques were developed for production of MgO powder compacts from high-purity carbonate precursors. A reagent grade magnesium carbonate (~500 p.p.m. cation impurities) and a high-purity magnesium carbonate (~40 p.p.m. cation impurities) were examined for the effects of process variables, such as transformation temperature and suspension liquid. The following physical and chemical characteristics of the powders and powder compacts were monitored: phase composition, specific surface area, average crystallite size, particle size distribution, particle morphology, weight loss, and thermal changes during heating, sedimentation volume, compact quality, infrared absorbance, and cation impurity level.

From these experiments, a series of procedures were chosen for the clean-room production of an uncracked oxide compact from the high-purity precursor. Calcium and sodium contamination, as measured by emission spectroscopy, increased from 30 p.p.m. to 170 p.p.m during the transformation of the carbonate to the oxide. The levels of other metallic elements, except Ba and Sr, remained below the detection levels of emission spectroscopy.

The effects of the suspension media on the adsorbed species, the particle size distribution, and the compact formability were substantial.

Acknowledgements

We thank Drs. Ellen Tormey, David Gabbe, and John Blendell for helpful discussions. We also thank Dr. Walter Zamechek of the Union Carbide Corporation for performing the chemical analysis. We also thank Mrs. Toni Centorino and the staff of the M.I.T. Word Processing Center for preparing the manuscript. This research was sponsored by the National Science Foundation, Division of Materials Research, Contract #77-11685-DMR and by Department of Energy contract #DE-AC02-76ER02390.

References

1. C.A. Handwerker, Ph.D. Thesis, M.I.T., 1982.

2. A. Debska-Horecka, T. Pukas, and I. Stronski, "The Preparation of High-Purity MgO," J. Rad. Chem. 13, 21-30, 1973.

3. A. Hulanicki, "Complexation Reactions of Dithiocarbamates," Talanta 14, 1371-1392, 1967.

4. R.M. Dell and S.W. Weller, "The Thermal Decomposition of Nesquehonite $MgCO_3 \cdot 3H_2O$ and Magnesium Ammonium Carbonate $MgCO_3 \cdot (NH_4)_2CO_3 \cdot 4H_2O$," T. Farad. Soc. 55, 2203-2220, 1969.

5. St. Malinowski, S. Szczepanska, and J. Sloczynski, "Magnesium Oxide as a Catalyst," J. Catal. 7, 67-75, 1967.

6. F. Pepe and F.S. Stone, "The Decomposition of Isopropanol over MgO and CoO-MgO Solid Solutions," J. Catal. 56, 160-168, 1979.

7. R.O. Kagel and R.G. Greenler, "Infrared Study of the Adsorption and of Methyl Alcohol on Magnesium Oxide," J. Chem. Phys. 49, 1638-1647, 1969.

8. A.J. Tench, D. Giles, and J.F.J. Kibblewhite, "Adsorption and Desorption of Methyl Alcohol in Magnesium Oxide," T. Farad. Soc. 67, 854-863, 1971.

9. N. Takezawa and H. Kobayachi, "Hydrogen Transfer Reaction between Ethanol and Acetone over Magnesium Oxide," J. Catal. 73, 120-127, 1982.

10. N. Ya. Turova, B.A. Popovkin, and A.V. Novoselova, "X-ray Diffraction and Crystallo-Optical Investigation of the Methylates and Ethylates of Magnesium and the Alkaline Earth Metals," Izv. Akad. Nauk SSSR Neorg. Mat. 3, 1435-1441, 1967.

11. N. Ya. Turova and A.V. Novoselova, "A Tensiometric Study of Alcohol-Solvated Alkoxides of Magnesium and the Alkaline Earth Elements, "Izv. Akad. Nauk SSSR Neorg. Mat. 4, 934-940, 1968.

12. Handbook of Chemistry and Physics, Ed. R. C. Weast, CRC Press, Cleveland, Ohio, 1973, F40-41.

13. N. Ya. Turova and A.V. Novoselova, "Solubility of Magnesium and Alkaline Earth Metal Alcoholates in Alcohols," Izv. Akad. Nauk SSSR Neorg. Mat. 3, 351-360, 1967.

14. R.C. Rossi and R.M. Fulrath, "Final Stage Densification in Vacuum Hot-Pressing of Al_2O_3," J. Amer. Ceram. Soc. 48, 558, 1965.

DEWATERING OF YTTRIUM HYDROXYNITRATE

PRECIPITATES BY ORGANIC LIQUIDS

M. D. Rasmussen, M. Akinc and O. Hunter, Jr.
Ames Laboratory USDOE†, Iowa State University
Ames, Iowa 50011
USA

The role of individual steps in dewatering of the yttrium hydroxynitrate precipitate by organic liquids in relation to powder morphology and sinterability of the oxide is studied. It was found that washing of the precipitate with toluene results in one or two orders of magnitude reduction in the particle size. A mechanism was offered for the effect of toluene wash. Variation of electrical double layer potential as a function of distance for different liquids is constructed to explain the effects of acetone-toluene-acetone treatment on the particle size and sinterability of the precursor.

†Work was done at the Ames Laboratory which is operated for the U. S. Department of Energy by Iowa State University under contract no. W-7405-ENG-82. This article was supported by the Director of Energy Research, Office of Basic Energy Sciences, Division of Materials Sciences.

Introduction

In recent years, considerable attention has been given to high temperature refractory materials in an effort to achieve greater efficiencies through higher operating temperatures. Rare-earth oxides and their solid solutions are candidates for higher temperature energy applications where conventional refractories fail. In addition to energy applications, optical and electrical properties of yttria, zirconia and rare-earth oxides seem to be more attractive than presently used materials. Electrical and optical applications as well as severe environments mandate low porosity (greater than 95% of theoretical density) materials. Production of low porosity sintered materials reproducibly depends on the characteristics of the oxide powder. Traditionally, the starting point for ceramic processing has been the calcined oxide powder; efforts to improve the materials have concentrated on compaction and sintering. However, recently an increasing amount of study has dealt with the preparation of the oxide powder including the precipitation, drying and calcining steps (1-3).

A common technique for obtaining oxide powder is to precipitate the cations in the hydroxide form which is then calcined to form an oxide. This process is usually carried out by the sol-gel method which has been used extensively in the production of ceramic nuclear fuels (4) and some refractory oxides (5,6). These methods either produce porous material (6) or require strict control of the variables in tedious procedures involving gelling, peptizing, aging, and dehydration (7). The acetone-toluene-acetone (ATA) method of oxide preparation described by Dole et al. (1,8) is significant in the sense that: 1) it is applicable to a very broad range of oxides, 2) the procedure is easy, and 3) control of variables during drying, calcining and/or sintering is not critical.

The object of this work was to understand the way the ATA washing steps affect the powder characteristics in relation to its sintering behavior.

Experimental

Preparation of Powders

Approximately 25g of yttrium oxide* was dissolved in 10 volume percent aqueous nitric acid (Baker reagent grade) to obtain 0.5M $Y(NO_3)_3$ and 0.01M HNO_3. This acidic cation solution was added to excess 3M NH_4OH solution at a rate of 20 ml per minute. The solution was constantly stirred during precipitation forming a viscous suspension with a final pH=10.5. The suspension was suction filtered in a sintered glass Buchner funnel. A gelatinous precipitate was obtained. This cake was re-suspended in a dilute NH_4OH solution (pH=10.5) and was then filtered again. The precipitate was split into two portions. The first portion was dried in an open beaker in an air convection oven at 110°C for overnight. This is a conventional method for drying the precipitate and will be referred to as oven dried (OD). The second portion of the precipitate was dewatered by the method described by Dole et al. (1,8). In the ATA dewatering method the cake was suspended in approximately 800 ml of acetone and then filtered. The same procedure was repeated with toluene and finally with acetone again. After the second acetone wash, the precipitate was transferred to a hot mortar and stirred with a pestle to evaporate the remaining acetone. This drying method will

*Obtained from Rare Earth Separation Group, Ames Laboratory USDOE, Iowa State University, Ames, Iowa. A typical analysis showed 99.9% Y_2O_3.

be referred to as the ATA method.

The powders dried by both the OD and ATA methods were calcined at 1000°C for 2 hours and then passed through a 200 mesh sieve; it was necessary to grind the OD powder in an agate mortar. The powders were dry pressed to 37 MPa (∿5400 psi) in a 9.5mm diameter tungsten carbide die which had been lubricated with stearic acid. The resultant 1g pellets were then isostatically pressed to 207 MPa (∿30000 psi), fired in a tungsten vacuum furnace at a rate of 10°C/minute to a peak temperature of 1900°C and soaked for 1 hour. Sintered densities were measured using the water displacement method and compared with the theoretical density of 5.031 g/cm^3. For the sake of clarity, a processing flowchart showing the basic steps of the OD and ATA methods is given in Figure 1. A small portion of precipitate was taken after each step and dried at 110°C in order to observe the morphological changes caused by each processing step. For electron microscopy studies, powders were dispersed in distilled water; a drop of dispersion was then placed on a carbon film supported on a copper grid. The grid was then transferred to an oven at 90°C and dried for 5 minutes. The specimens were coated with 5nm of gold. Observations were made on a scanning electron microscope (JEOL 100CX).

Specific surface areas of various powders were determined by nitrogen adsorption at its boiling temperature using a Micromeritics 2400D Sorptometer.

Sintered specimens were polished and chemically etched. Observations were made by an optical microscope.

Results

Properties of the Precipitate

As the acidic cation solution is added to aqueous ammonium hydroxide, precipitation starts immediately. Near the end of precipitation, the suspension becomes viscous and difficult to stir. At room temperature, a very small amount of precipitate settled in 12 hours. Filtering of this precipitate resulted in a gelatinous white cake. During the ATA treatment, after the material is suspended in acetone and filtered, the precipitate retains its gelatinous characteristics and exhibits some translucency. At this step it is believed that acetone, being miscible with water, replaces most of the H_2O in the colloidal system and provides a suitable medium for the toluene wash since acetone is soluble in toluene. The precipitate changes to flocs of particles when it is dispersed in toluene; these flocs settle out of the liquid phase very easily. The toluene wash removes most of the acetone-water solution from the medium and leaves colloidal particles in an organic phase; these particles flocculate readily. After the toluene wash, further filtering was very fast and the volume of the precipitate was reduced to a fraction of the original volume of the paste. This paste lacked any gel quality and it is very friable when partially dried. The most drastic changes take place at this step. The last acetone wash does not change the quality of the precipitate but replaces the toluene with the more volatile acetone. During evaporation of the acetone, the powder becomes very fine and friable. Chemical analyses of the powders show the composition is approximately $Y(OH)_{2.5}(NO_3)_{0.5}$, which is similar to the composition reported in reference 9.

The other portion of the precipitate was dried using the OD method; the material was glassy in appearance, difficult to grind, and sintered poorly. The tap density of this powder was significantly higher than ATA

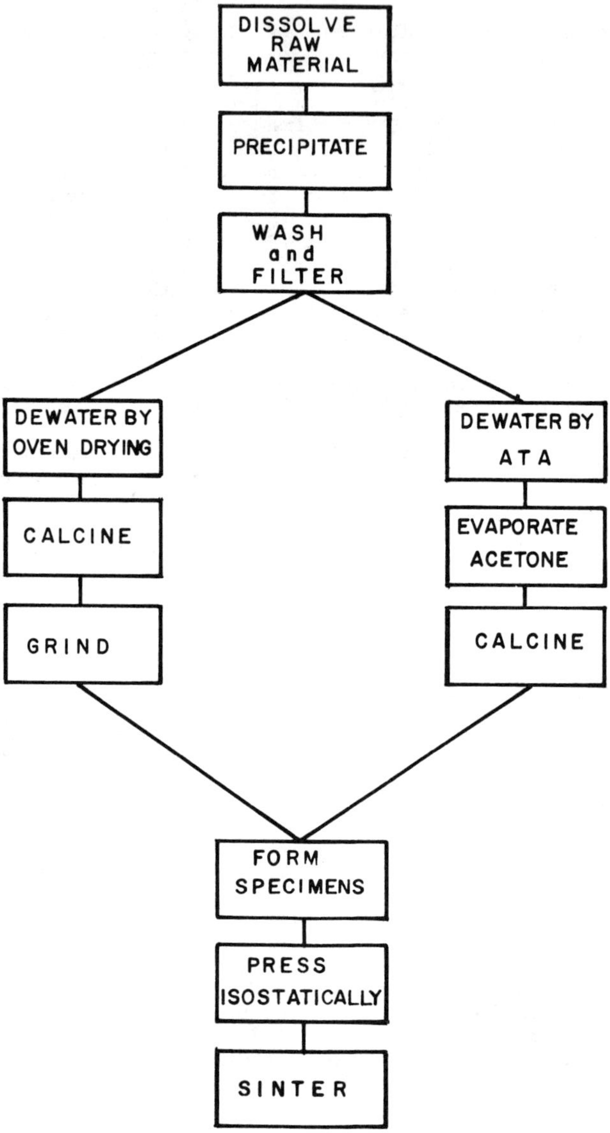

Figure 1 - Processing flowchart for yttrium oxide dewatered by oven drying and acetone-toluene-acetone methods.

prepared powder. Optical micrographs of sintered yttria pellets made from powder prepared by the OD and ATA methods are shown in Figure 2.

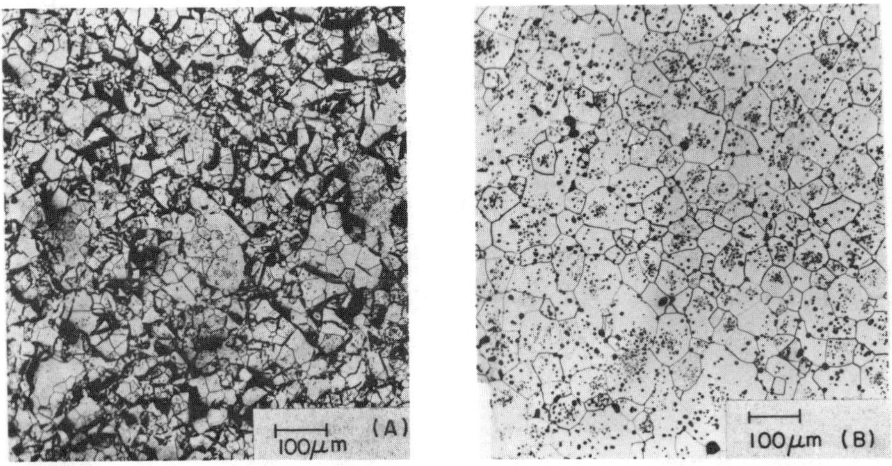

Figure 2 - Optical micrographs of sintered yttrium oxide specimens, polished and etched. A) Oven dried (∿80% theoretical density), B) ATA treated (∿97% theoretical density).

Morphological Changes

Scanning electron microscopy was found to be the most effective tool in determining how the wash steps were affecting the morphology of the material. Figure 3 shows the morphology of the powders after each step. The initial precipitate is given in Figure 3a. These results are analogous to the work on rapid dried erbium hydroxide reported by Dole et al. (1). The precipitated particles have irregular shapes with conchoidal fracture surfaces. At magnifications as high as 50000X, particles show no indication that they are an aggregation of smaller particles. This morphology is dissimilar to that reported by Holcombe et al. (9) who reported very small plates agglomerated in an edge-to-face arrangement. Later it was observed that the size and shape of the powder can be greatly altered when the ammonium hydroxide is added to the acidic cation solution, i.e., the direction of the precipitation is reversed (10). Figure 3b shows the morphology of the precipitate after the first acetone wash. The morphology remains more or less the same; it appears to exhibit glassy fracture with no definite shape. The average particle size of the original precipitate and the acetone-washed material is approximately 100 μm and no surface details were observed at magnifications up to 50000X. Figure 3c shows the precipitate after the toluene wash. There is a dramatic change in the size and shape of the particles as seen in the micrograph. The toluene wash reduced the particle size one or two orders of magnitude to an average particle size of approximately 1 μm. The final acetone wash does not change the size or shape significantly; however, it disperses the material and facilitates drying. Figure 3d shows an agglomerate of Y_2O_3 crystals after calcination. Individual crystals are cubic in shape and approximately 20-40nm in size. These primary particles are single crystals as indicated by electron diffraction. This observation is

Figure 3 - Micrographs (SEM) indicating morphological changes during ATA treatment, a) precipitated material (this micrograph corresponds to oven dried material), b) after first acetone wash, c) after toluene wash, d) ATA treated material after calcining.

supported by x-ray diffraction crystal size measurements by the line broadening technique which indicates a crystal size of 30-40nm. It appears that these individual crystals are aggregated to form secondary particles of about 100nm in size; these in turn agglomerate to form multimicron size particles. The secondary particles have very little porosity in the range of the primary particle size, but agglomerates are relatively porous giving rise to lower tap density.

Table I shows BET specific surface areas of powders obtained by N_2 adsorption together with calculated equivalent spherical diameters using true densities of yttrium hydroxynitrate and yttrium oxide for uncalcined and calcined powders respectively. Also in the table, values of particle size determined by microscopy are given.

Table I. Specific Surface Area and Particle Size Variations During Processing of Yttria

Processing Stage	S_{BET}, m^2/g	d^*_{BET}, nm	d_{MIC}, nm
Before ATA	1.8	$1x10^3$	$1x10^5$
After ATA	197	10	$1x10^3$
Calcined	13.9	80	100†

*Density of uncalcined powders is 3.12 and of calcined is 5.03 g/cc.

†Secondary particles; aggregates of primary particles.

The discrepancy between particle diameter calculated from nitrogen adsorption and microscopy is too large to be due to the shape factor or uncertainty in the density of the uncalcined materials. This is probably due to the intraparticle porosity of the uncalcined powders. These pores must be smaller than the microscope resolution (10nm) yet accessible for nitrogen adsorption. The calcined material gives comparable particle diameter values for adsorption and microscopy implying that the secondary particles are fully dense.

Discussion

Experimental observations and results so far can be explained if one considers the electric double layer at the precipitate-liquid interface. During the precipitation, the pH of the solution changes from 11.8 to 10.5. The entire precipitation process takes place in a highly concentrated solution (the ionic strength is approximately 0.25 during the entire process). The solubility products of $Y(OH)_3$ and $Y(OH)_{2.5}\,Cl_{0.5}$ are reported as 10^{-26} and 10^{-22}, respectively (11,12). The solubility of $Y(OH)_{2.5}\,(NO_3)_{0.5}$ ought to be similar to that of $Y(OH)_{2.5}\,Cl_{0.5}$. Conservatively, one might assume the solubility product of yttrium hydroxynitrate to be no larger than 10^{-20}. In fact, saturated solution of precipitate revealed a nitrate concentration of $2x10^{-3}M$ and a pH of 7.5. Solubility product based on these concentrations and charge neutrality requirement gives a solubility product of $1.5x10^{-20}$. At the beginning of the precipitation, the concentration of hydroxide ions is approximately $1x10^{-2}M$. The addition of one drop of the cation solution would produce yttrium and nitrate concentrations of $[Y^{3+}] \cong 5x10^{-5}$ and $[NO_3^-]=3x10^{-4}M$. These values together with the hydroxide concentration correspond to an ion product of

$$IP = [Y^{3+}][OH^-]^{2.5}[NO_3^-]^{0.5}$$

$$= (5x10^{-5})(1x10^{-2})^{2.5}(3x10^{-4})^{0.5} = 8.7x10^{-12} \qquad (1)$$

This value is about nine orders of magnitude larger than the presumed solubility product of yttrium hydroxynitrate. Thus, instantaneous supersaturation is extremely high. As precipitation nears completion, although the concentration of the hydroxide ions has decreased to $3x10^{-4}$, it was more than offset by the increase in nitrate ion concentration from $3x10^{-4}$ to about 0.25. Therefore, during the entire process of precipitation the condition of extremely high supersaturation remains. Under these conditions the nucleation is rapid; as a result an extremely fine colloidal precipitate forms. Similar observations were made by Rankin and Burney (13) in the precipitation of plutonium oxalate.

Precipitated particles are positively charged (10). A potential develops between the surface of the particles and the bulk solution. According to Debye-Huckel approximation variation of double layer potential as a function of distance is given by (14)

$$\psi = \psi_o \exp(-\kappa x) \qquad (2)$$

where ψ_o is the surface potential, x is the distance from the surface and $1/\kappa$ is known as double layer thickness and κ is equal to

$$\kappa = \left(\frac{8\pi e^2 N\ I}{1000\varepsilon kT}\right)^{\frac{1}{2}} \qquad (3)$$

where e = electronic charge, N = Avogadro number, ε = dielectric constant of the medium, k = Boltzmann constant, T = absolute temperature and I = ionic strength of the solution (for monovalent ions, ionic strength is equal to the concentration of the electrolyte). As the equations imply, the double layer potential at any given distance from the surface is proportional to the surface potential, ψ_o, but also it is an exponential function of ionic strength and the dielectric constant of the medium.

The surface potential of yttrium hydroxynitrate precipitate was determined in an aqueous solution by using electrokinetic measurements, and was found to be +60mV. The variation of double layer potential as a function of distance was constructed by using the known values of surface potential, ionic strength and dielectric constant of the medium.

The filtered material has a surface potential of +60mV in an aqueous environment, however, the double layer is compressed because the concentration is high ($[NO_3^-]$ and $[NH_4^+]$ are about 0.25M). This situation is represented by curve 1 in Figure 4. The precipitate flocculates to some degree and filtering is relatively easy. Drying of the precipitate in an oven probably involves a complicated process. The medium is alkaline, containing a high concentration of nitrate, hydroxide and ammonium ions. As the temperature of the paste increases, the pH of the medium decreases because

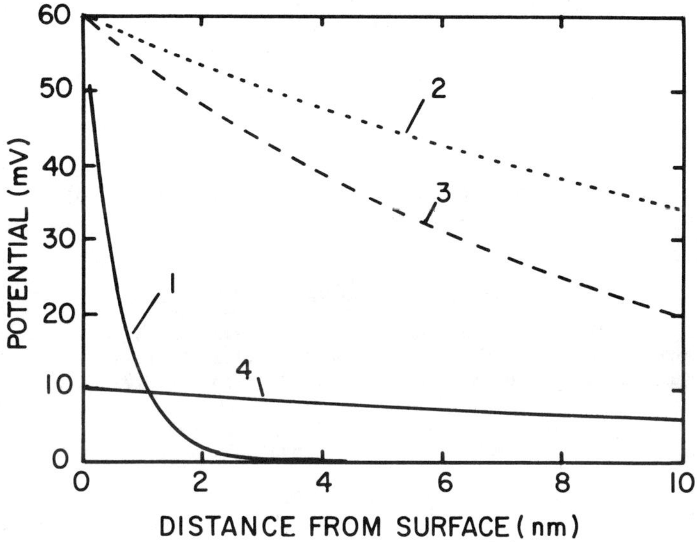

Figure 4 - Variation of double layer potential as a function of distance from the surface. 1) Precipitate in mother liquor, 2) Aqueous wash (pH=10.5), 3) Acetone wash, 4) Toluene wash.

of the evaporation of ammonia. This decreasing pH of the medium causes a corresponding increase in solubility of the precipitate. Furthermore, it is expected that increasing temperature will increase the solubility of the precipitate drastically. However, as water evaporates the solution becomes supersaturated and secondary precipitation takes place. This is a rather complicated phenomenon since temperature increase, change in pH and evaporation of H_2O take place simultaneously; no detailed analysis is available yet. However, it is conceivable that in such a process the individual colloidal particles are bonded by the secondary precipitate to form a hard, glassy material. Instead of drying the precipitate in an oven, if it is washed with an aqueous ammonium hydroxide solution, essentially all of the nitrate ions will be removed and the ionic strength will be limited to concentration of hydroxide and ammonium ions ($\sim 3x10^{-4}$M). This causes the electrical double layer to expand as shown by curve 2 in Figure 4. The expanded double layer causes the precipitate to swell and become more gelatinous; filtering becomes difficult and upon drying a hard glassy material is formed as in the previous case.

The effect of acetone washing on the properties of the precipitate should be similar to water washing except that the concentration of ions may be slightly lower in acetone, hence reducing the ionic strength. However, at the same time there is a compensating effect, namely the dielectric constant of the liquid. The dielectric constant of acetone is approximately one-third that of water. Variation of double layer potential as a function of distance is given in curve 3, Figure 4. After this material has been washed with toluene, the concentration of counter ions would be negligibly low, consequently surface charge and surface potential would be greatly reduced. In addition, the dielectric constant of toluene is extremely low thus potential drops faster than either water or acetone of identical ionic strength. The situation is represented in curve 4 of Figure 4 for a presumed 10mV (probably overestimated) surface potential. It is not surprising to see that the precipitate settled out of the toluene immediately in the form of fine powders, that it filtered easily, and that the toluene caused at least a two-fold decrease in the volume of the precipitate.

In summary, the ATA process prevents the dissolution-precipitation that takes place during conventional oven drying. This leaves the powder in a finely-divided, highly-sinterable form.

References

1. S. L. Dole, R. W. Scheidecker, L. E. Shiers, M. F. Berard, and O. Hunter, Jr., "Technique for Preparing Highly-Sinterable Oxide Powders," Materials Science and Engineering, 32 (1978) pp. 277-281.

2. K. Haberko, "Characteristics and Sintering Behaviour of Zirconia Ultrafine Powders," Ceramurgia International, 5 (4) (1979) pp. 148-154.

3. D. M. Ibrahim and H. W. Hennicke, "Preparation of Lead Zirconate by a Sol-Gel Method: Part II. Behavior of the Prepared Powders During Firing," Transactions British Ceramic Society, 80 (1) (1981) pp. 23-26.

4. ______________________, "Sol-Gel Processes for Ceramic Nuclear Fuels," Panel Proceedings Series, International Atomic Energy Agency, Vienna, 1968.

5. L. M. Sharygin, V. M. Galkin, and A. P. Shtin, "Pore Structure of Hydrated Zirconium Dioxide Made by the Sol-Gel Method," Izvestiya Akademii Nauk SSSR, Neorganiches kie Materialy, 15 (11) (1979) pp. 2051-2054.

6. B. E. Yoldas, "Alumina Sol Preparation from Alkoxides," American Ceramic Society Bulletin, 54 (3) (1975) pp. 289-290.

7. ______________________, "Sol-Gel Processes for Ceramic Nuclear Fuels," pp. 53-60, Panel Proceedings Series, International Atomic Energy Agency, Vienna, 1968.

8. M. F. Berard, O. Hunter, L. E. Shiers, S. L. Dole, and R. W. Scheidecker, Process for Preparing Active Oxide Powders, U. S. Patent #4140771 issued February 20, 1979.

9. C. E. Holcombe, C. C. Edwards, and D. A. Carpenter, "New Yttria Plasters," Report Y-2104, Union Carbide Oak Ridge Y-12 Plant, Oak Ridge, Tennessee, January 1978.

10. M. D. Rasmussen and M. Akinc, paper to be published.

11. V. B. Spivakouskii and L. P. Moisa, "Composition, Activity Products and Free Energies of Formation of Scandium, Yttrium, Lanthanum and Lanthanide Hydroxide Chlorides and Hydroxides," Russian Journal of Inorganic Chemistry, 22 (5) (1977) pp. 643-645.

12. O. A. Songina, Rare Metals, 3rd ed., p. 184; translated from Russian, Israel Program for Scientific Translations, Jerusalem, 1970.

13. D. T. Rankin and G. A. Burney, "Particle Size of $^{238}PuO_2$ obtained by Oxalate Precipitation and Calcination," American Ceramic Socity Bulletin, 54 (12) (1975) pp. 1061-1065.

14. P. C. Heimenz, Principles of Surface and Colloid Chemistry, pp. 367-372; Marcel Dekker, Inc., New York, N.Y., 1977.

AMORPHOUS Ni- AND Fe-B-Si POWDERS VIA THE P&WA RSR PROCESS

R. J. Patterson, II, D. L. Ledwith and J. C. Dwyer

Pratt & Whitney Aircraft Group
Government Products Division
P. O. Box 2691
West Palm Beach, Florida 33402 U.S.A.

Amorphous alloys in the Ni- and Fe-B-Si systems have been produced in the form of gas-quenched powder by the P&WA RSR powder process. Unlike powder produced by the spark erosion technique, these iron base powders show little variation in the magnetic Curie transition with particle size. We have, however, observed substantial variation with particle size of the thermal stability of some of the alloys produced to date.

An overview of the powder process is presented as is the typical response observed in particle size distribution, SEM, TEM, X-ray diffraction, and differential scanning calorimetry.

Introduction

Alloys in the Fe-B-Si and Ni-B-Si systems have been produced in the form of amorphous powder with the Pratt & Whitney Aircraft "rapid solidification rate" (RSR) process, in which rotary atomization converts a molten alloy into droplets which are rapidly quenched to the solid state by cross-flowing gas of high thermal conductivity (1). Most prior work reported on these and related metal-metalloid systems has been conducted on thin noncrystalline ribbon (2,3). More recently, powders have been produced by a variety of techniques including spark erosion (4,5), electrostatic atomization (6), water atomization (7), and gas atomization with water quenching (8).

Commercial interest in the metallic glasses stems from the unique mechanical, chemical, and physical properties obtained for a variety of the alloys (9-11). In the latter area, the low electromagnetic losses obtained with some of the alloys (12,13) make them potentially suitable for application to high-efficiency devices such as transformers and motors. Another aspect of growing interest is the use of the amorphous state as a precursor to new crystalline materials (14). Since most of the current alloys crystallize below 700°C, conventionally applied consolidation techniques such as extrusion, HIP, or sintering must be modified; recent success in the consolidation of amorphous materials (15-17) suggests that complex shapes may be fabricated from amorphous powders, though many questions remain regarding the effect of consolidation on properties.

Beyond the possible commercial aspects, the amorphous RSR powder is also of technical interest and work in this area can be expected to contribute to our knowledge of the effect of process parameters on particle solidification. In this process, the rate of quenching depends on a number of parameters including particle size, gas type, relative velocity, and the temperature difference between the particles and the gas (18). In general, the quench rate is not sufficiently high to prevent recalescence should the latent heat of fusion be released by the formation and growth of crystalline nuclei in the liquid. Thus, recalescence should lead to continued crystalline growth and the suppression of the formation of the amorphous solid, unless the growth rate is large enough that crystalline growth is not possible (19,20). This is not likely to be the case, however, since heat transfer studies (18,21,22) suggest that the velocity of the solid-liquid interface will decrease as solidification of the droplets proceeds. It is possible, however, that segregation occurring during such crystalline growth, combined with the heat transfer considerations, would lead to the presence of small volume fractions of an amorphous solid within otherwise crystalline particles, as has been observed by Kelly and Vander Sande (23) and Adam and Bourdeau (24). Thus, in the case of gas-quenched particles, at least, one might expect to find particles which are either substantially amorphous or substantially crystalline.

In this paper, we present some of the results obtained on RSR-processed powders of alloys in the Fe- and Ni-B-Si systems.

Experimental Techniques

Powder Production

Nine to fourteen kilogram ingots of one nickel base and several iron base alloys were prepared by vacuum induction melting using elemental stock of 98+ to 99.95+ percent purity. These ingots were then remelted and converted to powder with the RSR machine. In the case of the iron base alloys, only helium was used for quenching, and the powder conversion was accomplished prior to the addition in 1980 of a gas recirculation system to the RSR process. After the recirculator was added, we changed from the iron base alloys to Ni-17B-8Si (aim chemistry, atomic percent) to prevent iron contamination of nickel base superalloys which are also made in the rig, and nitrogen and argon were also used for quenching. For all runs, an 8.25 cm diameter atomizer operating at 400 revolutions per second was used. Post-production handling was done in air since no consolidation studies were planned for these early runs, where the objective was to obtain increased understanding of processing effects.

Powder Characterization

All lots of powder were sieved with an ATM Model L3P sonic sifter equipped with precision electroformed sieves to provide not only size distributions for the runs, but also narrow (either five or ten micrometer), precisely sized samples for further analysis.

A limited amount of characterization has been done using scanning and transmission electron microscopy (SEM, TEM), x-ray diffraction, and wet chemical analysis. The majority of the characterization has been conducted with differential scanning calorimetry (DSC) to obtain information on the variation of the magnetic Curie transition (the Ni-base alloy is non-magnetic), time-to-crystallization as a function of temperature, activation energy of crystallization, and possibly on the glass transition as functions of alloy chemistry and particle size. A DuPont 1090 thermal analysis system was used for this purpose, with aluminum pans serving both as standard and as sample container. Typically, 45 milligram samples were used except in cases where the energy of crystallization was quite large, where as little as 3 milligrams of powder was used. A variety of scan rates and isothermal conditions were used for these studies, conditions depending on the type of information being sought.

Results and Discussion

Powder Production

Wet chemical analysis of the earliest runs, not reported here, indicated that boron was being lost from the alloys. Processing was modified so that the iron or nickel would be induction melted and outgassed under vacuum, with the boron and silicon additions after the system had been backfilled with argon. During powder conversion, heating to within 50°C of the solidus was accomplished in vacuum, with the balance of the melting and conversion being conducted under the gas to be used for quenching. Subsequent chemical analyses have indicated that the resulting powders have been within an atomic percent of the aim chemistry.

These metal-metalloid alloys have been somewhat more difficult to atomize than other, more conventional iron, nickel, or aluminum alloys we have processed. For the latter alloys, the size distribution (by weight) is typically normal with a mean of about 88 micrometers. We have found that most alloys have a "temperature window" outside of which atomization is generally abnormal, leading to a log-normal distribution with a somewhat larger mean size. For the glass forming alloys, a similar log-normal distribution has been obtained with a mean typically near 125 micrometers. During these runs, a great deal of the material leaves the atomizer prematurely, being thrown upward toward the manifold carrying the quench gas. Impact with the manifold results in secondary atomization during such runs, and if the droplets thus produced are sufficiently large, they may subsequently be gas atomized. Thus, if the alloy behaves abnormally, a variety of atomization modes may occur, and conditions affecting quench rate and solidification can also be expected to vary widely. It was thus not too surprising to find that the quantity of amorphous powder from a given alloy and size fraction would vary from run to run: As the atomization conditions have approached those normally associated with this process, the quantity of amorphous powder in a given size fraction has increased. Unfortunately, our alloy studies have been varied enough, and our rig time limited enough, that the appropriate "window" for these alloys has not yet been defined.

Powder Characterization

In performing DSC, one of our hypotheses has been that the quantity of heat evolved during crystallization will be reasonably proportional to the volume fraction of amorphous material present. Since the technique cannot determine "how amorphous" a given material is, the results are relative, and since other factors such as residual stress may affect the energy released, the results are not expected to be exact. Similarly, the method cannot determine whether, as mentioned earlier, the particles are either completely amorphous or completely crystalline, or whether they each contain both amorphous and crystalline regions. Such factors require the correlation of results obtained by other techniques such as SEM, TEM, and X-ray diffraction, which we will attempt to accomplish by the consideration of several sets of data, some of which is presented in Figures 1-5.

In Figure 1, the results of the precision sieve analysis of two runs of Ni-17B-8Si are compared to the distribution typically obtained for the nickel base superalloy IN100 with the same atomizer size and speed. Other runs with the same master heat produced different distributions in which the yield correlated well with the general appearance of the atomization during a given run obtained by high-speed film and closed circuit video.

For these runs, the distribution of the amorphous fraction obtained by DSC is shown in Figure 2. For a given quench gas, the relative energy released by a given size fraction was found to vary with the qualitative "quality of atomization" obtained from the films and TV, and with the more quantitative results obtained by the sieve analysis. Thus, as the yield of powder below a given size fraction was found to increase from one run to another, the relative energy released upon crystallization in DSC was also found to increase, for a given quench gas. For all gases, nitrogen consistently resulted in a lower relative yield of amorphous material than either argon or helium, while the worst size distributions obtained with helium were

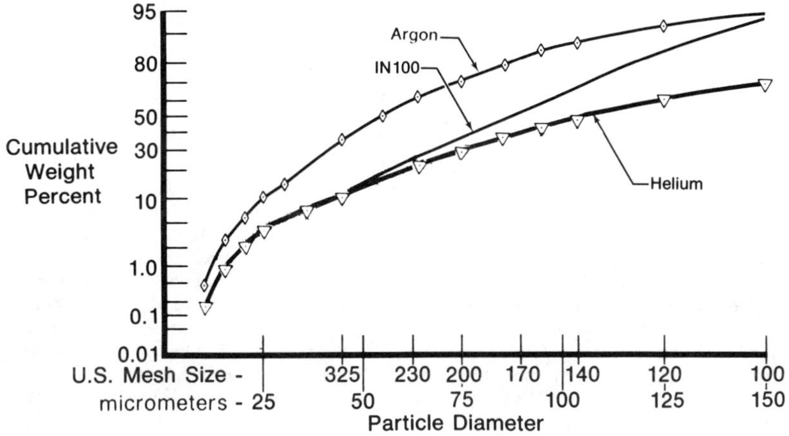

Figure 1 - Size distributions obtained for two runs of Ni-17B-8Si, helium and argon quenched, compared with the typical distribution obtained for the nickel base superalloy IN100. Both runs displayed abnormal atomization behavior, as discussed in the text.

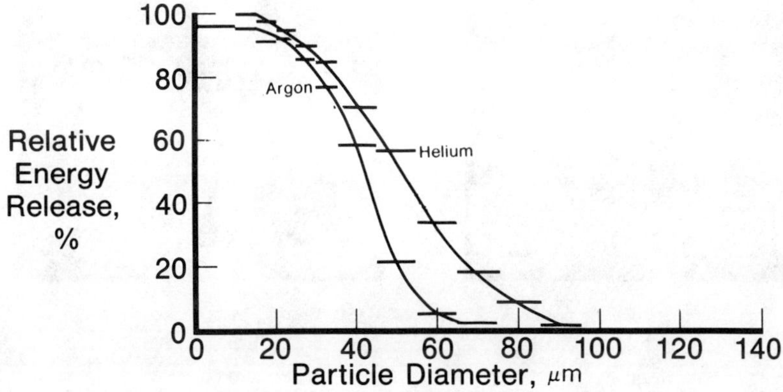

Figure 2 - Distribution of the amorphous fraction for the runs of Figure 1, based on relative energy released during crystallization. Data from other runs with this alloy indicate that improved atomization behavior leads to higher yields of the amorphous fraction. Horizontal bars represent the range of each size fraction.

found to yield less amorphous material, below 60 micrometers, than the best of the size distributions obtained with argon. Since helium has the highest thermal conductivity and argon the lowest, among the three, the performance of nitrogen was somewhat peculiar but might be caused by the lower purity of the gas. In any case, the results indicate the significance of atomization as a factor in the production of amorphous powder.

The general appearance of two fractions obtained from the runs presented in Figures 1 and 2 is shown in the scanning micrographs of Figure 3. We wondered whether the crystallinity of a "slightly amorphous" fraction would be apparent in the appearance of the powder, and that such is the case can be seen from Figure 3 (b). We were also curious to know whether the -15 micrometer argon-quenched powder might not be fully amorphous, with the slight difference in energy release from the data obtained for helium (Figure 2) being due to other causes, such as stress release. As seen in Figure 3 (a), there is some visual indication that some of the particles are crystalline. Note also that the surface features seen in the substantially amorphous fractions are typically smaller than one micrometer; what few features are observed generally consist of either debris picked up during handling or fine "splat" produced during abnormal atomization.

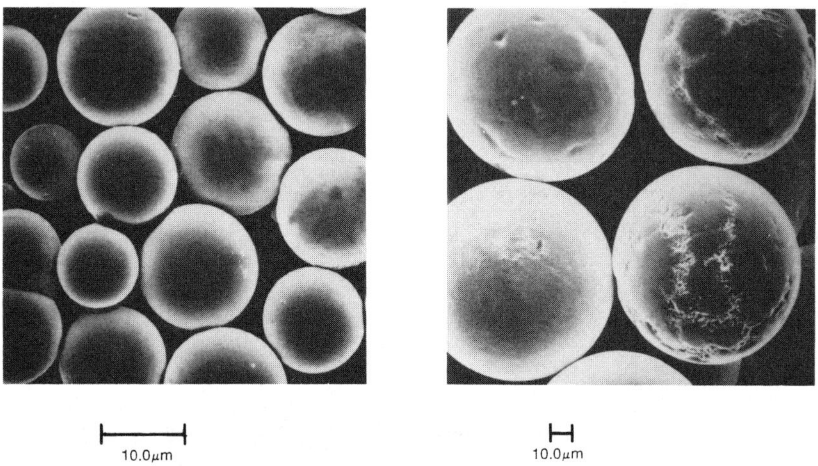

(a) (b)

Figure 3 - Scanning micrographs of two size fractions from the runs of Figures 1 and 2. (a) The -15 + 10 μm fraction from the argon-quenched run. Some of the surface features are the result of processing the sample, others are debris, while others are indications of crystallinity. (b) The -85 + 75 μm fraction of the helium-quenched run; most surface features are due to crystallinity.

X-ray diffraction was also used to check for the presence of crystalline material in the "highly amorphous" size fractions and to analyze the crystalline phases present; diffraction patterns taken with CuKα radiation at a rate of 1° per minute are shown in Figure 4. In Figure 4 (a), taken on the -15+10 micrometer argon-quenched sample shown in Figures 1, 2, and 3 (a), crystalline peaks can be seen riding the amorphous pattern. These peaks were analyzed from the data shown in Figure 4 (b), taken on a larger helium-quenched size fraction, and were found to be from only two phases: Ni_3B and Ni_3Si. These are the only phases expected from the aim chemistry. While we have not yet made extensive use of X-ray diffraction, the results from the several samples studied correlate well with the relative data obtained by DSC, although the presence of the amorphous "hump" becomes quite difficult to detect as the relative quantity of crystalline material increases.

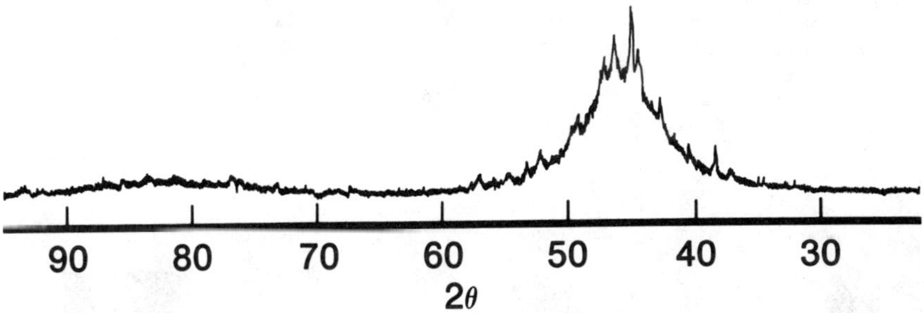

(a)

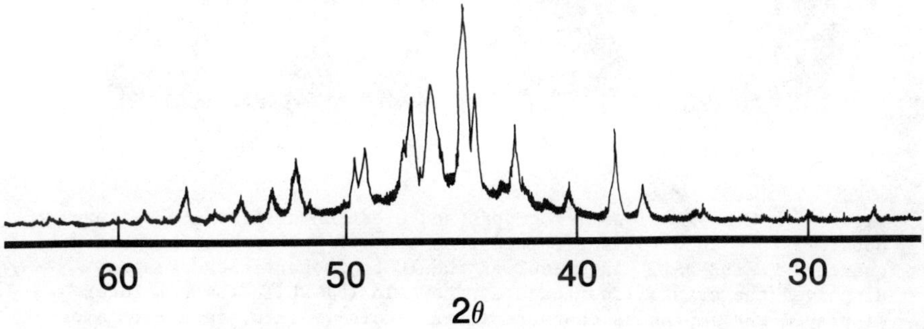

(b)

Figure 4 - X-ray diffraction patterns obtained from the Ni-17B-8Si runs of Figures 1-3. (a) The -15 + 10 μm fraction from the argon run. (b) The -55 + 45 μm fraction from the helium run. See text.

While these data are all consistent with the view that DSC can provide a relative distribution for the noncrystalline material, they do not attack the problem of whether the particles are either amorphous or crystalline, or whether a particle contains regions of both. To approach this problem, we turned to transmission microscopy, and micrographs obtained from one size fraction of one of the iron base alloys are presented, with the associated electron diffraction patterns, in Figure 5. Thin foils prepared from several of the particles in each of two size fractions were examined. In the finer fraction (-25 micrometer) none of the particles studied showed evidence of crystallinity, while in the larger fraction from which the micrographs were obtained, about 60 percent of the particles showed the crystalline structure observed in Figure 5 (b), while 40 percent displayed the features characteristic of amorphous material: A featureless bright field image, a "grainy" dark field image, and diffuse diffraction rings. DSC data similar to that presented for the nickel alloy suggested that this fraction was about 45% "as amorphous" as the finest size fraction (-10 micrometer) studied in DSC. Considering that relatively few (10-20) particles were examined in TEM for each fraction, and that even the finest fraction from this early run likely contained some crystalline particles, the correlation of the two results is good. Also, this preliminary TEM work suggests that the particles are either substantially amorphous or substantially crystalline, though conclusions to this effect would require significantly more data.

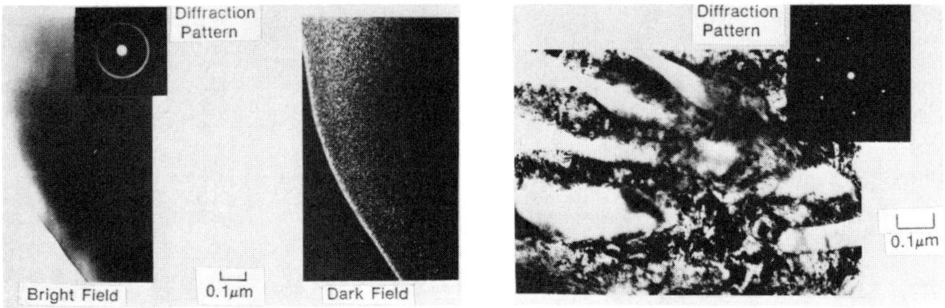

(a) (b)

Figure 5 - Transmission micrographs and electron diffraction patterns obtained from thin foils prepared from the -65 + 55 μm fraction of He-quenched Fe-16B-5Si. About 60% of the 10 to 20 particles examined displayed the crystalline features shown in (b) while the remainder displayed the amorphous characteristics of (a). In no case were amorphous and crystalline regions found in the same particle.

Two other points are worth mentioning. First, no glass transition was found for Ni-17B-8Si under any set of conditions used in DSC—this result was expected from the work reported by Donald and Davies (3). Second, in the study of the thermal stability of this alloy by the isothermal technique, several samples which did not crystallize during the time allowed (the longer runs, up to eight hours, were done overnight or on the weekend) were rescanned and were found to release the same energy as samples which had not been exposed to elevated temperatures but which were scanned at the same rate. Whether this implies that factors (beyond the volume fraction of amorphous material) which might affect the energy released during crystallization do not show up at these low times and temperatures (in the vicinity of 400°C), or whether it implies that the energy associated with such factors is small relative to the energy of crystallization, is not yet known.

Iron base alloys. In addition to the general characterization noted in the preceding paragraphs, the iron base alloys were also evaluated for magnetic Curie transition and thermal stability. A typical thermogram is shown in Figure 6. The large positive peaks in the vicinity of 550°C are the crystallization exotherms; the small endotherm seen near 400°C is enlarged in Figure 7, and is the transition from the ferromagnetic state (below the transition) to the paramagnetic state.

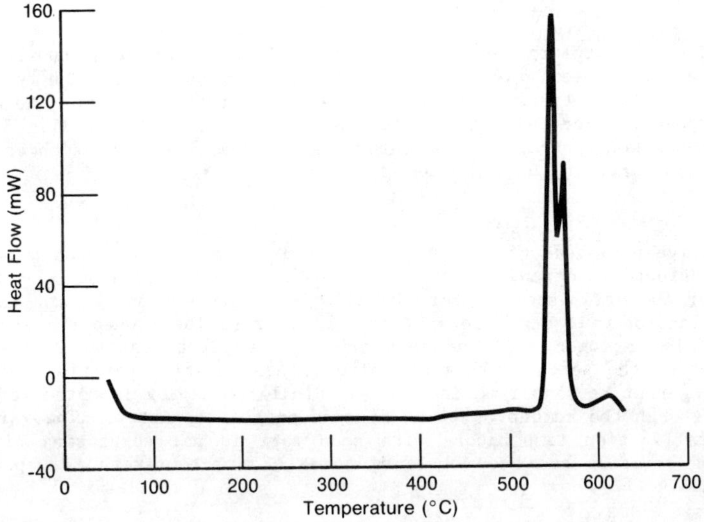

Figure 6 - DSC thermogram for Fe-16B-5Si at a scan rate of 50°C/min. The relative separation, amplitude, and temperatures of the peaks near 550°C vary with scanning rate and alloy chemistry.

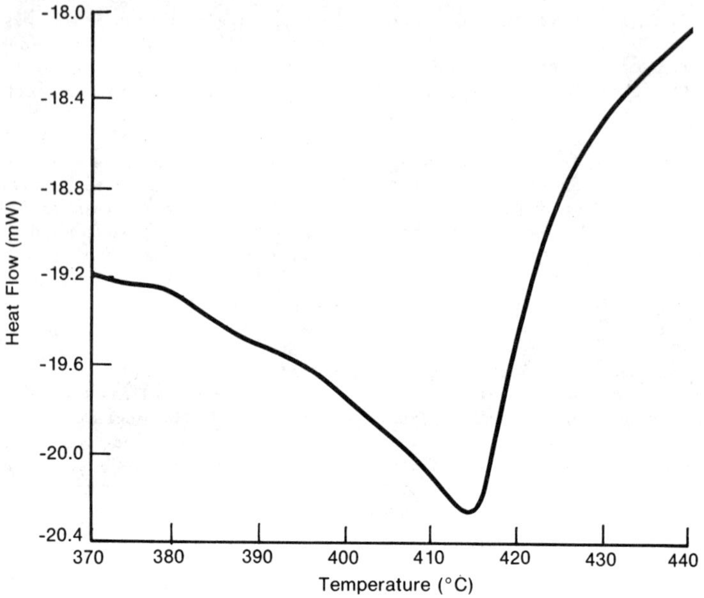

Figure 7 - Enlargement of a portion of Figure 6, showing the endotherm characteristic of the magnetic Curie transition. The temperature of the transition will vary with chemistry and prior thermal exposure. For this alloy, the transitions for the crystalline phases would occur at temperatures above the crystallization exotherms seen in the previous Figure.

The relative amplitude of the peaks and the temperatures at which they occur have been found to depend not only on alloy chemistry, but also on scan rate and prior thermal history. For the alloy of Figures 6 and 7, for example, crystallization will begin several tens of degrees lower when the scan rate is 20°C/min or lower, and the amplitude of the first peak will be smaller than that of the second. The peaks also change relative position, becoming closer together as scan rate increases. Similar behavior is noted for other alloys, though the number of peaks depends on the chemistry. The variation of crystallization temperature with scan rate is not surprising since an Arrhenius relation has been observed for the time-temperature response of these alloys (12).

As for the amorphous ribbon, we have found that the Curie transition varies in temperature not only with chemistry but also with prior thermal history (25). Contrary to reports on powder produced by the spark erosion technique (4, 5), however, we have found very little variation in the Curie transition with particle size, and the data we have obtained are generally within a degree or two of data in the literature for melt-spun ribbon of the same chemistry. We have, after much analysis, been able to detect a slight variation with size in some alloys; for Fe-16B-5Si, the variation is shown in Figure 8. On the chance that such variation might be a function of chemistry, we also ran Fe-10B-15Si where we obtained the transition at 434 $\pm$ 2°C for particle sizes ranging from below 10 to above 100 micrometers diameter. This value correlates well with data reported for "as-produced" ribbon of similar chemistry (2, 5).

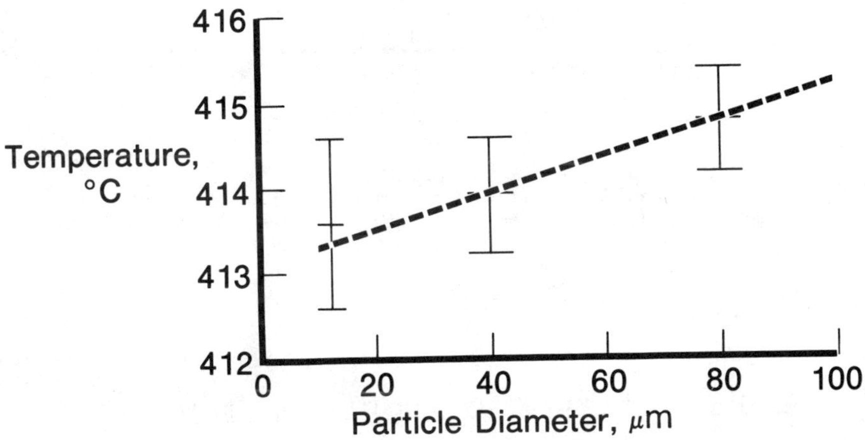

Figure 8 - Variation in the Curie transition (Figure 7) with particle size, for as-produced powder. The horizontal width of the bars represents the nominal range of each fraction, while the vertical extent provides the mean $\pm$ 2 σ (for a minimum of 7 data points).

Since we had found little variation in the Curie transition with particle size for any of the Fe-B-Si alloys studied to date, we were quite surprised to find substantial variation with particle size of both thermal stability and activation energy for crystallization for several of the alloys for which these factors have been determined. Using the relation (12):

$$t_x = t_o \exp(\Delta E_x / KT) \qquad (1)$$

where t is time, T is temperature, K is the Boltzman constant, and ΔE_x is the activation evergy, one can obtain the activation energy from isothermal data by plotting the time to crystallize, t_x, versus 1/T, and evaluating the slope. Such data are shown in Figure 9 for two particle sizes of Ni-17B-8Si, with the activation energies so calculated for three size fractions shown in Table I.

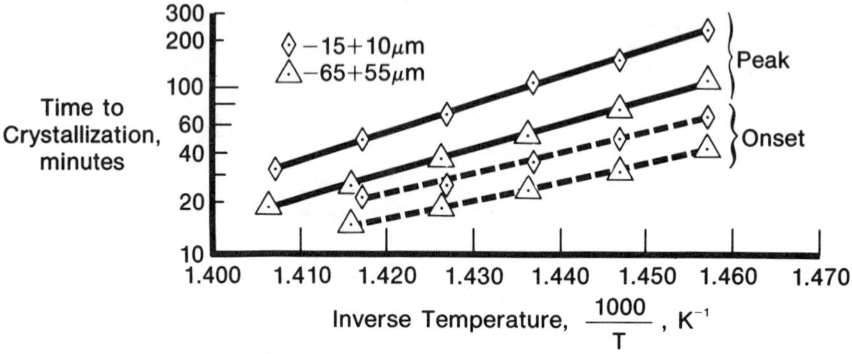

Figure 9 - The thermal stability of two size fractions of as-produced Ni-17B-8Si, using the isothermal technique. Samples were run at 50°C/minute up to the temperature of interest. Data for the -45 + 35 μm fraction fall between the curves shown.

Table I - Activation Energy for Crystallization of Ni-17B-8Si

Size Fraction	ΔE_x, eV	
	Onset	Peak
-15 + 10 μm	2.68	3.48
-45 + 35 μm	2.28	2.95
-65 + 55 μm	2.28	2.87

The data become more difficult to evaluate when the exotherm consists of two or more overlapping peaks, so not all of the iron base alloys have been tested. Difficulty also occurs because the energy is released over progressively longer periods of time as lower isothermal temperatures are used, making onset difficult to determine for size fractions where a reasonably large percentage of the material is crystalline. For higher temperatures, crystallization occurs earlier in time, but one reaches a point at which the time required to take the sample from room temperature to the isothermal condition becomes a substantial fraction of the elapsed time to crystallization, making interpretation difficult. Another technique for determining the activation energy is generally known as the Kissinger method (26-28), and it uses the variation in peak temperature with scan rate.

Trends obtained with this technique are similar to those observed from the Arrhenius relation for those cases evaluated, though typically the activation energies obtained were about 10%-15% higher for the former method. As with the Arrhenius relation, difficulties can be encountered when peaks overlap. Several possible causes for the variation with particle size remain to be evaluated.

Finally, we have not observed glass transition temperatures for any of the Fe-B-Si alloys, though some were expected. Whether this is due to our technique or to the modest volume fractions of amorphous material produced to date remains to be determined.

Conclusions

A great deal of work remains to be done in the production and characterization of gas-quenched amorphous powder. Yield can be expected to rise in the future as we acquire the melt practise appropriate to these alloys. Too, recent results at higher atomization speeds with the more conventional alloys suggest that it should be possible to produce amorphous powder in large quantity.

Acknowledgements

The authors are indebted to Prof. H. L. Fraser for the transmission microscopy, and to D. C. Williams for the extensive calorimetry performed on the nickel base alloy. We also thank H. R. Nesor and J. A. Umbach for the SEM and X-ray diffraction, respectively, and J. B. Moore, R. L. Athey and A. R. Cox for their support.

References

A. R. Mehrabian, B. H. Kear and M. Cohen, eds., Rapid Solidification Processing, Principles and Technologies; Claitor's Publishing Division, Baton Rouge, La., 1978.

B. R. Mehrabian, B. H. Kear and M. Cohen, eds., Rapid Solidification Processing, Principles and Technologies II; Claitor's Publishing Division, Baton Rouge, La., 1980.

1. M. R. Glickstein, R. J. Patterson, II and N. E. Shockey, "Rapid Solidification Effects of Micron-Size Droplets," pp. 46-63 in Ref. A.

2. F. E. Luborsky, J. J. Becker, J. L. Walter and H. H. Liebermann, "Formation and Magnetic Properties of Fe-B-Si Amorphous Alloys," IEEE Trans. Magn. MAG-15 (3) (1979) pp. 1146-1149.

3. I. W. Donald and H. A. Davies, "The Influence of Composition on the Formation and Stability of Ni-Si-B Metallic Glasses," J. Mat. Sci. 15 (1980) pp. 2754-2760.

4. T. Yamaguchi and K. Narita, "Magnetothermal Characteristics of Amorphous Fe-Si-B Powder Produced by Spark Erosion," IEEE Trans. Magn. MAG-13(5) (1977) pp. 1621-1623.

5. A. E. Berkowitz and J. L. Walter, "Preparation of Amorphous Metal Powders by Spark Erosion," pp. 294-305 in Ref. B.

6. J. Perel, J. F. Mahoney, B. E. Kalensher, K. E. Vickers and R. Mehrabian, "Electrohydrodynamic Generation of Submicron Particles for Rapid Solidification," pp. 258-269 in Ref. A.

7. T. Yamaguchi and K. Narita, "Preparation of Amorphous Magnetic Powder by Atomization Technique," Appl. Phys. Lett. 33(5) (1978) pp. 468-470.

8. S. A. Miller and R. J. Murphy, "The Production and Consolidation of Amorphous Metal Powder," pp. 385-392 in Ref. B.

9. T. Masumoto and K. Hashimoto, "Chemical Properties of Amorphous Metals," pp. 215-233 in Annual Review of Materials Science Vol. 8; Annual Reviews, Inc., Palo Alto, CA, 1978.

10. P. Duwez, "Structure and Properties of Glassy Metals," pp. 83-117 in Annual Review of Materials Science Vol. 6; Annual Reviews, Inc., Palo Alto, CA, 1976.

11. B. C. Giessen and C. N. J. Wagner, "Structure and Properties of Noncrystalline Metallic Alloys Produced by Rapid Quenching of Liquid Alloys," pp. 633-695 in Liquid Metals, Chemistry and Physics, S. Z. Beer, ed.; Marcel Dekker, Inc., New York, NY, 1972.

12. F. E. Luborsky, "Amorphous Ferromagnets," pp. 451-529 in Ferromagnetic Materials, Vol. 1, E. P. Wohlfarth, ed.; North Holland, Amsterdam, 1980.

13. J. J. Gilman, "Metallic Glasses," Science 208 (1980) pp. 856-861.

14. J. L. Walter, "Amorphous Metals as Starting Materials for Crystallization," pp. 79-92 in Ref. B.

15. C. F. Cline, J. Mahler, M. Finger, W. Kuhl and R. Hopper, "Explosive Fabrication of Rapidly Solidified Alloys," pp. 380-387 in Ref. A

16. H. H. Liebermann, "Warm-Consolidation of Glassy Alloy Ribbons," pp. 393-398 in Ref. B.

17. D. G. Morris, "Dynamic Compaction of Metallic Glasses and Microcrystalline Materials," pp. 372-384 in Ref. B.

18. R. J. Patterson, II, "The Modeling of Rapidly Solidified Powders" to be published in Rapidly Solidified Amorphous and Crystalline Alloys, B. H. Kear and B. C. Giessen, eds.; Elsevier North-Holland, New York, NY, 1982.

19. W. J. Boettinger, "Growth Kinetic Limitations During Rapid Solidification," ibid.

20. W. J. Boettinger, F. S. Biancaniello, G. M. Kalonji and J. W. Cahn, "Eutectic Solidification and the Formation of Metallic Glasses," pp. 50-55 in Ref. B.

21. W. N. Gill, J. Y. Jang and J. C. Mollendorf, "Rapid Solidification of Subcooled Small Metallic Drops," Chem. Eng. Commun. 12 (1981) pp. 3-131.

22. C. G. Levi and R. Mehrabian, "Heat Flow during Rapid Solidification of Undercooled Metal Droplets," Met. Trans 13A (1982) pp. 221-234.

23. T. F. Kelly and J. B.Vander Sande, "A STEM Analysis of Two Rapidly Solidified Stainless Steels," pp. 100-111 in Ref. B.

24. C. M. Adam and R. G. Bourdeau, "Transition and Refractory Element Additions to Rapidly Solidified Aluminum Alloys," pp. 246-259 in Ref. B.

25. C. D. Graham, Jr. and T. Egami, "Magnetic Properties of Amorphous Alloys," pp. 423-457 in Annual Review of Materials Science, Vol. 8; Annual Reviews, Inc., Palo Alto, CA, 1978.

26. H. E. Kissinger, "Reaction Kinetics in Differential Thermal Analysis," Analytical Chemistry 29(11) (1957) pp. 1702-1706.

27. H. E. Kissinger, "Variation of Peak Temeprature with Heating Rate in Differential Thermal Analysis," J. Research N.B.S. 57(4) (1956) pp. 217-221.

28. B. G. Lewis, H. A. Davies and K. S. Ward, "The Crystallization of $(Fe_{100-x}Ni_x)_{83}B_{17}$ Amorphous Alloys," pp. 325-336 in Rapidly Quenched Metals III, Vol. 1, B. Cantor, ed.; The Metals Society, London, 1978.

CHARACTERIZATION OF AN EXTERNAL GAS ATOMIZING NOZZLE

P. E. Hugo* and R. M. German

Materials Engineering Department
Rensselaer Polytechnic Institute
Troy, New York 12181

*now at Special Metals Corporation
Middle Settlement Road
New Hartford, New York 13413

A study of the gas atomization of pure tin has been conducted using nitrogen and argon as the atomizing gases with an annular nozzle design. Study variables included the melt superheat (which changes the viscosity and solidification time), atomizing gas pressure, and the gas flow rate (adjusted by the nozzle gap opening). Both particle size and shape were monitored as functions of the atomization conditions. A first effort is reported for characterizing the performance of an annular gas atomization nozzle in terms of the energy transfer from the expanding gas. Comparison of the data to earlier models shows this to be a superior representation for annular gas atomization.

Introduction

As interest in powder metallurgy expands, it becomes evident that many fundamental processes are not well understood. An increased understanding will allow us to construct mathematical descriptions to aid in optimization. This study characterizes a gas atomization system based on the operating parameters of pressure, gap size, gas and superheat. A unique aspect of this study is the use of an annular gas atomization nozzle. Although there is an abundance of literature on atomization (both gas and water), (1-9) little experimental attention has been given to annular systems. However, in practice the annular systems are used extensively in the production of high quality alloy powders. Thus, this paper describes the operation and variable relationships applicable to an annular gas atomization nozzle.

Background

Efforts to create mathematical descriptions of the atomization process are not new. Much of the earlier work has attempted to relate the particle size to melt stream disturbances while some recent studies have resorted to polynomial data fitting techniques. This section summarizes both approaches with more complete reviews being available in references 8-14.

The Rayleigh relation for a liquid stream breakup gives the droplet radius r as a linear function of the stream radius R,

$$r = 1.88R \qquad [1]$$

Castleman (15) has used this relation to describe water stream breakup by high velocity air jets and provided relations for ligament disintegration and spheriodization. Subsequently, several studies have treated atomization as a process of liquid sheet formation, wave formation in the sheet giving ligaments and subsequent droplet formation based on velocity and thickness calculations. (4,13-18)

Other approaches have relied on two jet dimensional analysis (19) and phenomenological (2,3,20) approaches to predict particle size as a function of atomization parameters. Analysis of this body of literature shows a general mathematical formulation based on liquid sheet breakup. Mathematical modeling for atomization proves difficult since there is a large list of variables that must be considered, including material parameters, gas characteristics, nozzle geometry, flow rates, pressures, viscosities, etc. (6) Thus apparatus differences as well as differences in metals are influencial on the mathematical models. Hence, equations which apply to one apparatus do not as a rule apply to other designs. Furthermore, in spite of the literature on stream breakup physics, there is a general inability to predict particle size as a function of operating parameters. This is especially true for annular gas atomization nozzles. Of course one difficulty is the distributed nature of the powders; hence, our approach has been to analyze both the mean and distribution about the mean as functions of the atomization conditions.

Experimental Plan

An annular gas atomization nozzle was designed for pressures up to 35 MPa to produce fine metal powders. The nozzle is shown schematically in Figure 1. It is constructed from brass with a 21 cm diameter and is 8 cm in height, with a variable gas discharge gap. The feed tube had a 3 mm inner diameter. As can be seen in Figure 1, this is a confined, external, annular type nozzle. (11)

The study matrix included four variables— annular gap opening, superheat, pressure and gas at the levels shown in Table I. All possible combin-

Table I. Experimental Test Matrix - 48 Experiments

Variable	Settings	Levels
Gap Opening, μm	50, 100, 150, 200	(4 levels)
Pressure, MPa	1.0, 1.7, 2.4	(3 levels)
Superheat, °C	100, 150	(2 levels)
Gas	argon, nitrogen	(2 levels)

ations of the four variables were used to generate 48 experimental runs, one of which was unsuccessful. Three sets of repeat runs tested the system reproducibility. In one set of replications, the nozzle was disassembled, cleaned and reassembled to ensure the gap size was retrievable.

Atomized droplets were collected in a 0.3 m diameter by 1.8 m height cylindrical chamber with gas exit ports. Tin with a purity of 99.86% (analysis shown in Table II) was used as the metal. The tin was melted under

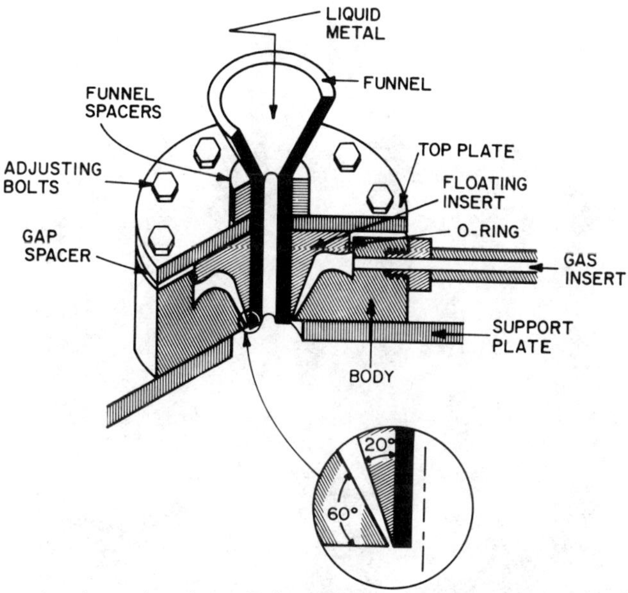

Fig. 1: A schematic view of the RPI atomization nozzle showing the key elements.

Table II. Chemical Analysis of the Tin

Element	Wt. %
Al	0.0003
Ag	0.0006
B	0.0004
Cu	0.0005
Fe	0.004
In	0.04
Mg	0.001
Na	0.0001
Pb	0.03
Sb	0.06
Si	0.002

an argon gas cover in 175 g batches external to the nozzle assembly. When the metal was properly heated, the argon shield was removed and the tin poured into the running atomizer. The metal head was not adjusted, thus the melt atomized at its own rate and the total atomization time recorded to calculate the siphoning rate. The resulting powder was collected and screen analyzed for size using standard techniques. Some variation in the funnel length occurred inadvertently and this is thought to be the major source of error in the experiments.

Results and Discussion

Particle Shape

Thompson (7) has previously established that particle shape at a given size was common among atomization runs. As Figure 2 shows this commonality

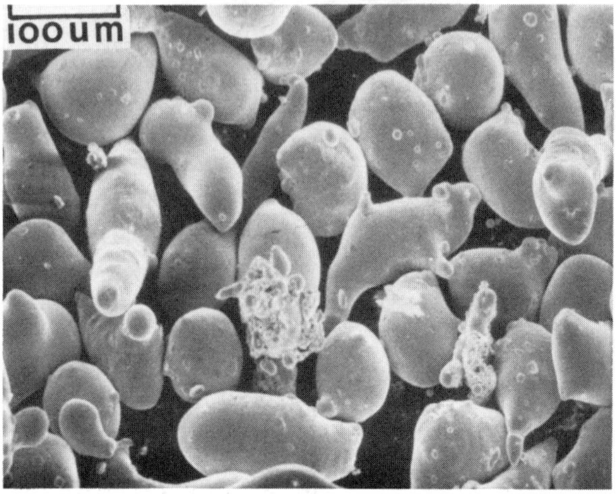

a

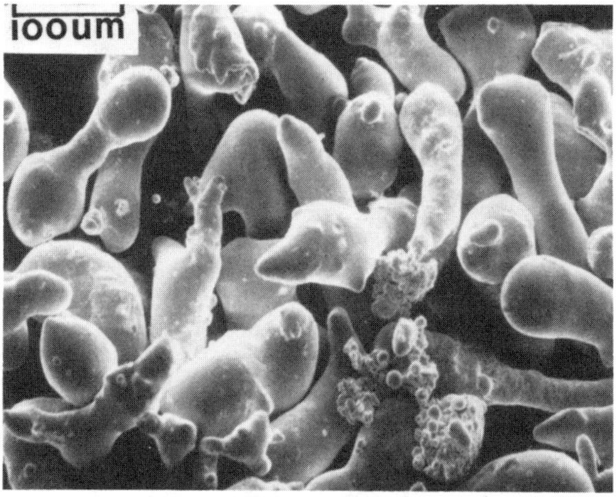

b

Fig. 2: Comparison of particle shapes from different atomization runs. Both are -200+270 mesh fractions (74 to 53μm); a) 1.7 MPa argon, 150°C superheat, 100μm gap, b) 1.7 MPa argon, 150°C superheat, 200μm gap.

of shape is true for the tin powder generated in this study. Figure 3 shows a typical variation in particle shape with size for powder atomized with 1.7 MPa argon at a 100°C superheat and 200 μm gap opening. The coarse powder is flake shaped, resulting from droplet collision with the chamber walls before solidification. Figure 4 schematically summarizes the shape changes with size for the tin powders. At particle sizes of approximately 400 μm the particles transform to an ellipsoidal shape. Thus, we can assume for particles below 400 μm, solidification occurred during free-fall in the chamber. At progressively finer particle sizes the particle shape is predominantly ligamental until the maximum dimension falls below approximately 40 μm. At particle sizes below 100 μm there are clusters or agglomerates of fine particles, probably resulting from turbulence near the nozzle. Finally, with the finer particle sizes there are several occurrences of satellite particles. The satellite particles result from collisions between solid, fine particles and unsolidified larger particles away from the atomization zone. The finer particles quench most rapidly during free-fall, but they also decelerate rapidly. (9-11) Alternatively, the coarser particles solidify slower but maintain their exit velocities, hence collisions occur during free-fall. Consequently the larger, semi-liquid particles pick up highly adherent solid fine satellites. These shapes are expected to be dependent on the amount of superheat, chamber size and turbulence near the atomization nozzle.

Size Analysis

The raw size data were in the form of the weight of each mesh size fraction for a given run. The data were transformed into cumulative weight percent after which a least squares fit of cumulative weight percent versus log particle size was generated on a log-normal coordinate system. Figure 5 shows an example log-normal plot, showing the mean particle size, slope, intercept, and correlation coefficient. Typically the correlation coefficient for the log-normal data fit exceeded 0.99 (41 out of 56 cases). Appendix I lists the atomization results in detail.

Two-dimensional plots between the operating parameters and the slopes or intercepts of the log-normal plots were reviewed to suggest variable inter-relations. Most of the two-dimensional graphs showed no significant trends, with the exception of slope versus pressure and slope versus flow rate. The trend is for the size distribution slope to increase as either flow rate or pressure increase. Another important observation is that the size distribution is narrower as the average particle size gets smaller as shown in Figure 6. These trends being the only easily visible ones, suggests that pressure and its relation to flow rate is the most important parameter (a significance level of .998 as will be shown later). Based on scatter in the superheat plots, superheat is treated as the least significant factor (a significance level of .394 also will be shown later). In contrast, Rao and Mehrotra (12) showed superheat had some effect on particle size of pure tin; increases in superheat gave slight size decreases.

Statistical Analysis

Standard computer based statistical packages were used to perform linear regressions to determine statistical significance for each of the variables. Since the computer treats each value equally, there was only one representation from each set of operating parameters (data from multiple runs was averaged before being entered). The operating parameters for each run were coded and entered with the particle diameters to perform a regression of each set of variables against one another. From this program the level of significance for each of the operating parameters was determined as fol-

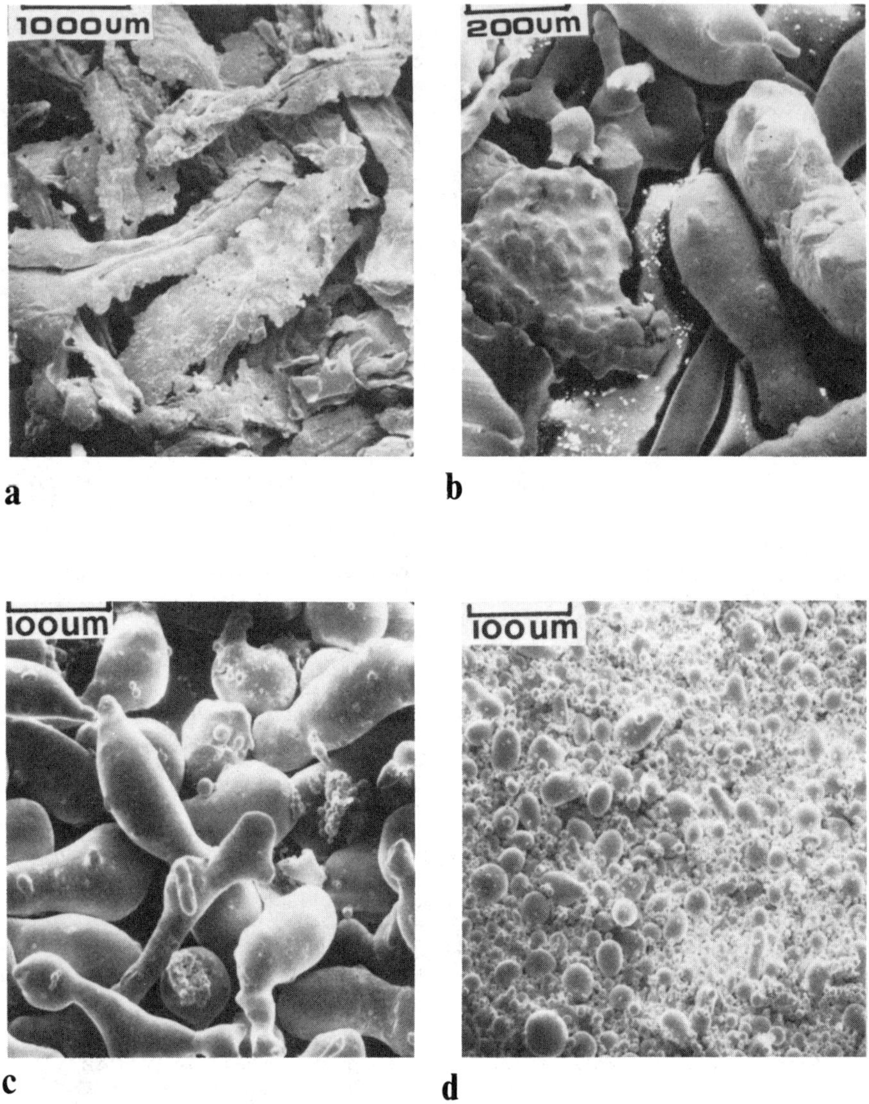

Fig. 3: Scanning electron micrographs showing the shapes of various size fractions for tin atomized with 1.7 MPa argon, 100°C superheat and a 200μm gap. a) -20+45 mesh (840 to 350μm), b) -60+80 mesh (250 to 177μm), c) -200+270 mesh (74 to 53μm), and d) -400 mesh (below 37μm).

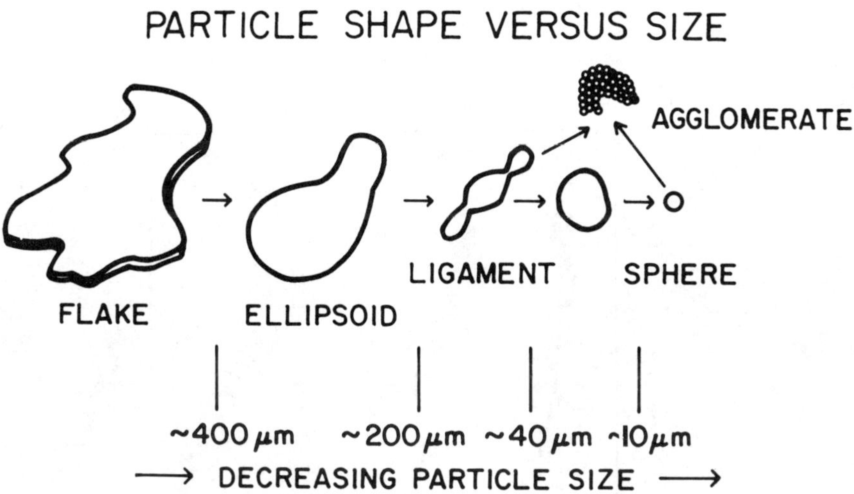

Fig. 4: Schematic diagram of the particle shape changes with particle size as observed in this study.

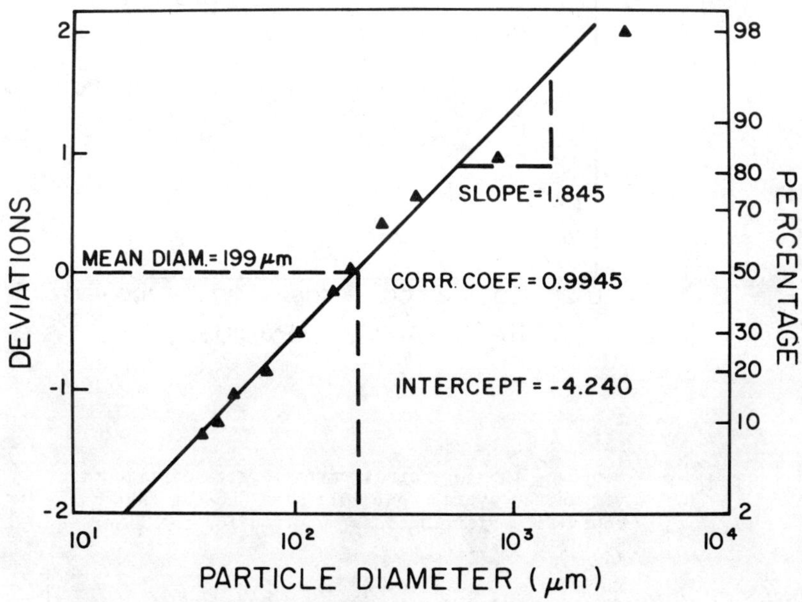

Fig. 5: Example log-normal particle size distribution for atomized tin powder fabricated by argon atomization at 1.7 MPa, 150°C superheat and 100μm gap.

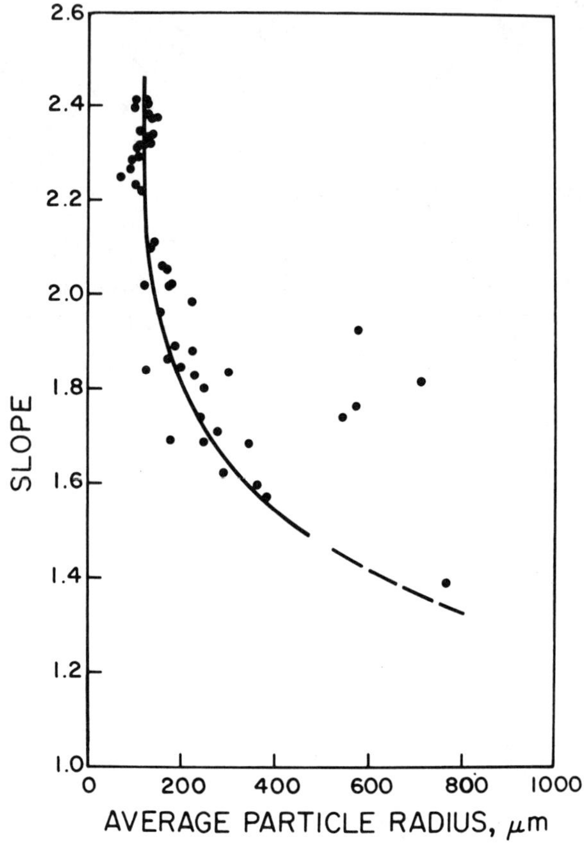

Fig. 6: Slope of the log-normal particle size distribution versus the average particle size showing that narrower size distributions occurred for the finer particle sizes.

lows: superheat (as expected was very low) = 0.394, pressure (very high) = 0.998, gap size (highly significant) = 0.990, and gas = 0.950.

More information was still needed regarding the interactions of the parameters and their exponential level (linear, quadratic or cubic). Further regression analysis was used to generate the significance level of the interactions and exponential levels. This information gives some feeling of how the individual parameters and the interactions between them will affect the final equation. Pressure was found to be the most significant parameter accounting for 55% of the variance in the data, with 92.4% being linear, the rest quadratic. Gap size was the next highest significance, accounting for 23% of the variance with 87.2% being linear, 9.2% quadratic, 2.4% cubic, and 1.2% undeterminable. The only other important parameter was found to be the interaction between pressure and gap size accounting for 9% of the variance. All other parameters, gas, superheat, and the interactions between superheat and pressure, gas and gap size, superheat and gap size, pressure and gap size, gas pressure and gap size, superheat gas and gap size, and superheat pressure and gap size, were found to be of low significance levels and thus not dependable.

Based on this information it was believed that the equation was a linear relationship of gap size and pressure with some small dependence on the gas. An operational form for the equation was developed next, based on the physics of atomization, after which the computer was used again to generate the best fit. The next section deals with the development of the physical model, and its correlation with the data.

Physical Model

Based on the statistical work, superheat only accounted for 1.5% of the variance. Therefore, it was not considered a significant factor in the mathematical model. Superheat was used only as an experimental operating parameter. This is not to say that in all atomization the superheat is not significant, but in our atomization of pure tin, and possibly other pure metals, it does not play a significant role. When dealing with alloy systems, superheat may be a more important parameter.

Neglecting superheat, it was assumed that all the energy input into the system was contained in the kinetic energy of the gas, thus:

$$\text{Kinetic Energy in Gas} = E_{gas} = \frac{1}{2} m_g V^2 \qquad [2]$$

with m_g = mass of gas

and V = velocity of gas.

The maximum velocity of the gas can be calculated at the point where it passes through the opening of the nozzle; assuming the gas to be incompressible:

$$V = \frac{\text{Flow Rate (STP)}}{\text{Area of Opening}} = \frac{F}{2\pi R_2 G} \qquad [3]$$

where R_2 = annulus radius

G = gap size, and

F = gas flow rate (at STP) in volume/time.

This approximation is made since the solution of the compressible flow problem requires knowledge of boundary conditions, choking conditions, and complex two-phase flow which has not been determined for the present nozzle geometry. Flow rate is a complex function of pressure, gap size and gas; it is most sensitive to pressure thus the high statistical significance. The velocity at the time of impact with the melt stream will not be the same as when it passes through the nozzle. (8) Allowance was made for this in terms of an efficiency constant, C which compensates for both the non-perfect energy transfer and the loss of gas velocity prior to collision with the melt stream.

The energy from the gas is transferred to the melt stream, thus creating surface area:

$$E_{gas} = C\gamma\Delta A \qquad [4]$$

where γ = surface energy, and

ΔA = change in surface area.

The initial shape of the melt was a cylinder, and the final surface area that of the powder. For simplicity it will be assumed that the particles are spherical, thus:

$$\Delta A = A_{final} - A_{initial} \qquad [5]$$

or

$$\Delta A = 4\pi\bar{r}^2 n - 4\pi R_1 H \qquad [6]$$

with $\bar{r}$ = average particle size,

n = number of particles,

R_1 = melt stream radius, and

H = melt stream height.

It can be shown that use of an average particle radius is valid for log-normal powders even though there actually is a range of particle sizes and shapes. Assuming that the initial surface area (the melt column) is small, it can be ignored when compared with the surface area of the powder, giving;

$$\Delta A = 4\pi\bar{r}^2 n. \qquad [7]$$

However, during atomization, volume of the metal is conserved. If the melt stream is allowed to fall undisturbed it will eventually break apart into ligaments which will speroidize. Using Rayleigh's relationship and conservation of volume gives,

$$1.88 \left(\frac{4}{3}\right) \pi R_1^3 = \frac{4}{3} \pi \bar{r}^3 n \qquad [8]$$

or

$$n = \frac{1.88\, R_1^3}{\bar{r}^3}. \qquad [9]$$

Substituting n back into the expression for ΔA yields:

$$\Delta A = \frac{7.25\ \pi R_1^3}{\bar{r}} \qquad [9]$$

And substituting ΔA back into the energy transfer expression yields:

$$\frac{1}{2} mg \left(\frac{F}{2\pi R_2 G}\right)^2 = \frac{C(7.52\ \pi R_1^3)}{\bar{r}} \qquad [10]$$

Collecting constants and solving for $\bar{r}$, gives

$$\bar{r} = C_1 \left(\frac{G}{F}\right)^2 . \quad [11]$$

To determine the nature of the relationship between $\bar{r}$ and $(G/F)^2$ a plot of $\bar{r}$ vs. $(G/F)^2$ was constructed. As shown by the plot in Figure 7 the relationship between $\bar{r}$ and (G/F) is log-log. The above expression shows a physical representation of the relationship between the mean particle size and the ratio of the gap size to the gas flow rate.

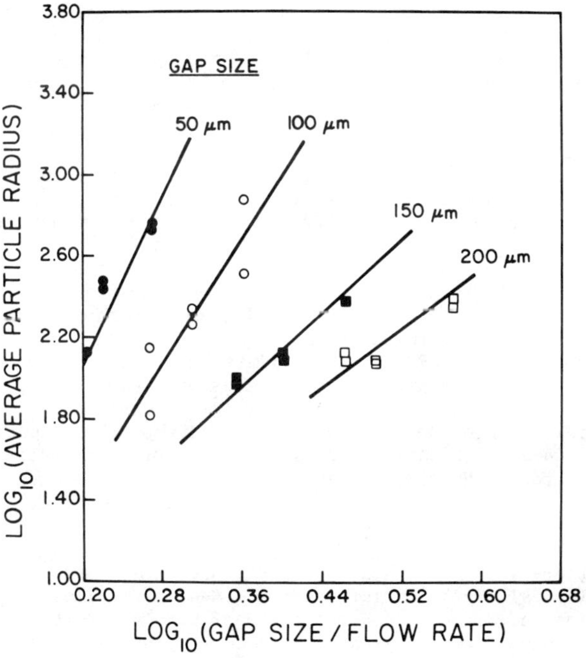

Fig. 7: Average particle size versus the gap size to flow rate ratio (log-log basis) for four gap openings.

A few of the assumptions require highlighting. In development of the model assuming all particles were spherical was not accurate; however, the smaller the particle size gets the more spherical the shape as demonstrated in Figure 3. The predominant particle shape was ligamental so some shape factor to compensate for this might give a small change in the results. Satellites and agglomerates were also ignored in development of the model; these particles tend to skew the size distribution data to larger sizes. Neglecting the initial surface area of the melt stream when compared with the surface area of the powder was a safe assumption as the surface area of the powder will be about 700 times that of the melt stream before atomization.

To determine the validity of this mathematical model a fit to the experimental data was performed as shown in the next section.

Computer Fit

The computer was used to generate the best fit of the mathematical model to the experimental data via regression analysis. The program optimizes the fit by adjusting the constants to minimize the variance in the equation,

$$\log \bar{r} = n \log (G/F) + k \qquad [12]$$

where n and k are the constants to be determined. To keep the units consistent, F was input in millimeters per second and $\bar{r}$ in millimeters. Because the computer considers each measurement equally, only 47 cases were used to avoid the skewing caused by multiple cases with common parameters. The computer determined n to be 0.807 and k to be 7.489. The correlation coefficient of predicted versus actual particle size was 0.890. It is fairly significant that the particle size dependence on the gap size to flow rate ratio is 0.807 as compared to the model value of 2.0. This discrepancy is attributed to the assumed simplified atomization model wherein we have ignored the energy for particle acceleration out of the nozzle, momentum conservation and the decay in gas velocity with distance. However, the concept of energy transfer from the gas to liquid has successfully guided this first analysis of annular, confined, external gas atomization. Based on these results, it is possible to tailor the atomization conditions for producing specific particle sizes. It should be noted that this same data was computer fit to the models presented by Small and Bruce, (2) Kishidaka, (21) and Tamura and Takeda. (19,22) These earlier models predict the particle size dependence on the specific flow properties of the atomizing fluid. The quality of fit was assessed through the correlation coefficient. The correlation coefficients ranged between 0.244 and 0.447 for these models. Thus, the model represented by Equation [12] is found to provide better agreement with data taken from an annular gas atomizer.

Conclusions

Although there is a great deal of literature on gas atomization, surprisingly little deals with annular systems. Annular nozzles are very common in production and are used on most types of metals. Particle shape follows a progression towards a sphere with decreasing particle size. In this study, flakes, ellipsoids, ligaments, spheres and agglomerates were observed at the different particle sizes.

This study has presented an energy transfer model for atomization based on data collected from an annular system. A correlation of 0.890 for the plot of predicted versus experimental particle size shows the model to be a good first representation of the system.

The statistical methods used to analyze the data proved invaluable. The regressions showed superheat to be an insignificant factor. They also showed pressure to be the most significant factor (0.998 significance level), gap size to also be significant (0.990 level), and the gas to be only mildly significant (0.950). The model developed was therefore based on pressure and gap size, using a flow rate term to incorporate the gas pressure. Regression techniques were used to best fit the mathematical model to the data. The results show that particle size can not be predicted with any accuracy using past multiple jet models. An analysis based on energy transfer has proven more successful.

Appendix I: Atomization Data

Run #	Gas	Gap Size mm	Superheat Degrees C	Pressure MPa	Gas Flow m^3/sec	Metal Flow kg/min	M.F/G.F	Slope Log 10	90% Slope	Intercept	90% Int.	Correlation	D Ave. µm
1	Nitrogen	.100	150.	1.0	.0038	NA	NA	1.573	0.08	-4.060	0.19	.9937	381
2	Nitrogen	.100	150.	1.0	.0038	NA	NA	1.392	0.08	-4.012	0.19	.9919	762
3	Nitrogen	.100	100.	1.7	.0084	NA	NA	1.890	0.07	-4.288	0.15	.9971	186
4	Nitrogen	.100	150.	1.7	.0084	NA	NA	1.879	0.08	-4.410	0.18	.9959	223
5	Nitrogen	.100	100.	2.4	.0161	NA	NA	2.108	0.07	-4.535	0.16	.9974	142
6	Nitrogen	.100	150.	2.4	.0161	NA	NA	2.247	0.26	-4.100	0.59	.9696	67
7	Argon	.100	100.	1.0	.0028	2.76	986.	1.629	0.05	-4.171	0.12	.9975	364
8	Argon	.100	150.	1.0	.0028	2.85	1018.	1.597	0.06	-4.085	0.12	.9973	361
9	Argon	.100	100.	1.7	.0066	4.80	727.	2.016	0.08	-4.183	0.17	.9967	119
10	Argon	.100	150.	1.7	.0066	4.15	629.	1.844	0.09	-4.240	0.20	.9945	199
11	Argon	.100	100.	2.4	.0136	3.54	260.	2.779	0.29	-5.625	0.64	.9767	106
12	Argon	.100	150.	2.4	.0136	3.70	272.	2.231	0.15	-4.467	0.33	.9899	101
13	Argon	.150	100.	1.0	.0040	4.20	1050.	2.064	0.05	-4.594	0.12	.9984	168
14	Argon	.150	150.	1.0	.0040	2.70	675.	1.687	0.08	-4.035	0.19	.9942	246
15	Argon	.150	100.	1.7	.0081	3.64	449.	2.402	0.13	-5.056	0.30	.9930	127
16	Argon	.150	150.	1.7	.0081	3.06	378.	2.059	0.09	-4.525	0.20	.9959	158
17	Argon	.150	100.	2.4	.0155	3.45	223.	2.725	0.28	-5.418	0.63	.9766	98
18	Argon	.150	150.	2.4	.0155	3.00	194.	1.218	0.24	-2.489	0.55	.9194	111
20	Nitrogen	.150	150.	1.0	.0048	3.40	708.	1.740	0.08	-4.144	0.18	.9954	241
21	Nitrogen	.150	100.	1.7	.0099	3.00	303.	2.381	0.15	-4.995	0.33	.9912	125
22	Nitrogen	.150	150.	1.7	.0099	3.14	317	2.093	0.10	-4.442	0.21	.9953	133
23	Nitrogen	.150	100.	2.4	.0167	2.90	174.	2.395	0.07	-4.762	0.16	.9980	97
24	Nitrogen	.150	150.	2.4	.0167	2.60	156.	2.846	0.29	-5.685	0.51	.9858	99
25	Nitrogen	.150	150.	1.7	.0081	2.50	309.	2.265	0.09	-4.401	0.21	.9963	88
26	Nitrogen	.150	150.	1.7	.0081	2.49	307.	2.310	0.11	-4.631	0.25	.9948	101
27	Argon	.050	100.	1.0	.0015	3.00	2000.	1.761	0.04	-4.855	0.09	.9987	571
28	Argon	.050	150.	1.0	.0015	3.78	2520.	1.815	0.05	-5.174	0.11	.9983	709
29	Argon	.050	100.	1.7	.0046	3.25	707.	1.983	0.05	-4.655	0.12	.9984	223
30	Argon	.050	150.	1.7	.0046	3.73	811.	1.623	0.07	-3.998	0.16	.9957	290

Appendix I continued

31	Argon	.050	100.	2.4	.0090	4.57	508.	1.862	0.08	-4.146	0.18	.9956	168
32	Argon	.050	150.	2.4	.0090	4.28	476.	1.691	0.07	-3.799	0.16	.9959	176
33	Argon	.050	100.	1.0	.0019	3.29	1732.	1.919	0.06	-5.297	0.14	.9977	575
34	Nitrogen	.050	150.	1.0	.0019	3.30	1737.	1.739	0.09	-4.755	0.20	.9937	542
35	Nitrogen	.050	100.	1.7	.0061	3.18	521.	1.709	0.06	-4.175	0.14	.9969	277
36	Nitrogen	.050	150.	1.7	.0061	4.23	693.	1.823	0.07	-4.521	0.15	.9971	302
37	Nitrogen	.050	100.	2.4	.0092	4.09	445.	2.806	0.37	-5.954	0.83	.9620	132
38	Nitrogen	.050	150.	2.4	.0092	4.84	526.	2.804	0.39	-6.007	0.88	.9580	138
39	Nitrogen	.050	100.	2.4	.0092	2.12	230.	2.382	0.15	-5.026	0.33	.9915	129
40	Nitrogen	.050	100.	2.4	.0092	3.67	399.	2.375	0.16	-5.034	0.36	.9898	132
41	Nitrogen	.200	100.	1.0	.0052	4.21	810.	1.802	0.07	-4.315	0.16	.9966	248
42	Nitrogen	.200	150.	1.0	.0052	3.60	692.	2.850	0.45	-6.327	1.01	.9477	166
43	Nitrogen	.200	100.	1.7	.0106	2.87	271.	2.412	0.11	-5.042	0.25	.9951	123
44	Nitrogen	.200	150.	1.7	.0106	4.35	410.	2.767	0.35	-5.768	0.79	.9649	122
45	Nitrogen	.200	100.	2.4	.0141	4.04	287.	2.318	0.19	-4.935	0.42	.9852	135
46	Nitrogen	.200	150.	2.4	.0141	4.82	342.	1.838	0.12	-3.848	0.26	.9907	124
47	Argon	.200	100.	1.0	.0059	3.28	556.	2.501	0.08	-5.284	0.18	.9977	130
48	Argon	.200	150.	1.0	.0059	3.29	558.	2.015	0.07	-4.521	0.16	.9970	175
49	Argon	.200	100.	1.7	.0084	4.89	582.	2.178	0.12	-4.688	0.28	.9927	142
50	Argon	.200	150.	1.7	.0084	2.91	346.	2.290	0.12	-4.637	0.27	.9939	106
51	Argon	.200	100.	2.4	.0104	4.65	447.	2.040	0.09	-4.447	0.19	.9959	151
52	Argon	.200	150.	2.4	.0104	2.47	238.	2.286	0.07	-4.458	0.16	.9977	89
53	Nitrogen	.200	150.	1.7	.0106	4.63	437.	2.322	0.18	-4.901	0.41	.9858	129
54	Nitrogen	.200	150.	1.7	.0106	3.04	287.	2.751	0.31	-5.590	0.69	.9725	108
55	Helium	.200	100.	1.7	.0106	5.26	496.	1.710	0.05	-4.036	0.12	.9977	229
56	Helium	.100	150.	1.7	.0106	3.01	284.	2.324	0.14	-4.822	0.32	.9913	119

References

1. A. Lawley, "Atomization of Specialty Alloy Powders", J. Metals, 33, No. 1 (1981) pp. 13-18.

2. S. Small and T. J. Bruce, "The Comparison of Characteristics of Water and Inert Gas Atomized Powders", Inter. J. Powder Met., 4, No. 3 (1968) pp. 7-17.

3. U. R. Date, G. S. Tendolkar, and M. N. Vartak, "A Statistical Interpretation of Metal Atomization", Inter. J. Powder Met., 3, No. 2 (1967) pp. 49-58.

4. J. A. Tallmadge, "Powder Production by Gas and Water Atomization of Liquid Metals", pp. 1-32 in Powder Metallurgy Processing, H. A. Kuhn and A. Lawley (eds.), Academic Press, N.Y., 1978.

5. E. Klar and W. M. Shafer, "High-Pressure Gas Atomization of Metals", pp. 57-68 in Powder Metallurgy for High Performance Applications, J. J. Burke and V. Weiss (eds.), Syracuse University Press, 1972.

6. P. U. Gummeson, "High-Pressure Water Atomization", pp. 27-55, ibid.

7. J. S. Thompson, "A Study of the Process Variables in the Production of Aluminum Powder by Atomization", J. Inst. Metals, 74 (1948) pp. 101-132.

8. J. B. See, J. C. Runkle and T. B. King, "The Disintegration of Liquid Lead Streams by Nitrogen Jets", Met. Trans., 4 (1973) pp. 2669-2673.

9. J. K. Beddow, The Production of Metal Powders by Atomization, Heyden, London, 1978.

10. S. P. Mehrotra, "Mathematical Modeling of Gas Atomization Process for Metal Powder Production", Powder Met. Inter., 13 (1981) pp. 80-84 and pp.

11. H. Schmitt, "Mathematical-Physical Considerations Regarding the Production of Metal Powders for Powder Metallurgy", Powder Met. Inter., 11 (1979) pp. 17-21 and pp. 68-71.

12. K. P. Rao and S. P. Mehrotra, "Effect of Process Variables on Atomization of Metals and Alloys", pp. 113-130 in Modern Developments in Powder Metallurgy, H. H. Hausner, H. W. Antes and G. D. Smith (eds.), Metal Powder Industries Federation, Princeton, 1981.

13. D. Bradley, "On the Atomization of Liquids by High-Velocity Gases", J. Phys. D: Appl. Phys., 6 (1973) pp. 1724-1736 and pp. 2267-2272.

14. N. Dombrowski and W. R. Jones, "The Aerodynamic Instability and Disintegration of Viscous Liquid Sheets", Chem. Eng. Sci., 18 (1963) pp. 203-214.

15. R. A. Castlemen, Jr., "The Mechanism of the Atomization of Liquids", National Bur. of Standards J. Res., 6 (1931) pp. 369-376.

16. O. S. Nichiporenko and Y. I. Naida, "Disintegration of a Metal Jet by a Gas Stream", Soviet Powder Met. Metal Ceram., 6 (1967) pp. 973-977.

17. A. F. Silaev, "Mechanism of Disintegration of Molten Metal Streams by a Gas Jet", Soviet Powder Met. Metal Ceram., 6 (1967) pp. 350-353.

18. See for example A. Lawley, "Preparation of Metal Powders", Ann. Rev. Mater. Sci., 8 (1978) pp. 49-71.

19. K. Tumura and T. Takeda, "Manufacture of Irregularly-Shaped Copper Powders by Gas Atomization", Trans. Nat. Res. Inst. Met. (Japan), 7 (1965) pp. 163-164.

20. O. S. Nichipanenko, "Role of Melt Viscosity in the Formation of Powder Particles During Atomization", Soviet Powder Met. Metal Ceram., 8 (1969) pp. 931-933.

21. H. Kishidaka, "Theory of Production of Atomized Iron Powder", Proc. Multidisciplinary Meeting on Sintered Metals and Magnetic Materials, Japan Society of Powder and Powder Metallurgy, Tokoyo, 1972.

22. K. Tamura and T. Takeda, "A Study of Production of Copper Powder by Atomization", Trans. Nat. Res. Inst. Met. (Japan), 5 (1965) pp. 82-86.

THE PREPARATION OF HIGH PURITY CALCIUM-NICKEL OXIDES OF PRECISION STOICHIOMETRY

E. F. Funkenbusch, P. Singh, and B. C. Cornilsen

Department of Chemistry & Chemical Engineering
Michigan Technological University
Houghton, Michigan 49931

The preparation of high purity powders with controlled cation and anion stoichiometries is prerequisite to the study of material properties. Solid solutions in the NiO-CaO system have been prepared using a liquid mix technique to achieve formation at low temperature, maximizing cation homogeneity. Characterization of the products using differential scanning calorimetry, laser Raman spectroscopy and, in particular, X-ray fluorescence will be discussed to demonstrate the precision attainable.

Introduction

CaO-NiO powders of precision cation composition (Ca/Ni ratio) and with controlled oxygen nonstoichiometry have been prepared. High precision is necessary in order that the detailed influence of composition and structure upon the other properties can be studied. The long range objective of this work is to understand this structure-property relationship, specifically the mechanical, magnetic, and electronic properties as a function of nonstoichiometry and dopant concentration. For example, diamagnetic calcium influences the magnetic order in antiferromagnetic NiO (1,2). Furthermore, the precision measurement of nonstoichiometry is prerequisite to a study of the influence of nonstoichiometry upon properties. The latter is a very subtle effect in NiO_{1+x}, with the oxygen excess being a maximum of 0.1% (i.e. $x \leq 0.001$) (3). The development of methods to directly and precisely measure nonstoichiometry is one of our objectives.

The immediate goal is to demonstrate the liquid mix technique (LMT) (4) as a means of achieving precision control of the cation composition in the NiO-CaO system (5). Careful characterization of LMT prepared powders provides evidence of the precision attainable using the LMT. Subtle variations in cation composition and oxygen nonstoichiometry can be monitored if controlled precisely enough. It will be seen that X-ray fluorescence (XRF) measurements confirm cation composition within the empirical limits of the analytical instrumentation (± 0.1%) and provide a measure of cation homogeneity. Differential scanning calorimetry (DSC) and laser Raman spectroscopy (LRS) are used to further confirm and characterize the cation composition. In turn, these studies can provide point defect structural information.

High cation homogeneity, i.e. nickel-calcium mixing at the atomic level, is another beneficial result of the LMT (4). Maximum cation homogeneity and the presence of undesirable second phases are ascertained. LRS observation of low level second phases will be discussed. In short, careful preparation and characterization of materials are required if one wishes to effectively study structure-property relationships. The preparation technique and characterization methods discussed are applicable to other oxide systems as well.

Experimental

Instrumentation and Materials

XRF data have been measured using a Philips 1L AXS automated X-ray wavelength spectrometer.* Nickel intensities have been counted using the K_β line and an operating voltage of 30 Kv at 10 ma. The K_β line was used for calcium counting at 25 Kv and 10 ma. The flow proportional counter and the LiF(200) analyzing crystal were used for both elements. Background points were taken and background corrections were made with a linear background. DSC and LRS data were collected using instruments and conditions described in the literature (1,2). "Puratronic" Johnson-Matthey†, high purity carbonates of calcium (56.03 w/o CaO) and nickel (hydrated and 57.71 w/o NiO) were used as starting materials. These each contained less

*Philips Electronic Instruments, 85 McKee Drive, Mahwah, NJ 07430
†Johnson Matthey Chemicals Limited, Orchard Road, Royston, Hertfordshire, SG8 5HF, England

than 10 ppm observed cation impurities. The cation compositions were determined in triplicate by thermal gravimetric decomposition under oxygen. Baker Analyzed reagent grade citric acid, monohydrate, was used. The oxygen partial pressures were controlled to define the oxygen nonstoichiometry using standard techniques (3).

The Liquid Mix Technique

The first step in this preparation technique involves formation of an aqueous solution containing both cations complexed in citric acid. Carbonates dissolve in citric acid solution via a simple acid-base reaction. Other solvents can be used if the cations of interest hydrolyze in water (4). The high precision of this method is attained in this step. For example, one can weigh a 50 gram sample to better than ± 0.5 mg precision on an analytical balance and obtain a final precision of ± 10 ppm. The relative precision of two sample compositions is greater than that obtained by chemical analysis. An excellent example of the inter-sample precision and absolute accuracy obtainable using this LMT has been given by Eror and Loehr (6). Overall accuracy depends upon the analyses for cation content of each carbonate and simply shifts the entire composition curve. This does not affect intersample precision.

The second step in the LMT procedure involves slow heating of the solution to remove solvent and to form a solution of increasing viscosity. The resulting "glassy" product is homogeneous and mixed at the atomic level. Oxidation of any remaining organics at a relatively low temperature, e.g. 900° C, completes the process. No precipitation should be extant during this process, as it would lead to drastic inhomogeneities. This fine grained powder can be further processed, for example, pressed and sintered under controlled oxygen partial pressures.

Three additional mixtures have been prepared for comparison with the LMT powders. These provide information about the influence of cation inhomogeneity and matrix effects upon the XRF analyses. Two are physical mixtures, a carbonate mixture (CM) and an oxide mixture (OM). The third was sintered at 900° C, after mixing the carbonates, to form a sintered oxide product (SOM). A Spex Mixer* was used to mix powders in glass using a plastic mixing ball.

Results

XRF analyses of the calcium and nickel compositions were carried out for each of the four sets of samples, LMT, SM, OM, and SOM (Figure 1). Figure 2 depicts the LMT nickel and calcium X-ray fluorescence peak intensities as a function of "as-weighed composition". The LMT nickel intensity as a function of nickel composition is linear. The calcium intensity is non-linear but the data are fit using a non-linear term which is proportional to the nickel concentration, indicating nickel absorption of the calcium fluorescence. Figure 1 compares the XRF results for the four different oxide sets. The scatter in the LMT data is less than that for the other three sample sets. This scatter is quantified in terms of the standard deviation (σ) of the composition for each data set. σ measures the deviation from the best fit linear least squares line for each set and is 0.6, 1.7, 5.0 and 7.1 for the LMT, CM, OM, and SOM sets, respectively. (The deviation of the line from zero for the y-intercept also is a measure

*Spex Industries, Inc., Box 798, Metuchen, NJ 08840

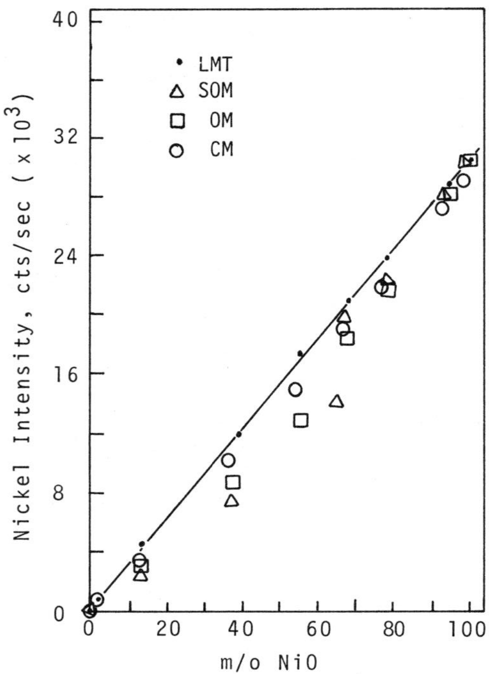

Figure 1 - Nickel X-ray fluorescence intensities as a function of composition and comparing the LMT results with those for the SOM, OM, and CM powders.

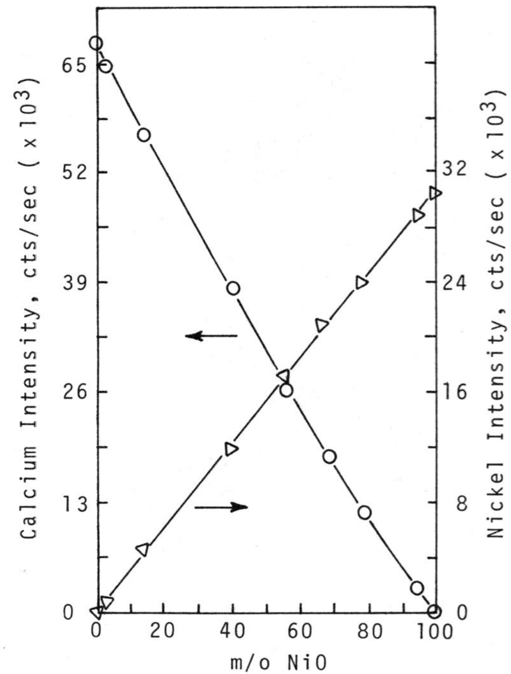

Figure 2 - XRF intensities in LMT oxides. The non-linear standard deviation in composition is 0.38 m/o for each curve (0-100%). The non-linear term for each cation is proportional to the intensity of the other cation.

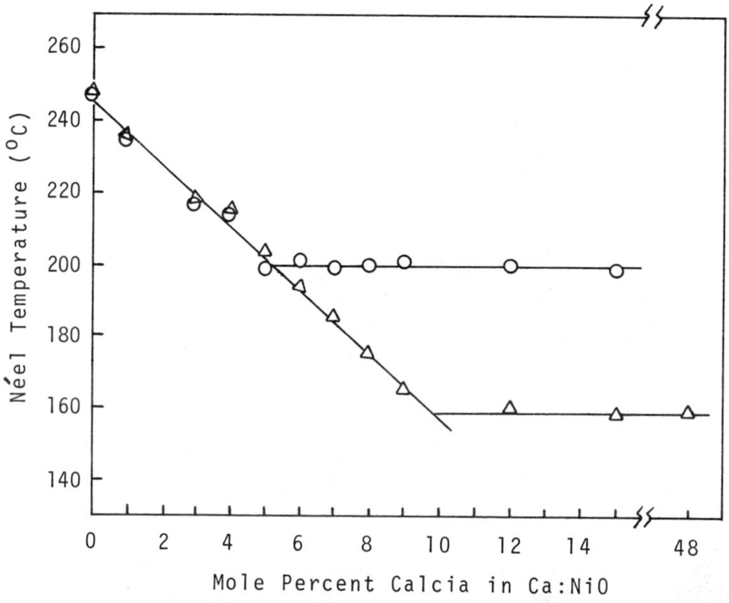

Figure 3 - The variation of the Néel temperature for Ca:NiO as a function of calcium concentration for samples quenched from 1300°C (○) and 1500°C (△).

of the precision and is closest to zero for the LMT sample.) The three non-LMT sample sets produce curves that consistently fall below the LMT line.

Discussion

These XRF data demonstrate the precision of the LMT preparation technique as compared with samples prepared via the three other methods. The deviations observed for the CM, OM, and SOM samples reflect Ca/Ni sample inhomogeneity and variable matrix effects. The latter include the influence of particle size and density differences upon absorption (e.g. absorption of calcium radiation by nickel); these lead to the systematic shift, i.e. drop, in the intensities for the non-LMT samples (Figure 1). The improvement in precision directly reflects the greater homogeneity of the Ca/Ni ratio throughout an LMT sample.

Once the composition is controlled, defined, and verified it is possible to consider the variation in properties which depend upon composition or even nonstoichiometry. DSC has been used to monitor the heat capacity and Néel temperature (T_N) of calcium-doped NiO (2). In the NiO-rich single phase region the Néel temperature drops linearly as predicted by an Ising model (1). In the two-phase region the cation activities are fixed and no further change in T_N occurs (see Figure 3). It is therefore possible to monitor second phase formation and the position of the solidus <u>if</u> the Ca/Ni ratio is known precisely. The decreasing T_N

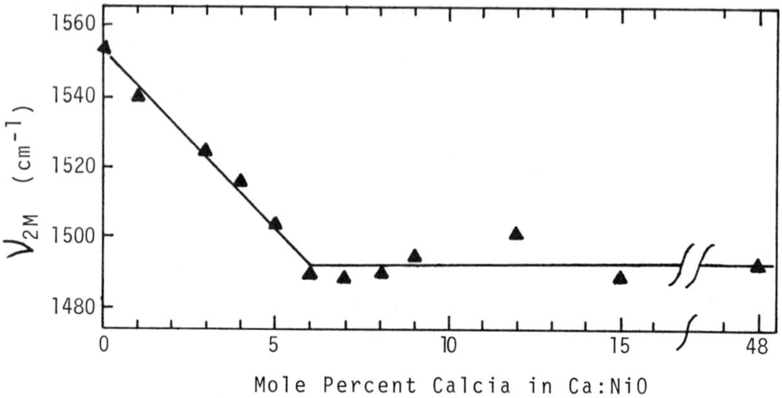

Figure 4 - The influence of calcium composition upon the Raman 2-magnon band, ν_{2M}, in Ca:NiO. Samples quenched from 1300°C.

reflects the magnetic disorder introduced by the diamagnetic calcium dopant.

Raman spectroscopy is a simple, nondestructive test which requires a minimum of sample. Raman spectra, which measure the inelastically scattered light from a sample, are complementary to infrared spectra. Disruption of the antiferromagnetic order in nickel oxide by calcium doping is reflected in the behavior of a two-magnon (8) Raman band in NiO (1). The wavenumber (or energy) of this band decreases linearly with calcium concentration (Figure 4) and this relationship can therefore be used to determine the calcium content of an unknown sample. A calcium-induced vibrational band is also observed, and its intensity is linearly related to the calcium concentration (7).

Occurrence of a CaO-rich second phase is reflected in NiO-rich DSC and LRS data by a decrease in calcium content. The appropriate calcium-dependent parameters, e.g. 2-magnon wavenumber or Néel temperature, can be monitored to follow precipitation of the calcia-rich second phase (Figure 5). This method can be applied to follow precipitation kinetics . Alternatively one can demonstrate the attainment of equilibrium. In the latter example, equilibrium was reached after ca. 60 hours. Because of the precision of the standards prepared using the LMT, together with the instrumental sensitivity, a small amount of second phase can be identified, e.g. 0.5 m/o calcia-rich precipitate is readily observed. These applications are only feasible if accurate standards with precisely known cation contents are available, as is possible using the LMT.

Conclusions

As empirically confirmed, precision control of cation stoichiometry is feasible using the LMT. Because the LMT allows homogeneous, atomic scale mixing, the diffusion distances are short and preparation temperatures and times can be minimized. The shorter times at high temperatures minimize

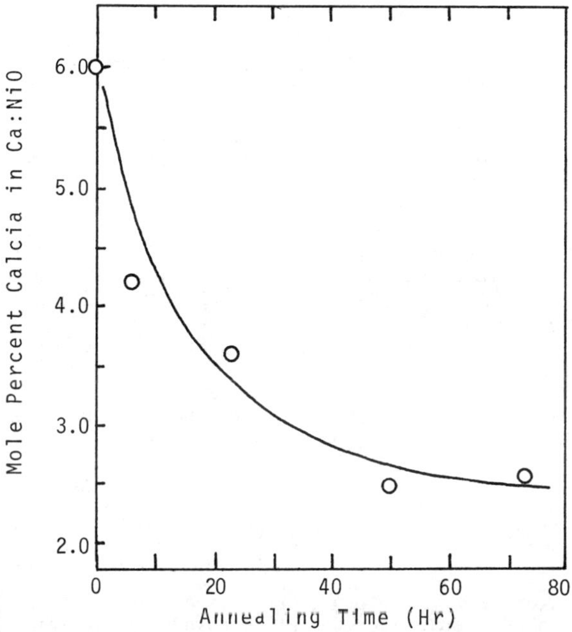

Figure 5 - Decrease in calcium composition in Ca:NiO as CaO-rich second phase precipitates during 1090°C anneal as measured using laser Raman spectroscopy.

impurity introduction at the high temperatures. A second common source of impurities is also avoided since no ballmilling is required in the LMT.

Precision control of cation composition is prerequisite to the observation of antiferromagnetic disorder and of nonstoichiometry, both oxygen nonstoichiometry and cation nonstoichiometry. The experimental techniques discussed have been used to confirm composition. Raman spectroscopy has been useful for monitoring low levels of second phase precipitation. DSC and LRS can be used to monitor the approach and attainment of equilibrium. Solvus compositions can be determined. Demonstration of the latter results depends upon the ability to prepare precise standard samples using the LMT.

To study physical properties which may be influenced by stoichiometry it is necessary to control both the cation ratio and oxygen nonstoichiometry. As demonstrated, high precision attainable from the LMT allows one to examine subtle physical property changes, including those which may be dependent upon composition and/or the point defect structure.

References

1. E. F. Funkenbusch and B. C. Cornilsen, "Two-Magnon Raman Scattering in Calcium Doped NiO," Solid State Commun., 40 (6) (1981) pp. 707-710.

2. V. Lorprayoon, W. M. Lee, and B. C. Cornilsen, "A DSC Study of the Heat Capacity and Phase Transition in Calcium-Doped Nickel Oxides," pp. 83-88 in Proceedings of the North American Thermal Analysis Society, Volume 11, NATAS, New Orleans, LA, 1981.

3. C. M. Osburn and R. W. Vest, "Defect Structure and Electrical Properties of NiO-I, High Temperature," J. Phys. Chem. Solids, 32 (6) (1971) pp. 1331-1342; "Cation Self-diffusion in Monocrystalline Nickel Oxide Under Oxygen Partial Pressure Close to the Pressure of Dissociation of the Oxide," J. Phys. Chem. Solids, 42 (1981) pp. 705-712; and references therein.

4. M. P. Pechini, "Method of Preparing Lead and Alkaline Earth Titanates and Niobates and Coating Method Using the Same to Form a Capacitor," U.S. Patent 3,330,697, July 11, 1967.

5. D. E. Smith, T. Y. Tien, and L. H. Van Vlack, "The System NiO-CaO," J. Am. Ceram. Soc., 52 (8) (1969) pp. 459-60.

6. N. G. Eror, Jr. and T. M. Loehr, "Precision Determination of Stoichiometry and Disorder in Multicomponent Compounds by Vibrational Spectroscopy," J. Solid State Chem., 12 (3/4) (1975) pp. 319-323.

7. E. F. Funkenbusch and B. C. Cornilsen, "Raman Spectroscopic Determination of Calcium Solubility Limits and Cation Composition in Calcium-doped NiO," Commun. Am. Ceram. Soc., (in press).

8. A "2-magnon" band is the overtone of a transition between two spin states in antiferromagnetic NiO.

Acknowledgments

We wish to thank Prof. D. E. Mikkola and Prof. W. I. Rose for helpful X-ray fluorescence discussions. We gratefully acknowledge the National Science Foundation for support of this research.

Consolidation

DIRECT POWDER ROLLING OF

HIGH-STRENGTH ALUMINUM ALLOY STRIP

D.H. Ro and M.W. Toaz

Imperial Clevite Inc., Technology Center
Cleveland, Ohio 44108
U.S.A.

This paper describes a process to form high-strength aluminum alloy strip directly from powder rolling. While high-strength powder metallurgy (P/M) aluminum alloy strip is not currently available, the probable conventional processing to form strip will include costly and complex steps. For example, isostatic compaction, preheat evacuation, and vacuum hot compaction could be used to form a billet which is subsequently multi-pass rolled to a final gauge desired. Thus, the development of a simpler method to produce an aluminum alloy strip would result in a significant reduction in cost and time. In the present work, X7091 strip was produced directly from an air-atomized powder to establish its applicability as an aerospace material. Experimental data will be discussed in terms of the microstructure and mechanical properties of the strip.

---

*This work was supported by the U.S. Air Force, Wright Aeronautical Laboratories under Contract No F33615-80-C-5161.

Introduction

This paper addresses a currently unmet need for low-cost, high-strength aluminum alloy sheet/strip for aircraft structures. Extensive research under government sponsorship has recently demonstrated that rapidly solidified (RS) aluminum powder metallurgy (P/M) alloys have superior engineering properties to their commercial ingot metallurgy (I/M) counterparts.[1] These properties include static and fatigue strength, fracture toughness, stress-to-rupture and creep strength, exfoliation and stress corrosion cracking resistance, and fabricability. Unfortunately, an expensive and complex experimental or developmental process is currently employed for new alloys such as X7090, X7091, MR 64 and 1519B to provide useable mill products with superior properties. This process results in projected P/M product costs of 50 to 300 percent over those of conventional I/M products. Widespread commercial implementation of aluminum P/M alloys appears likely only when favorable cost/benefit ratios are established.

While high-strength aluminum P/M alloy sheet is not commercially available at present, the probable processing steps to produce such products include isostatic compaction, encapsulation, vacuum preheat, vacuum hot compaction, and scalping to form a billet which is subsequently multi-pass rolled to the final gauge (Fig. 1). The process we will describe had been developed initially under company, and then Air Force Sponsorship.[3] It has the capability of producing economical sheet from a wide variety of current and future alloys by eliminating many of the labor - and capital - intensive steps currently used.

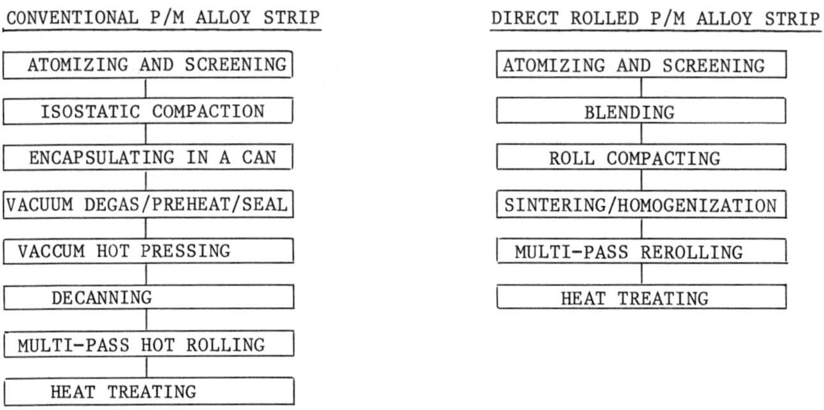

Figure 1: Comparison of the conventional process and Imperial Clevite Process for production of P/M alloy strip.

The basic process has been used commercially for over 6 years to produce a fatigue-resistant engine bearing material from RS aluminum-lead base powders.[2] The process sequence employed to produce high-strength aluminum strip is illustrated in Figure 1. The essence of the process, as applied to high-strength alloys, is that alloy powder particles are directly roll-compacted to form strip or sheet of near-final thickness. Subsequent heat treatment and rolling produces material which has properties equivalent to those of P/M forgings and extrusions based on billets produced by canning, evacuating/heating, and hot compacting, followed by decanning.

The direct powder rolling process can utilize a variety of high-strength, prealloyed aluminum alloys. However, previous IRAD work has led us to conclude that more effective densification and better properties can be achieved using a blended master alloy technique. The essence of this technique is that a highly enriched master alloy powder is blended with virtually pure aluminum powder to yield the nominal composition of the chosen alloy. The blended powder mixture is sized and roll compacted into a porous "green" strip. The "green" strip is homogenized and densified by sintering. The sintered strip then goes through multi-pass re-rolling steps to a final thickness. The final treatment is to solutionize and age the strip to raise the properties to the target potential of the alloy.

Experimental Procedures

All of the powder alloys used in this study were produced by an air atomization process. Three master alloy (Ma) powders with varying ratios of enrichment and one filler alloy (Fa) powder to be used as a common blending agent were produced. The composition of the alloys is given in Table I.

Table I: Composition of Experimental Powders

Alloy Designation	Composition - Weight percent					
	Cu	Mg	Zn	Co	Fe	Al
Ma 2	1.68	2.87	7.83	0.38	0.08	87.16
Ma 4	1.88	3.26	8.95	0.38	0.08	85.45
Ma 5	2.12	3.40	8.88	0.38	0.08	85.14
Fa 1	-	-	-	0.40	-	99.60

Master alloy powders with varying amounts of enrichment (between 20% and 50%) were blended with filler alloy powders in appropriate ratios to produce a mixture with the overall composition required for X7091. Limited experiments were conducted using three different fractions of Ma 5 powder blend in an attempt to find the optimum powder size to roll compaction. Five blended powder lots were prepared for roll compaction as shown in Table II.

Atomized aluminum powders were characterized by particle shape and by the determination of dendrite arm spacing using optical and scanning electron microscopy. The explosibility of powder materials was evaluated in accordance with the Bureau of Mines' Method (Report 5624).

The thickness of the "green" strip ranged from 1.5 mm to 2 mm. Process and quality control procedures included monitoring the powder rolling process parameters, documenting dimensions and metallographic examination of the rolled strip. The roll-compacted, "green" strip was sintered at temperatures between 510°C and 540°C for up to 12 hours in nitrogen. Sintered strips were then reduced in thickness by cold rolling. Intermediate annealing at 340°C for a half hour was used when required. Reduction in thickness ranged up to 85%.

Rectangular specimens of 1.27 cm x 10-15 cm (1/2 in x 4-6 in) with a 2.54 cm (1 in.) gauge length were cut in the longitudinal and/or transverse direction from thermomechanically processed strip for tension testing. The specimens were solution treated at 480°C for one to two hours and aged at 120°C for 24 hours. The tension tests were performed

Table II: Blended Powder Lots Prepared for Roll Compaction

Blended Powder Identification	Constituents*
8-1014-80	2 parts (Ma 5, -40)** + 1 part (Fa 1, -40)
10-1215-80	2 parts (Ma 5, -60/+100) + 1 part (Fa 1,-60/+100)
11-1215-80	2 parts (Ma 5, -100) + 1 part (Fa 1, -100)
12-122-81	5 parts (Ma 2, -100) + 1 part (Fa 1, -100)
13-122-81	5 parts (Ma 4, -100) + 2 parts (Fa 1, -100)
	*Powder Designation
	Ma 2: 20% enriched master alloy
	Ma 4: 40% enriched master alloy
	Ma 5: 50% enriched master alloy
	Fa 1: Filler alloy (99.6% Al + 0.4% Co)

** Numbers immediately following the powder designation indicate the particle size in mesh.

at room temperature in accordance with ASTM standard method E 8-79. Similar gauge (1.3mm) I/M 7075 alloy sheet was also tested for control purposes. Strip density was determined in accordance with ASTM standard method C 693-74. The grain size was determined by the mean intercept method (ASTM Standard E 112).

Atomized, screened and blended X7091 powder was roll compacted directly into "green" strip using a modified, two-wide Stanat rolling mill. A total of ten roll compaction tests were performed (Table III)*.

Table III: Roll-Compaction Conditions

Blended Powder Identification	Roll Gap,** mm	Mill Speed, rpm	Strip Thickness, mm	Surface Crack
8-1014-80	1.02	1.75	1.93	None
8-1014-80	1.52	1.75	1.93	None
8-1014-80	0.51	1.75	1.91	None
10-1215-80	0.51	1.6	1.96	None
11-1215-80	0.51	2.0	1.91	Moderate
11-1215-80	0.00	2.3	1.73	Moderate
11-1215-80	-0.51	1.8	1.83	None
12-122-81	-0.51	2.0	1.75	Severe
12-122-81	-1.02	1.6	1.78	Severe
13-122-81	-0.76	2.0	1.78	Slight

Note: * Our previous IRAD work has shown that fully pre-alloyed X7091 powder was not readily rollable.

** A negative roll gap denotes the presence of mutual roll pressure before roll-compaction.

Results and Discussion

Powder Characterization

Physical properties of the powders are given in Table IV.

Table IV: Physical Properties of Experimental Powders

		Alloy Designation			
		Ma 2	Ma 4	Ma 5	Fa 1
Sieve Analysis in Mesh (wt.%)	+40	0.89	1.62	0.96	1.20
	-40/+60	29.03	22.74	16.05	25.22
	-60/+100	36.66	32.00	32.92	34.95
	-100/+140	15.70	17.26	21.03	17.47
	-140/+200	9.91	12.94	15.25	11.39
	-200/+270	3.67	5.55	6.10	4.25
	-270/+325	2.15	3.76	3.90	2.66
	-325	1.96	4.01	3.76	2.71
Apparent Density (gm/cm^3)		1.15	1.14	1.14	1.17
Flow Rate (sec/50gm)		107	108	100	102

The morphology of the master and filler alloy powders was largely irregular with many elongated particles as shown in Figure 2. This irregularity of shape encouraged particle interlocking which provided adequate "green" strength to withstand the subsequent strip handling procedures.

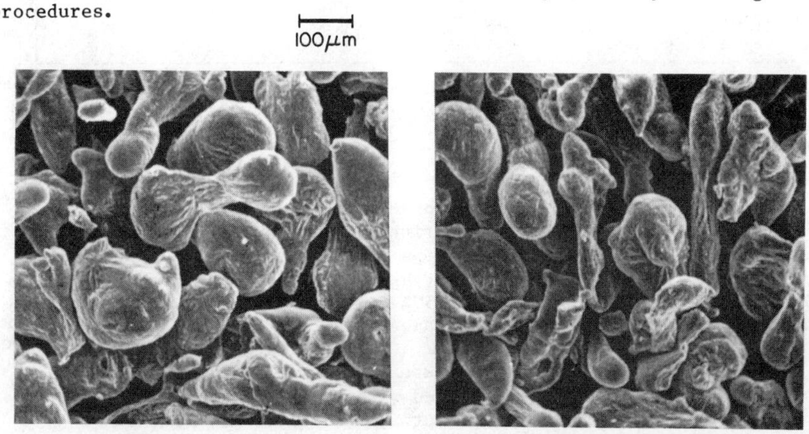

Master Alloy Filler Alloy

Figure 2: Scanning electron micrographs of representative powder particles of master and filler alloys.

Dendrite arm spacings of representative powder particles were determined to establish the approximate cooling rate of the atomized powder. The dendrite spacings of an average particle for each powder fraction were correlated with the cooling rate following a relationship shown by Flemings(4) and Dean et al(5). The results are summarized in Table V.

Table V: Variation of Cooling Rate with Dendrite Arm Spacings

Mesh Size	Particle Size Range (μm)	Dendrite Spacing (μm)	Cooling Rate (°C/sec.)
-40/+60	420/250	3.2	3.0×10^3
-100/+140	149/105	2.0	1.0×10^4
-200/+270	74/53	1.5	3.0×10^4
-270/+325	53/44	1.5	3.0×10^4
-325	< 44	1.1	8.0×10^4

The use of fine powder particles would appear to be advantageous since, in general, they have smaller dendrite spacings and consequently higher cooling rates as shown in Table V. However, extreme care must be exercised in handling such powders because of the higher explosibility risk. Table VI shows the minimum "explosive" concentration (MEC) for selected heats in comparison with pure aluminum and magnesium powders. The lower the MEC value, the greater the explosibility of the powder.

Table VI: Minimum Explosive Concentrations (MEC) for Various Heats

Alloy Designation	Mesh Size	Minimum Explosive Conc. (oz/cu ft)	-200 Mesh in Weight Percent of a Total Heat
Ma 5-1	-200	0.129	15.8
Ma 5-2	-200	0.081	20.8
Ma 5-3	-200	0.182	8.6
Fa 1-1	-200	0.145	9.4
Pure Aluminum	-200	0.045	-
Pure Magnesium	-200	0.020	-

The MEC values in Table IV appear to correlate well with the amount of fine powder particles present in each heat.

Roll Compaction

Numerous surface cracks were observed in "green" strips produced from powders employing Ma 2 (20% enriched master alloy powder). Slight surface cracks were observed in strips produced from the blended powder containing Ma 4 (40% enriched master alloy powder). Finally, strips from the blended powder employing Ma 5 (50% enriched master alloy powder) exhibited little or no visible surface cracks.

Metallographic samples were taken at the end of the rolling test in transverse cross-sections, and etched with an acid mixture of HNO_3, HF and CrO_3. This etch revealed the outline of individual powder particles, and the internal structure of the alloy particles. Figure 3-1 shows the typical microstructure of as-rolled strip employing Ma 2 powder. This strip exhibited excessive surface cracks during rolling. The microstructure in the etched condition (Figure 3) clearly revealed that the path of internal cracking was along particle boundaries. In contrast, typical microstructures obtained from "green" strip employing Ma 5 powder were crack-free (Figure 3-2). It was evident that a large number of master alloy powder particles were encapsulated by soft filler alloy powder particles (lighter background). In general, the presence of surface cracks in as-rolled strip correlated well with the presence of internal defects. Thus, the presence of cracks caused rejection of the strip for evaluation of mechanical properties.

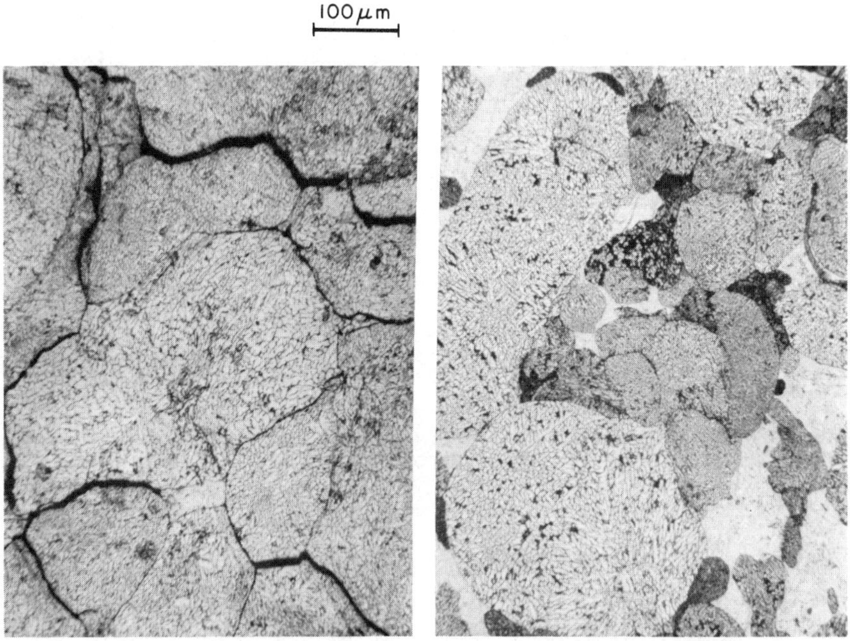

3-1: Typical internal cracks along particle boundaries observed in "green" strip

3-2: Typical crack-free microstructure observed in "green" strip employing Ma 5 masteralloy powder

Figure 3: Representative microstructures of "green" strip

There appears to be a minimum amount of soft filler alloy powder necessary to encapsulate the hard master alloy powder particles without causing surface cracks during rolling. This hypothesis gained some support from the trend that the severity of surface cracking increased with decreasing filler alloy content in blended powders. This behavior explained why no attempts were made to roll fully pre-alloyed X7091 powder.

As-rolled "green" strips exhibited a relatively uniform density ranging from 94 to 97% of the theoretical density of a X7091 alloy depending on blended lots. Furthermore, the density variation between two edges of a strip was within 2% in most cases. Density results are shown in Table VII.

Table VII: Results of Density Measurements on "Green" Strip

Blended Powder Lot No	Density, g/cm^3			Average density* in % theoretical
	Edge 1	Middle	Edge 2	
8-1014-80	2.68	2.74	2.70	96.7
8-1014-80	2.75	2.73	2.69	97.2
8-1014-80	2.71	2.55	2.67	94.5

* Theoretical density of a X7091 alloy was calculated to be 2.80 g/cm^3.

Table VIII: Results of Tension Test at Room Temperature

ROLLING TEST NUMBER	BLENDED POWDER COMPOSITION	POWDER SIZE (MESH)	TENSILE PROPERTIES							
			% REDUCTION BY RE-ROLLING	LONGITUDINAL			% REDUCTION BY RE-ROLLING	TRANSVERSE		
				UTS MNm^{-2}	YS MNm^{-2}	% ELONG.		UTS MNm^{-2}	YS MN^{-2}	% ELONG
1	2/3(Ma5)+1/3(Fal)	-40	75	580	526	12.1	-	-	-	-
2	2/3(Ma5)+1/3(Fal)	-60/+100	75	555	506	11.2	75	529	518	2.4
3	2/3(Ma5)+1/3(Fal)	-100	75	600	540	9.8	75	574	498	5.6
4	5/6(Ma2)+1/6(Fal)	-100	75	507	449	10.2	75	485	434	2.3
			85	567	512	9.5	85	572	514	6.1
5	5/7(Ma4)+2/7(Fal)	-100	75	521	474	10.9	75	533	499	3.6
			85	507	434	10.2	85	501	448	8.2
I/M 7075-T73			-	471	399	15.5	-	478	399	15.0

Room Temperature Tensile Properties

Five of the strips produced during roll-compaction tests were selected for tensile tests at room temperature. Result of tension tests are shown in Table VIII. Room temperature tensile data of a control alloy, 7075-T73, were also included for comparison.

Tensile test results were, in general, in the expected range of values with the exception of elongation which was quite low. Tensile strength and ductility increased with increasing amounts of cold work by re-rolling up to approximately 75% reduction in thickness. Transverse ductility was poor in most cases and was significantly improved after 85% reduction. Decreasing the average powder size appeared to yield improved strength. A better combination of strength and ductility was obtained from -100 mesh powder (Table IX).

Table IX: Tensile Properties of a X7091-T6 Strip as a Function of Particle Size

Tensile Properties	-100 Mesh	Particle Size -60/+100 Mesh	-40 Mesh
UTS, MNm^{-2}	600	559	580
YS, MNm^{-2}	538	504	524
% Elongation	10	11	12

In Table X compares the tensile properties of various P/M products of a similar type with conventional I/M sheet product. An examination of the properties of direct-rolled X7091 strip indicated that the properties obtained from the direct rolling experiments to date compared favorably with those generated by other consolidation techniques. This strongly suggests that with further refinement in the process, a much superior material could be developed.

Table X: Tensile Properties of P/M and I/M Aluminum Products

Products	Direction	UTS (MNm^{-2})	YS (MNm^{-2})	El.(%)
ICI(1) X7091-T6 Strip	Longitudinal	641	572	13.4
	Transverse	593	545	8.0
Alcoa (2) X7091-T7E69 Forgings	Longitudinal	572	531	15
	Transverse	538	489	3.0/5.0
Reynolds (3) X7091-T7511X Extrusions	Longitudinal	593	538	7.0
	Transverse	-	-	-
Reference(4) I/M 7075-T6 Sheet	Longitudinal	572	524	11.0
	Transverse	-	-	-

Notes: (1) Data from AF Contract No. F 33615-80-C-5161
(2) Data from AF Contract No. F 33615-74-C-5077
(3) Data from AF Contract No. F 33615-79-C-5161
(4) Data from Aerospace Structural Metals Handbook, 1973

Microstructural Analysis

Optical metallography was performed in an attempt to correlate the microstructure and the mechanical properties. Figure 4 shows the typical microstructure of strips produced from a powder blend (8-1014-80) at various levels of reduction in thickness. As-sintered strip (0%) showed the presence of porosity and more or less defined prior particle boundaries. At 35% reduction, the strip appeared to reach full density. However, prior particle boundaries were still apparent and grains of 50 μm in size are frequently observed. As the amount of cold work increased up to 75%, the average grain size became smaller and more uniform, which is believed to have contributed to a better combination of strength and ductility.

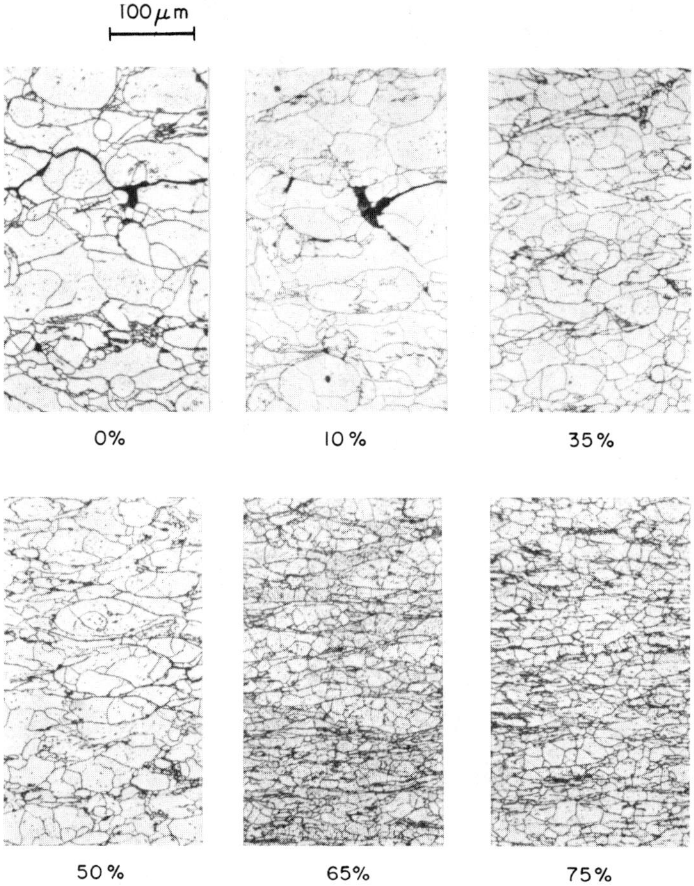

Figure 4: Representative transverse microstructure of powder-rolled X7091-T6 strip with varying amounts of cold work.

Electron probe micro analysis (EPMA) was performed on a strip which had been re-rolled to a 75% reduction. Figure 5 shows a backscattered electron image of an area containing small and large particles (light) as well as dark regions. Elemental images of oxygen, magnesium, silicon, cobalt, iron, zinc and copper were obtained to show their distribution (Figure 5 (b) - (h)). Zinc and copper appeared to be uniformly distributed. Large particles (marked A) appeared to be Co_2Al_9 intermetallic compound. A separate semi-quantitative microprobe analysis of ten particles (4$\sim$9 μm in size) showed that a particle consisted of 13.7% Co, 0.4% Zn, 82.1% Al, 3.0% Fe, 0.7% Cu, and 0.1% Si (in atomic percent). A sign of iron contamination in Co_2Al_9 particles was noteworthy. For improved static and fatigue properties along with fracture toughness, such large Co_2Al_9 particles and impurities such as iron and silicon are not desirable and must be avoided.

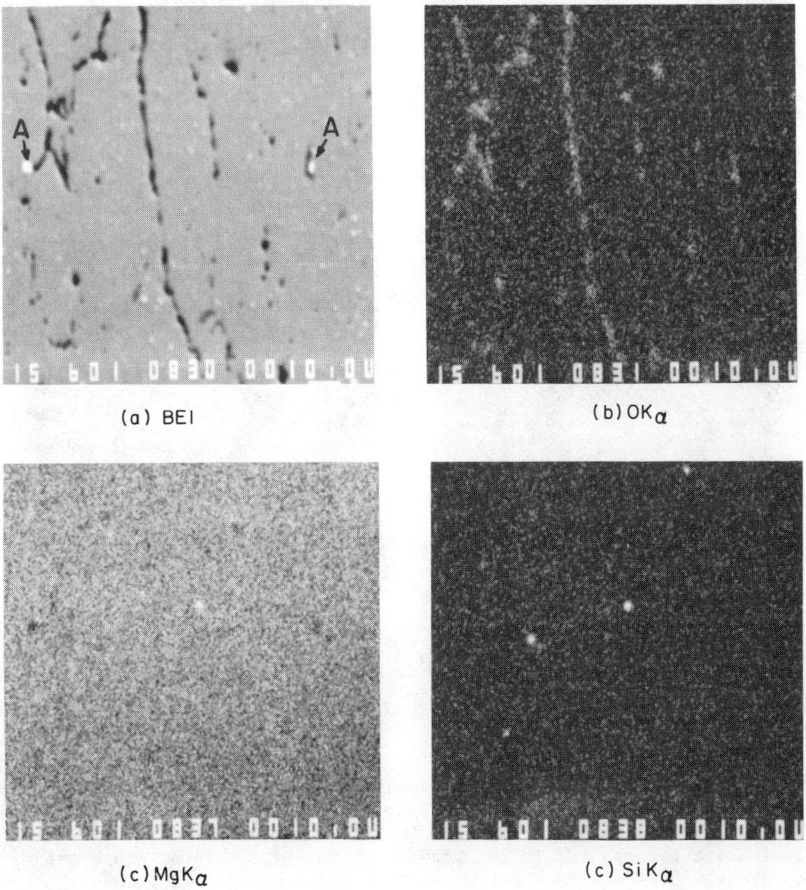

Figure 5: Electron probe microanalysis of a X7091-T6 strip
a) back scattered electron micrograph;
b) O Kα ; c) Mg Kα ; and d) Si Kα

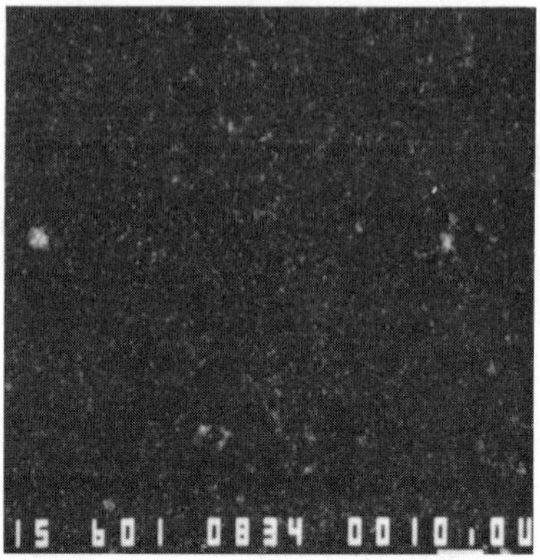

(e) CoK_{α}

(f) FeK_{α}

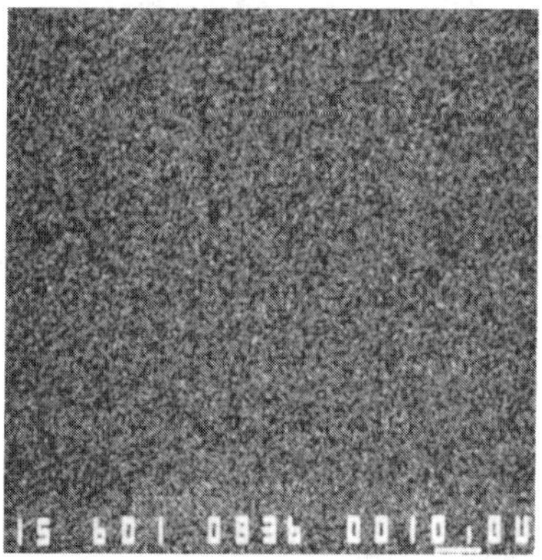

(g) ZnK_{α}

(h) CuK_{α}

Figure 5: Electron probe microanalysis of a X7091-T6 strip.
e) Co K_{α}; f) Fe K_{α}
g) Zn K_{α}; and h) Cu K_{α}

Dark regions in Figure 5(a) appeared to be related to magnesium, silicon, and oxygen images, suggesting the presence of residual compounds of mixed oxides and/or silicates. The possibility of the presence of aluminum-rich compounds in the form of oxides or silicates could not be ruled out. Line scan analysis of a similar area shown in Figure 5 indicated that magnesium and oxygen were highly concentrated at grain boundaries (Figure 6). At present we can only speculate that tenacious films of oxides formed on powder surfaces during air atomization may not have been uniformly broken up, even after thermomechanical procesing, leaving scale at powder particle boundaries.

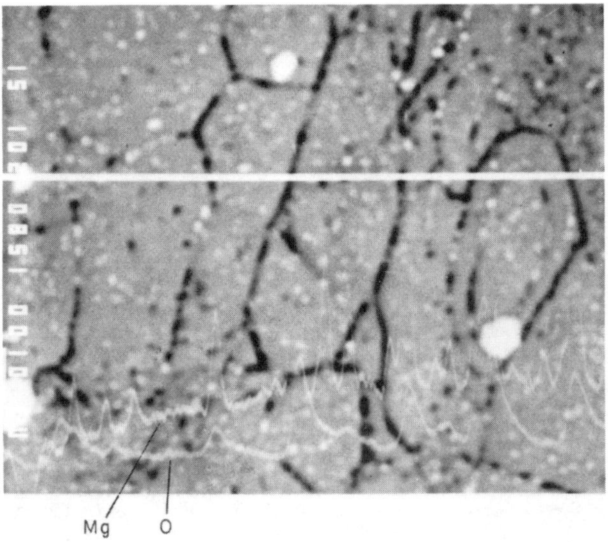

Figure 6: Line scan analysis for magnesium and oxygen.

The fracture appearance of an X7091 strip was examined by scanning electron microscopy. Figure 7 (a) shows a typical fracture surface of a strip which had been rerolled to 75% reduction, heat treated to the T6 condition, and tensile tested along the rolling direction. While the majority of the fracture surface exhibited well-developed dimples (marked A), some areas contained shallow, poorly developed dimples (marked B). Parts of the fracture appeared to have occurred by intergranular rupture evidenced by a dark arrow. Figure 7 (b) shows small, flat, shallow dimples at a higher magnification which are marked B in Figure 7(a).

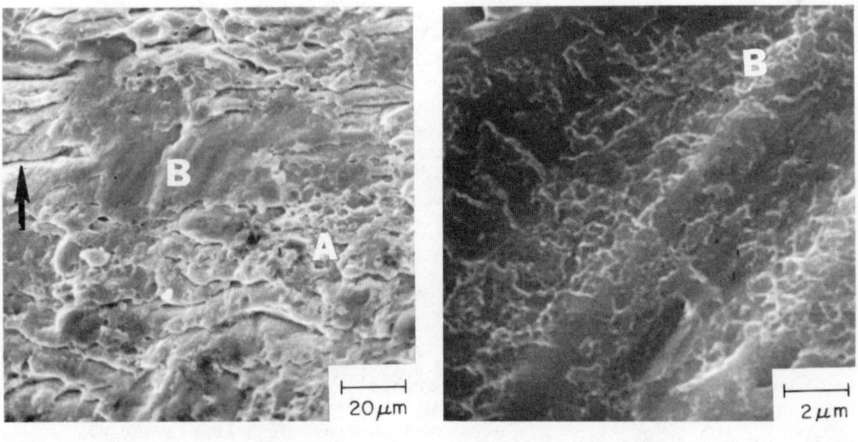

Figure 7: Fracture appearance of an X7091-T6 strip

Table XI: Results of Grain Size Measurements

Reduction in Thickness (%)		Avg. Count	ASTM No.	Grain Size (mm)	Photo Mag.
35	Long.	169	6.7	.031	100X
	Transv.	129	8.0	.020	200X
50	Long.	141	6.4	.036	100X
	Transv.	126	8.0	.020	200X
65	Long.	166	8.8	.015	200X
	Transv.	172	8.9	.014	200X
75	Long.	163	8.7	.016	200X
	Transv.	155	8.6	.017	200X

Grain size was determined on strips made from a powder blend (8-1014-80) at various reduction levels. The grain intercept counts and ASTM grain size number are given in Table XI. Average grain size appears to decrease with increasing reduction in thickness from 50% to 65% approximately by a factor of two. No further reduction in grain size was observed after 65% reduction in thickness.

Conclusions

The following conclusions can be drawn based on the experimental results of a direct-rolled X7091 powder alloy in this study.

(a) Local variations in the alloy composition due to varying ratios of blending between master and filler alloy powders as well as powder size exert a major influence on roll compactability.

(b) A -100 mesh, 50% enriched master alloy powder blended with a -100 mesh filler alloy powder at two to one ratio appears to produce strip with a satisfactory combination of strength and ductility.

(c) The room temperature tensile strength and ductility increase with increasing amounts of cold work to 75% reduction in thickness.

(d) The low transverse ductility of powder-rolled strip appears to be related to a poor interparticle bonding.

(e) The direct powder rolling process has the potential to produce high-strength aluminum strip/sheet for airframe applications. When fully developed, this process should lend itself to application to a wide variety of high-strength, low density and high temperature aluminum alloys.

Acknowledgments

This research is based in part on the work of the powder rolling process developed initially under the Imperial Clevite's IRAD program, and then the Air Force Materials Laboratory's sponsorship (AF Contract No. F 33615-80-C-5161). The authors wish to thank Imperial Clevite Inc. for permission to publish this paper. K. Geiger, J.J. Kysela and R. Stutzman assisted with the experimental part of the work.

References

(1) Rapidly Solidified (RS) Aluminum Alloys - Status and Prospects, NMAB-368, MDA-903-78-C-0038, National Materials Advisory Board, Washington, D.C., 1981.

(2) D.H. Ro, "Direct Rolling of Aluminum Powder Metal Strip,"Interim Report, Contract F 33615-80-C-5161, March 31, 1981.

(3) M.L. MacKay, "Innovation in Powder Metallurgy: An Engine Bearing Material," Metal Progress, June, 1977.

(4) M.C. Flemings, Proceedings of the 12th Army Materials Research Conference, p. 235, Syracuse, N.Y. 1966.

(5) W.A. Dean and R.E. Spear, ibid, p. 268.

GAS GUN COMPACTION OF TITANIUM SPONGE

P. R. Frausto, A. H. Clauer, and A. G. Jackson

Battelle
Columbus Laboratories
505 King Avenue
Columbus, Ohio 43201
USA

The gas gun utilizes a high-velocity ram to produce a compact of a powder material. The effect of various ram velocities has been studied in terms of the density achievable in chloride-derived Ti sponge. Densities obtained by this method ranged from 64 to 96%. The highest density was acheived using a velocity of 186 m/sec and having the powder in a mild vacuum. Various energy-absorber designs were tried in order to determine their effect on the product.

Introduction

The objective of this research was to investigate the effects of gas gun dynamic compaction on titanium sponge fine powder with the intent to maximize the resulting density of titanium compacts within the energy constraints of the gun itself.

Dynamic compaction as a possible means of producing high density P/M materials has been carried on for more than 20 years. Although explosives have been used from the inception of dynamic compaction, the use of a gas gun to acheive high shock pressures has received increased attention only within the last 10 years. Gas gun dynamic compaction is a high energy rate consolidation process which incorporates shock waves as a means of compacting a powder column.

The use of shock waves can result in high density materials which do not require an additional sintering step to increase density and mechanical properties (1-5). This results from the shock waves causing significant particle deformation, sliding, shearing, etc., which can lead to highly localized heating of the particle surfaces to the extent that particle surface melting results. The particle surface melt acts as a lubricant facilitating the compaction process and as a weldment upon solidification. It was determined early in the investigation that Battelle's gas gun, which was modified for this program, did not have the capability of producing the projectile velocities needed to explore the high energy regimes where particle surface melting is observed. The thrust of the work then was to optimize the process within the constraints posed by the gun. Even without particle surface melting, dynamic compaction offers several advantages:

1) Lower energy requirements than conventional P/M consolidation processes for the same degree of green densification (in the high density regime).[6,8]
2) Less shrinkage during sintering and lower sintering temperature for a dynamically compacted part vs. a conventionally processed one.[8,9,11]
3) High resulting mechanical and physical properties.[5,7,8]

Material

The powder used in these experiments was RMI Titanium Sponge Fines (-270 mesh) made by a chloride reduction process. This powder is highly irregular and contains a large amount of internal porosity (Figure 1).

Equipment

The design of Battelle's gas gun assembly is shown in Figure 2. The guide barrel was not evacuated during firing. Instead, openings were machined in the side of the guide barrel to allow the air to escape during acceleration of the projectile. Figure 3 shows the projectile and die assembly in greater detail. The projectile consisted of a 6061 aluminum cup, a hardened 4340 steel support, and a nylon impact head. Aluminum was chosen for the cup because of its low mass. The steel support, which screwed into the cup, was selected for its strength and impact resistance. Nylon was selected for the impact head so that if there was misalignment during firing, the die would not be severely damaged. The nylon impact head was made to slip-fit onto the steel support.

The dies, machined from 4340 tool steel,consisted of a collar and base plug. This assembly was supported at the end of the guide barrel in an

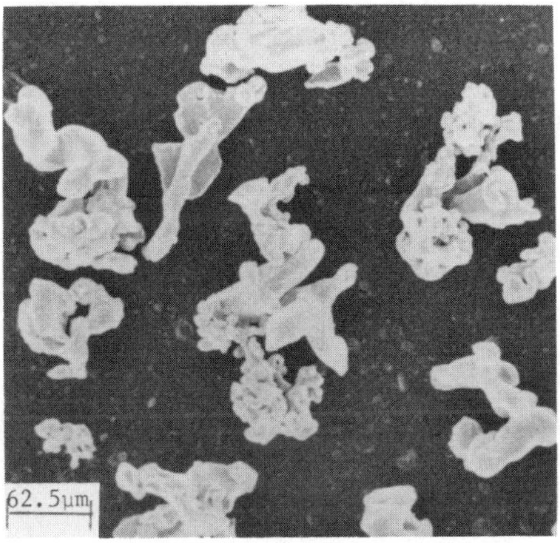

(a) RMI Titanium Sponge Fines Powder

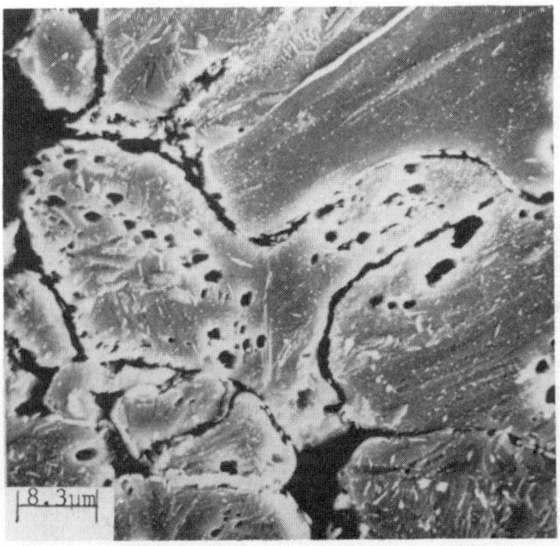

(b) Internal Porosity of RMI Titanium Sponge Fines

FIGURE 1. RMI TITANIUM SPONGE FINES POWDER CHARACTERISTICS

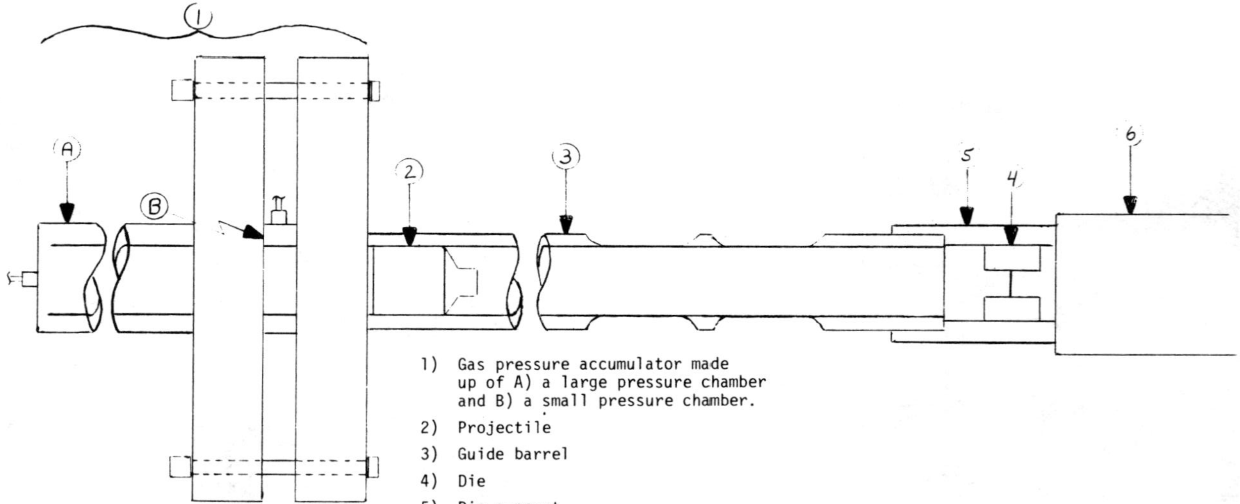

1) Gas pressure accumulator made up of A) a large pressure chamber and B) a small pressure chamber.
2) Projectile
3) Guide barrel
4) Die
5) Die support
6) Momentum trap system

FIGURE 2. GAS GUN ASSEMBLY

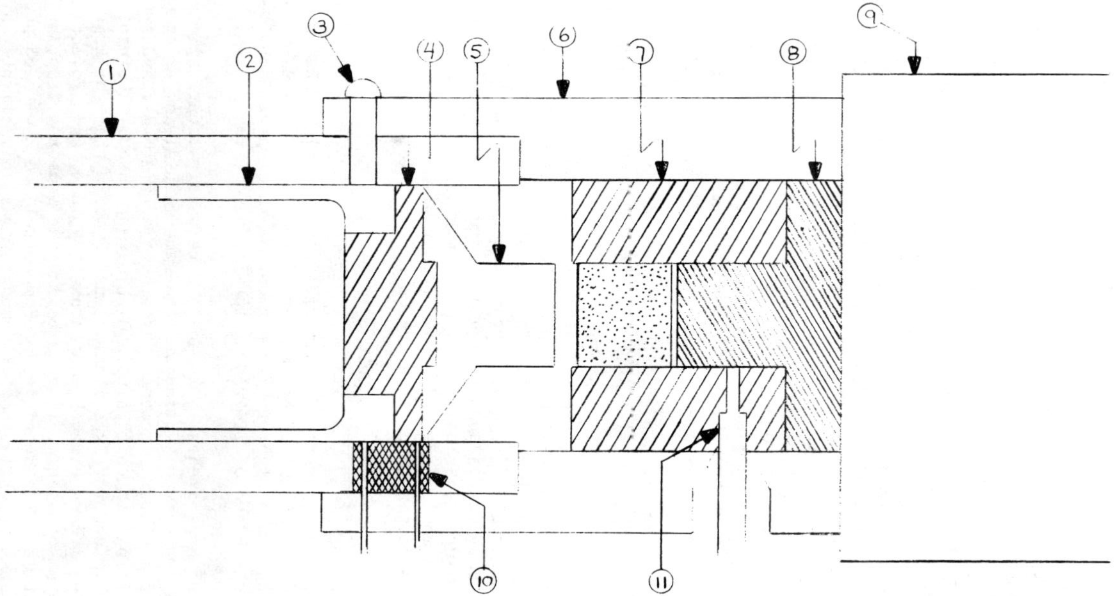

1) Guide barrel
2) 6061 aluminum cup
3) Ground pin
4) 4340 steel support
5) Nylon impact head
6) Aluminum support sleeve
7) Die collar
8) Die base plug
9) Momentum trap
10) Velocity pins
11) Vacuum port

FIGURE 3. PROECTILE, DIE, AND MOMENTUM TRAP ASSEMBLY

aluminum sleeve. A tapered hole was machined near the back of the die collar for evacuation of the die.

A momentum trap was coupled to the die to inhibit reflected shock waves from passing back through the sample. Use of the momentum trap is important because tensile waves which travel back into the compact can damage it. This trap consisted of an array of plates of which the number, mass, and length of the plates could be varied. The momentum trap can also keep the net movement of the die to a minimum.

Procedure

Prior to each experiment, the guide barrel and projectile were cleaned and lubricated with a fluorcarbon spray. The projectile was then placed in the barrel just outside the small pressure chamber. The pressure accumulator was readied by inserting two brass diaphrams of the same known burst pressure; one between the large and small chambers (back diaphram) and the other between the small chamber and guide barrel (front diaphram). The accumulator was then bolted together. The die cavity was lubricated, filled with powder, tamped down, and covered on its free side with a powder support. If an evacuated powder column was used, vacuum grease was applied to the critical areas and the die was evacuated through the vacuum port. Spacers could be placed in the die cavity so that the column length could be varied. For each experiment, an aluminum shim was placed behind the powder column to prevent the titanium from possibly bonding to the spacers or the base plug. (After each experiment, the titanium compact was bonded to the shim but could be removed with a small knife edge chisel and mallet.) The die was then placed on the aluminum support sleeve and the momentum trap assembled behind it.

Both chambers of the pressure accumulator were filled at relatively the same rates with nitrogen gas--the large chamber to slightly less than twice the pressure of the small chamber. To initiate firing the gas in the small chamber was quickly bled off causing the back diaphram to burst and the projectile to be accelerated down the guide barrel. The velocity of the projectile was measured just prior to impact using an electrical timing device. The projectile would make and break contact with 2 electrically charged pins and a ground pin which triggered a differential time indicator.

Results

The results of 12 experiments are summarized in Table 1. Final compact densities ranged from a low of 88 percent of theoretical density for the RMI Titanium powder to a high of 97.9 percent of theoretical density for an elemental powder blend of Ti-6w/o Al-16 w/o Sn.

The conditions of Experiments 1, 2, and 3 were nearly identical (see Table 1). There was no vacuum applied and the same initial powder column length (making use of a spacer), momentum trap, projectile mass, and type of support cover (a 1/16 inch thick aluminum shim) were used. The projectile velocity was $\sim$85 m/sec for Experiments 1 and 2 while that for Experiment 3 was 202.7 m/sec. Each time the momentum trap, die, and projectile were all driven into a receptacle, located behind the momentum trap, by the impact of the projectile. The densities of Compacts 1 and 2 were 88 percent of theoretical density while that of Compact 3 was only slightly higher at 92 percent of theoretical. Each of the first three compacts exhibited cracks running parallel to the plane of the compact, perpendicular to the direction of compaction. This was due to gas entrapment during compaction and also possibly to tensile waves which were reflected from the back surface of the base plug.

TABLE I. SUMMARY OF COMPACTION CONDITIONS AND RESULTS

Exper.	Material	Vacuum (μ)	Initial Column Length (cm)	Projectile Mass (kg)	Projectile Velocity (M/sec)	% Theoretical Density
1	Ti	No	1.1	.51	87.5	88.0
2	"	No	1.1	.51	83.2	88.0
3	"	No	1.1	.51	202.7	92.0
4a	Ti	∿500μ	2.5	.52	147.8	90.0
4b	"	No	2.5	.52	146.3	93.0
5	"	100	2.5	.52	140.8	92.0
6	Ti	200	2.5	.52	N/A	96.5
7	"	100	2.5	.52	186.2	95.8
8	"	150	2.5	.65	167.6	93.4
9	Ball Milled Ti	150	2.5	.52	185.9	94.0
10	Ball Milled Ti	200	1.1	.52	N/A	95.6
11	Ti	200	1.1	.52	189.3	95.5
12	Ti-6wt.% Al 16wt.% Sn	200	1.1	.52	187.5	97.9

For Experiment 4a the following changes were made. To prevent gas entrapment the die volume was evacuated. In order to pull a better vacuum on this volume and reduce fit-up problems between the support cover and the die, the aluminum shim powder support used in Experiments 1, 2, and 3 was replaced by a flexible 0.015 inch thick acetate disk. Also, the momentum trap mass, the nylon impact head length, and initial powder column length (by removing the spacer) were all increased in an attempt to increase the energy and reduce the reflected tensile waves delivered to the compact.

The evacuation caused the powder to be compacted to about three quarters of its original column length. It is probable that the projectile, traveling at 148 m/sec, lost energy due to friction as it moved down the die prior to impact. As in Experiments 1, 2, and 3 the projectile, die, and momentum trap were all driven into the receptacle. The resulting compact was 90 percent dense. This compact, like all subsequent compacts made, had a slightly concave fron surface.

To observe the effects of recompaction, the compact was surface ground, placed back into the die and recompacted without the use of a vacuum (Experiment 4b). The density of this compact was increased over the prior density to 93.2 percent theoretical.

Additional changes were made for Experiment 5. To investigate the effect of the support cover configuration on the attainable vacuum pressure, an acetate sheet was used which covered the entire die face rather than just the die cavity. No apparent effect on the vacuum was observed but this configuration was not as effective in supporting the powder column as the acetate disk configuration. The mass of the momentum trap was increased and the projectile velocity was 141 m/sec. This resulted in the projectile rebounding slightly up the guide tube while the die remained in the support sleeve. The momentum traps were again driven into the receptacle. The final density of this compact was 92 percent of theoretical.

For Experiment 6 the acetate disk support was used. The momentum trap configuration was not changed from Experiment 5, but the projectile velocity was increased to ∿180 m/sec. The projectile again rebounded back into the guide barrel but both the die and momentum trap were driven into the receptacle. The final density of this compact was 96.5 percent of theoretical. Experiment 7 was made under conditions similar to Experiment 5 except that the projectile velocity was 186 m/sec. The resulting compact was 95.8 percent dense. Higher projectile velocities were desired but because of safety considerations the gas gun pressure accumulator could not be subjected to pressures much greater than 2000 psi.

For Experiment 8 the masses of the impacter and the momentum trap were increased and the acetate disk cover was used as in Experiment 4. The projectile velocity was 168 m/sec and the final density of the compact was only 93.4 percent of theoretical. The projectile rebounded up the guide barrel, the die remained stationary, and the momentum trap was driven into the receptacle. The results of this experiment when compared to the others indicated that a lighter, higher speed projectile is more efficient than a heavier, slower one at approximately the same projectile kinetic energy. Based on these results, the mass of the projectile was reduced for the remaining experiments.

The RMI titanium used for Experiment 9 was ball milled for 9 hours in argon gas in an attempt to reduct its irregular shape. The initial packing density was increased, but the final compacted density (∿94 percent of theoretical) was actually lower than the previous results. Whatever advantages may have been gained by reducing the particle shape irregularity were diminished by a probable increase in the surface hardness of the ball milled powder and/or changes in the specific energy-mass relationship of the powder column, and die wall friction as described by El-Wakil and Davies(6). The ball milled titanium was again compacted in Experiment 10. Conditions remained the same as Experiment 9 except that the initial column length was decreased by placing a spacer in the die as in Experiments 1, 2, and 3. The final density of this compact was 95.6 percent of theoretical. In Experiment 11, as-received RMI Titanium was compacted under identical conditions as Experiment 10. The compact produced was 95.5 percent dense. From these last three experiments, it was deduced that at shorter column lengths, the effect of ball milling was small but for the larger initial column lengths, the use of the ball milled powder was detrimental to the compaction process.

Finally, an elemental blend of Ti-6w/o Al-16 w/o Sn was compacted under the same conditions as Experiments 10 and 11. Both the aluminum and tin powders were very fine (-325 mesh) in order to get a good mixing of these powders between the titanium particles. The use of these additions had a beneficial effect on the compaction process, producing a compact which was 97.9 percent dense. The aluminum and tin additions, being more ductile, were squeezed into voids which otherwise might not have been closed and thereby increased the density of the compact.

Optical micrographs of an 88 percent dense compact and a 95.6 percent dense compact are shown in Figures 4a and b. The higher density compact (Figure 4b) has less porosity and also exhibits pressing directionality with the particles being somewhat elongated perpendicular to the compaction direction.

Scanning electron microscopy revealed no evidence of particle surface melting in any of the specimens. The microstructure of the compact elemental powder blend (Figure 5) exhibited a large degree of deformation and flow of the aluminum and tin, but again no evidence of surface melting.

In any mechanical powder pressing operation, die wall friction plays an important role in the subsequent physical characteristics of the compacted specimen. Sano, et.al.(9) examined the effect of die wall friction during dynamic compaction, and found that the velocity of the shock wave passing through the powder column is lower near the die walls and higher near the center. The result is that an initially flat surface becomes concave. This effect could also lead to variation of the density with the compact.

Compacts 4 through 12 exhibited concave front surfaces which are attributed to frictional effects at the die wall and to the use of the acetate support covers. Compacts 1-3 did not have this characteristic. Density variations within the specimens were revealed by macro- and microhardness measurements and examination of fracture surfaces through the thickness of the compacted disks. Macrohardness measurements taken on the front and back surfaces of Compacts 6-11 revealed a hardness variation along the diameter. Hardness values near the centers of both the front and back surfaces were higher than near the edges. Microhardness measurements made on polished transverse sections through the thickness of the same specimens also showed values higher near the center than near the side edges. This spatial variation in hardness is most likely due to a corresponding variation in the local density through the compact.

For Compact 6-9, microhardness measurements taken in the direction of compaction on transverse sections through the compacts revealed the existence of a density gradient having maximum values near the front and back surfaces with minimum values near the center. This was confirmed metallo-

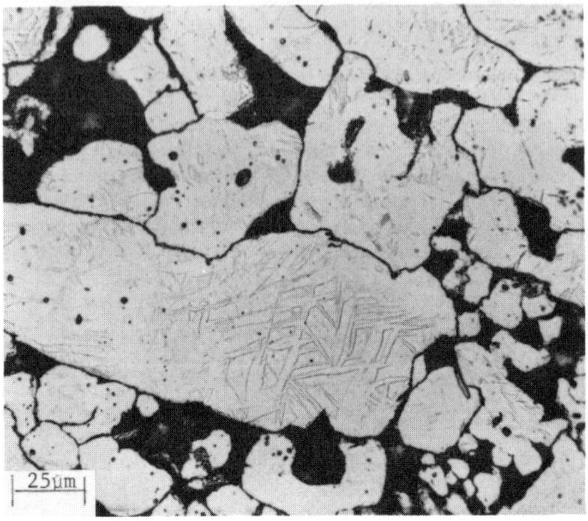

(a) 88% Dense Titanium Compact

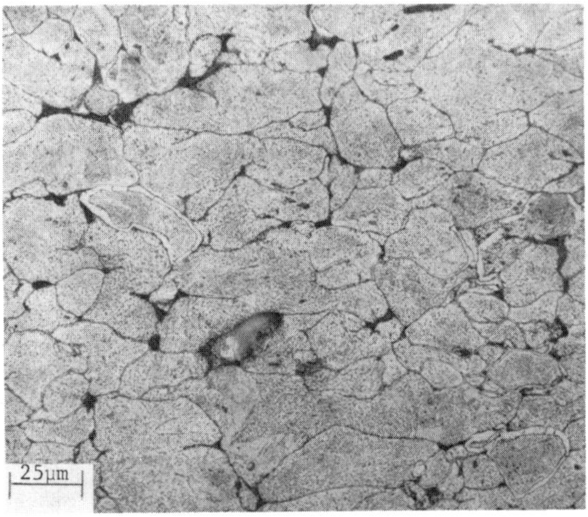

(b) 95.6% Dense Titanium Compact

FIGURE 4. MICROSTRUCTURES OF HIGH AND LOW DENSITY TITANIUM COMPACTS

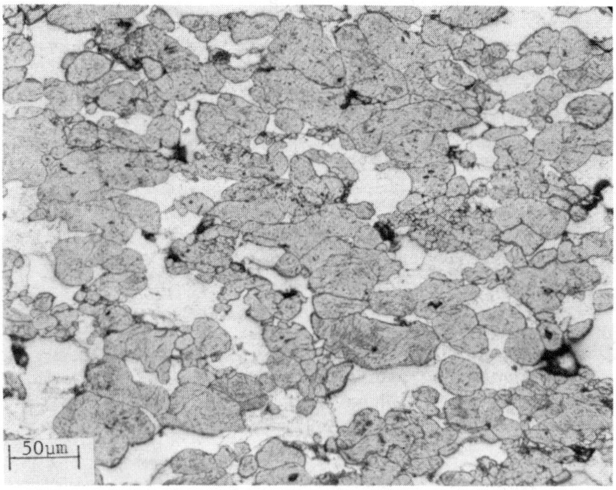

(a) Ti-6Al-16Sn

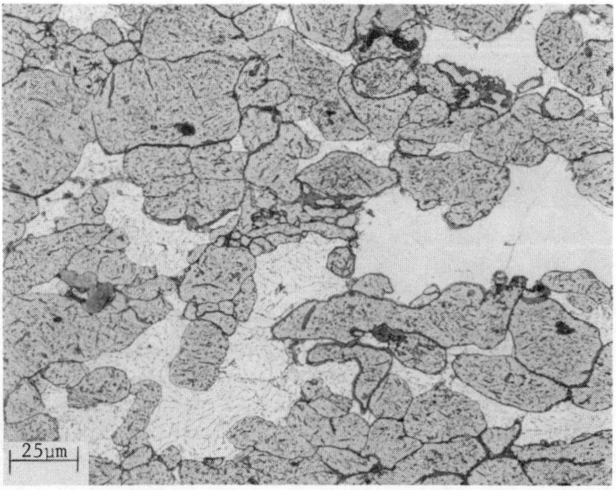

(b) Ti-6Al-16Sn

FIGURE 5. MICROSTRUCTURE OF Ti-6Al-16Sn 97.9% DENSE COMPACT

graphically as described later. Microhardness measurements taken on Compacts 10 and 11, which were of shorter initial column length, revealed similar microhardness/density variations as described above for Compacts 6-9 but to a lesser degree. Optical metallography did not confirm the density variations revealed by the hardness measurements, possibly due to smearing of the titanium during polishing.

Compacts 6-12 were then broken through their thickness and the resulting surfaces examined. Optically, on Compacts 6-9, the fracture surfaces exhibited three distinct regions. Figure 6 shows a typical example of these regions and high magnification shows these regions to be areas of high and low density (Figure 7). In each of the Compacts 6-9, the front and back high density zones are similar in thickness and appearance to each other. The presence of these regions shows that the initial powder column length was too long to obtain uniform compaction under the conditions used. The high and low density zones could conceivably be a result of compaction wave/release wave interactions, much like those described by Raybould(1).

A typical fracture surface of Compacts 10 and 11 is shown in Figure 8, while that of Compact 12 is shown in Figure 9. These fracture surfaces did not reveal noticeable density variations.

Concluding Comments

For the conditions employed in this investigation, the following comments can be made:

1) Evacuation of the powder column is necessary to produce a sound, crack-free compact.

2) A lightweight, high velocity projectile produces higher compacted densities than the use of a heavier, lower velocity one.

3) Shorter initial powder column lengths did not appear to produce higher density compacts as might be expected, but it did produce more uniform ones.

4) Powder shape modification by ball milling did not lead to increased final densities.

5) Frictional effects, and powder column lengths which are too long for the compaction conditions result in density variations within the compacted specimens.

Acknowledgement

This program was particially supported by Battelle's Office of Corpoorate Technical Development and the U.S. Air Force under contract F-33601-80MN591.

FIGURE 6. TYPICAL FRACTURE CROSS SECTION OF COMPACTS 6 TO 9

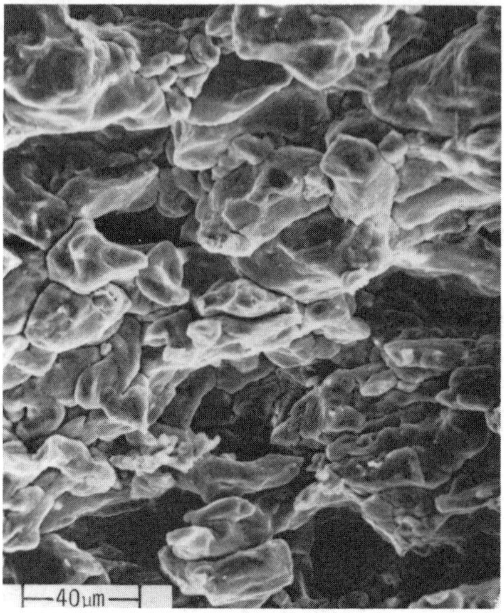

(a) Middle Zone

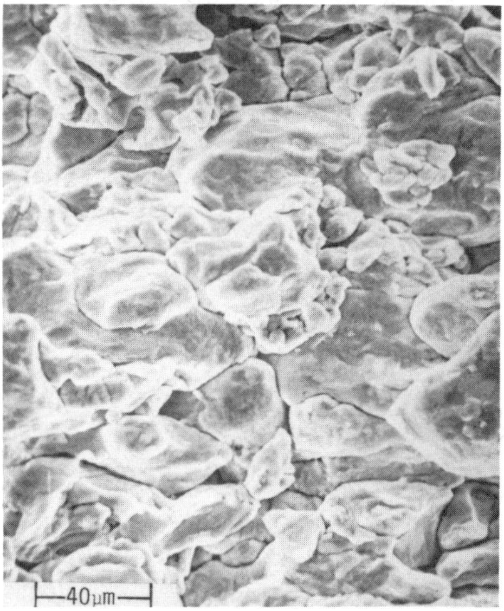

(b) Front Zone

FIGURE 7. FRACTURE SURFACES OF MIDDLE AND FRONT ZONES

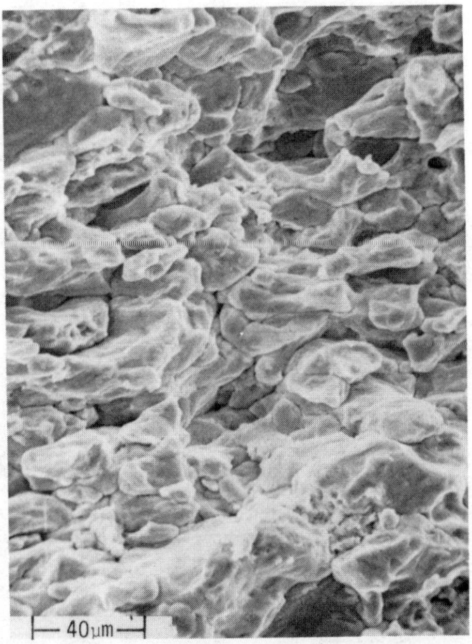

FIGURE 8. REPRESENTATIVE FRACTURE SURFACE OF COMPACTS 10 AND 11

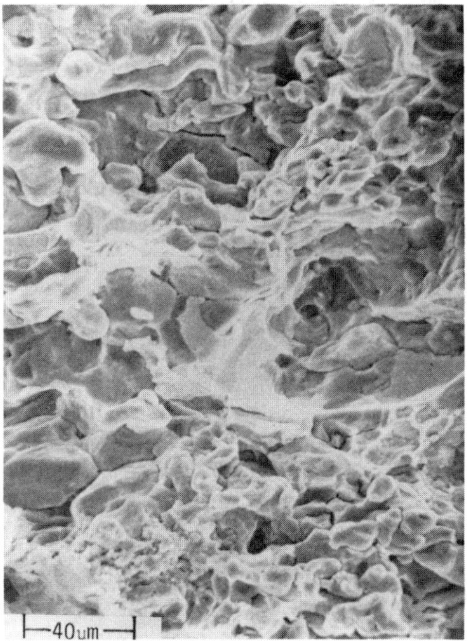

FIGURE 9. FRACTURE SURFACE TO Ti-6Al-16Sn COMPACT

References

1. D. Raybould, 7th International Conference of the International Association for the Advancement of High Pressure Science and Technology, Le Creusot France, July (1979).

2. D. Raybould, D. G. Morris, G. A. Cooper, Journal of Material Science, 14 (1979). Pg. 2523

3. D. Raybould, Powder Metallurgy, No. 1 (1980). Pg. 37.

4. D. G. Morris, Metal Science, (June, 1980). Pg. 215.

5. D. Raybould, Journal of Materials Science, 16 (1981). Pg. 589.

6. S. Elwakil, R. Davies, Proceedings of the 13th International M.T.D.R. Conference, Birmingham, Macmillan, London (1972).

7. P. A. Vityaz, O. V. Roman, 13th International MTDR Conference of Birmingham, Macmillan Press, London (1972).

8. R. M. Rusnak, International Journal of Powder Metallurgy and Powder Technology, 12, No. 2 (April 1976). Pg. 91.

9. Y. Sano, K. Miyagi, T. Hirose, International Journal of Powder Metallurgy and Powder Technology, 14, No. 4 (1978). Pgs. 291-298, 300-303

10. J. W. Hagemeyer, J. A. Regalbuto, International Journal of Powder Metallurgy 4, 3 (1968). Pg. 19

11. L. W. Garber, J. K. Beddow, Powder Metallurgy International, 6, No. 2 (1974). Pg. 78.

THE EFFECT OF PARTICLE SIZE AND STATE OF AGGREGATION

ON THE SINTERING OF ALUMINUM NITRIDE*

Glenn C. Culbertson and James P. Mathers

University of Maryland
College Park, Maryland 20742

A commercial AlN powder having a specific surface area of 3.9 m^2/g (0.5 um equivalent spherical diameter) was found to contain a large fraction of hard aggregates, ranging up to 200 um in size. The aggregates were composed of submicron crystallites which accounted for the relatively high specific surface area of the powder. In attempts to sinter this powder it was observed that the aggregates survived dry-pressing and prevented densification.

The same commercial powder was processed to yield a submicron powder and a coarse powder, both consisting of individual crystallites or small aggregates, and having approximately the same oxygen content. The submicron powder sintered to high density (>95% theoretical) while the coarse powder showed no densification, indicating the importance of particle size on the sintering behavior. The results also illustrated the danger in relying too heavily on surface area determinations to evaluate the potential of AlN powders to densify during sintering.

*This work was supported by NSF Award No. DMR80-01980.

Introduction

Aluminum Nitride has a number of attractive properties which make it a potentially important material for a number of advanced technologies. This has spurred interest in its sintering behavior. As with other covalently bonded ceramics, a good deal of difficulty has generally been encountered in attempts to sinter AlN to high density (ref. 1 & 2). This has been linked to the formation of an unsinterable microstructure (ref. 3) along the lines suggested by Greskovich and Rosolowski (ref. 4). Some recent investigations have reported that AlN powder of high specific surface area can be sintered to full density. However, this experience has not been universal as can be seen from Table I.

Table I. Previous Sintering Studies of Aluminum Nitride

Reference	Spec. Surface(dia.)	Oxygen Content	Rel. Density
Kuzenkova (ref 5)	37.0 m^2/g (0.05 um)	2.0 wt%	100.0%
Trontelj (ref 6)	30.0 (0.06)	-	85.0
Kuzenkova (ref 5)	26.0 (0.07)	2.0	100.0
Poluboyarinov (ref 7)	13.0 (0.14)	-	99.7
Prochazka (ref 3)	9.2 (0.20)	2.1	68.0
Kuzenkova (ref 5)	4.6 (0.36)	2.5	90.0

Experimental Procedure

A high purity AlN powder was obtained from a commercial source*. The chemical analysis supplied by the vendor is presented in Table II. As will be shown later, this powder contained a large fraction of large aggregates composed of submicron crystallites. In order to fairly measure the effect of particle size on the sintering behavior two powders of different average particle size (0.1-0.2 um vs. 1-2 um) were desired. Additionally it was desired that the two powders have roughly the same degree of spread in their particle size distribution, that the particles consist primarily of individual crystallites or small aggregates, and that the oxygen content of each powder be approximately the same. Powders which closely approached this goal were prepared from the commercial powder in the following manner.

The submicron powder was obtained through a combination of dry ball-milling and sedimentation. The commercial powder was milled for 120 hours in a linear polyethylene jar, using high purity Al_2O_3 media (99.5 wt%)**. The dry milling followed procedures which have proven useful in breaking-up aggregates of calcined Al_2O_3 (ref. 8). A high media to powder ratio was used (32:1 by weight) and 2.5 wt% of triethanolamine (TEA) was added as a grinding aid. Large aggregates which survived the milling process were removed from the powder by sedimentation in anhydrous acetone. A mixture containing 10 ml of acetone per gram of AlN, 10 cm deep, was allowed to stand undisturbed for 90 minutes. The AlN remaining in suspension was removed from the acetone by

*Cerac, Inc., Wis.
**Degussit AL 23, Degussit Co., Germ.

vacuum distillation to minimize exposure of this ultrafine powder to the atmosphere. The resulting powder, like all other powders used in this work, was contained in a closed jar, and stored in a desicator until use.

Table II. Chemical Analysis of Commercial AlN Powder

A-1120 Aluminum Nitride   Lot Number 3411
-200 mesh
Typically 99.8%
(N = 32%, C = 0.01-0.05%)

Specific Analysis: N = 33%, C = 150ppm

Spectrographic Analysis:

Cu	0.0005-0.005%	Mn	0.001%
Fe	0.001-0.01%	Mg	0.0005-0.005%
Si	0.0005-0.005%		

X-ray Diffraction Analysis:

AlN matches PDF 25-1133
Hexagonal
No extra lines observed

A coarse powder was prepared by heating the commercial powder in N_2 at 1950 C for one hour, to increase the grain size. This was followed by wet ball-milling in anhydrous acetone for 12 hours to break-up aggregates and reduce the particle size to the desired level. Again a linear polyethylene jar and high purity Al_2O_3 grinding media were used, with 8.3 g of media and 2 ml of acetone per gram of AlN.

Each of the three powders used in this study were characterized by scanning electron microscopy (SEM), X-ray diffraction, single point B.E.T. surface area measurement*, X-ray scattering particle size distribution**, and neutron activation analysis***.

Prior to cold-pressing each powder was granulated. Tannic acid, 0.5wt%, was added, as a binder, via an ethanol solution (0.005 g/ml). The resulting slurry was worked until dry with a high purity (99.5%) Al_2O_3 mortar and pestle. The product was sieved to yield granules between 150 and 500 um.

The granulated powders were dry-pressed in a cylindrical, single action stainless steel die of 1 cm diameter. Charges to the die were about 0.4 g which kept the height to diameter ratio of the pressed compact below 0.4. A thin coating of polyethylene glycol**** in methanol was applied to the die surfaces as a pressing lubricant. A pressure of 132 MPa was used in the pressing operation.

*Monosorb, Quantachrome Corp., N.Y.
**Sedigraph Particle Size Analyzer, Micromeritics Inst. Corp., GA.
***by Forest Burns of the Army Mechanics and Materials Research Ctr., Watertown, Mass.
****Carbowax 4000, Union Carbide

The sintering behavior was studied by examining the fracture surfaces of powder compacts at different stages of sintering, and by monitoring changes in the density and weight. The time-temperature sintering schedule is shown in Fig. 1, solid line. Compacts were fired to various points on this curve and then rapidly cooled as indicated by the dashed lines. The fracture surfaces were examined using an SEM after sputter coating with 100-200Å of Au-Pd.

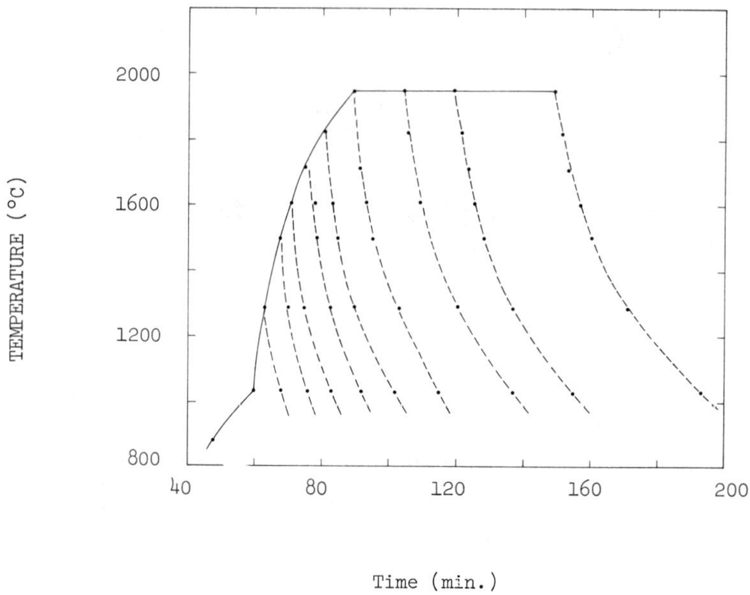

Figure 1 - Sintering Schedule

The powder compacts were fired in a graphite element resistance furnace.* An ultrapure grade N_2 atmosphere** at slightly above atmospheric pressure was used with a flow rate of 6 l/hr. A 1cm layer of AlN powder was used to support the samples within an open BN crucible. Compacts of each powder were fired together at each point of the time-temperature schedule.

Temperature was measured with an optical pyrometer sighted directly on the crucible contents with an accuracy of ± 20 C. The compacts were slowly heated, over one hour, to 1000 C and then rapidly heated to the desired temperature as indicated in Fig. 1.

*Astro 1000A, Astro Industries, CA
**Ultrapure Carrier Grade (99.999%), Air Products

Results and Discussion

The character of the as-received commercial powder is revealed in Fig. 2. As mentioned earlier this powder contained a substantial portion of hard aggregates ranging up to 200 um in diameter. The specific surface area of this powder was 3.9 m^2/g (0.47 um equivalent spherical diameter). The latter figure reflects the crystallite size of the powder, but it is clearly not representative of the particle size. The particle size distribution by X-ray scattering is shown in Fig. 3. The average particle size determined by this method was 15.5 um. Neutron activation analysis of this powder indicated an oxygen content of 2.35 wt%. If all the oxygen present is assumed to be present in a surface layer of Al_2O_3 the thickness of this layer would be 32 Å. Research by Slack (ref. 9) indicates that an oxide layer of 50-100 Å in thickness is formed on AlN after exposure to the atmosphere for 24 hours.

An SEM micrograph of the submicron powder is shown in Fig. 4. The largest aggregates present were about 2-3 um in size and a large fraction of the powder is clearly submicron. The specific surface area was 23 m^2/g (0.08 um), but the average particle size measured by X-ray scattering was 0.64 um. The entire particle size distribution is presented in Fig. 3. The oxygen content of the powder was 5.71 wt%, as determined by neutron activation analysis. This value was lower than had been expected since it was known that sizeable amounts of Al_2O_3 had been picked-up from the milling media. Apparently much of this contamination was removed in the sedimentation process which followed dry-milling. An X-ray diffraction pattern of the sediment showed a relatively strong Al_2O_3 {100} peak, while no such peak was present in the submicron powder which remained in suspension.

Figure 2 - As-received Powder (a) bar=100 um (b) bar= 0.1 um

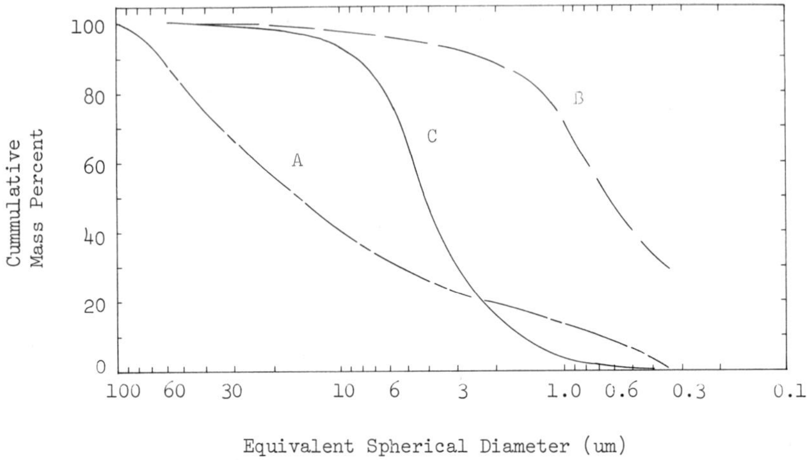

Figure 3 - Particle Size Distribution as Determined by X-ray Scattering (A-As-received, B-Submicron, C-Coarse).

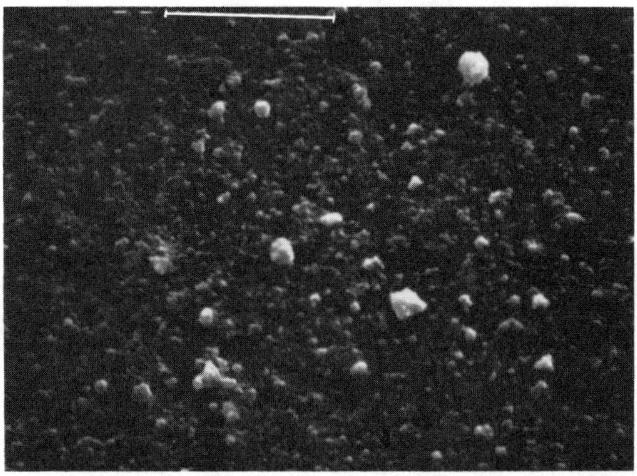

Figure 4 - Micrograph of Submicron Powder (bar-10 um)

Although an oxygen content of 5.71 wt% may seem rather high, it is surprisingly low considering the high surface area of the powder. If it is assumed that all of the oxygen is present as oxide surface contamination a thickness of only 14Å is calculated. Thus it appears that some degree of success was achieved in limiting reaction with the atmosphere.

The coarse powder which was prepared for sintering had a specific surface area of 1 m^2/g (1.84 um), however, SEM micrographs (Fig. 5) and X-ray scattering indicate a larger particle size. It is clear that few large aggregates are present. The entire particle size distribution is shown in Fig. 3, and the average particle size determined to be 4.2 um. The oxygen content of this powder was 5.90 wt%, very near that of the submicron powder. In this case however, much of the oxide content probably resulted from wear of the alumina media during wet milling.

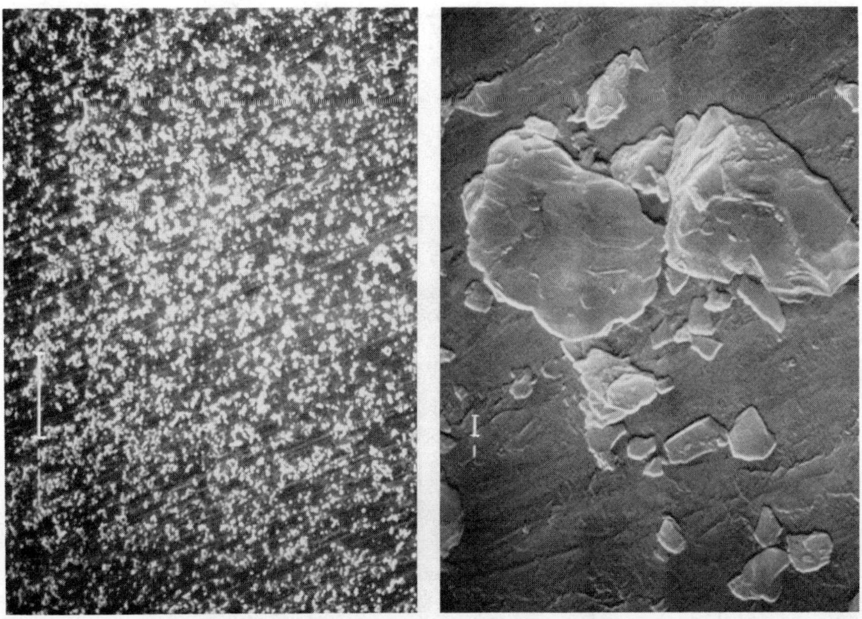

Figure 5 - Coarse Powder - Prepared by heat treatment and wet milling (a) bar=100 um, (b) bar=1 um.

As-received Powder

The results of the sintering study were somewhat surprising in that the development of a non-sinterable microstructure was not observed, except in the case of the as-received powder. The microstructure development during the sintering of this powder can be followed to some degree in Fig. 6. Each micrograph represents a fracture surface of a different sample making it difficult to follow the exact course of sintering. The large aggregates which characterize this powder are evident in Fig. 6(a). As sintering proceeds grain growth is evident both in the aggregates and the fines filling the voids between these aggregates. In the later stages of sintering the aggregates are difficult to distinguish from the remainder of the structure. Pore growth seems to accompany grain growth in both the aggregates and the fines until in the later stages of sintering it is difficult to distinguish the aggregates from the rest of the structure. The end result resembles the non-sinterable microstructure described by Greskovich and Rosolowski (ref. 4), and the mechanism of its formation seems similar in many respects to that proposed by these investigators. However, the fact that this structure was only observed with the as-received powder strongly suggests that its appearance in AlN is dependent in some manner on the presence of the large aggregates. It may be, for example, that the aggregates form a rigid network which impedes the normal densification process. Figure 7 shows that little overall densification occurs during sintering of the as-received powder and significant weight losses are incurred.

Submicron Powder

In contrast to the as-received powder the submicron powder prepared by dry-milling and sedimentation sintered to densities as high as 95.7% of the theoretical value for AlN (3.26 g/cm^3). Figure 7 shows that densification occurs fairly rapidly between 1550 and 1850 C, and that significant weight losses are observed despite the density achieved. The microstructure development during sintering can be seen in Fig. 8. Densification and grain growth occur uniformly throughout the process, and the final product shown in Fig. 8(d) is nearly fully dense. Most of the porosity in this material actually resulted from gross voids caused by incomplete breakdown of the granulated powder. Figure 9 shows a polished section of a sintered compact. On close examination the individual granules are visable. After completion of the sintering studies a single attempt was made to improve the granulation procedure in hopes of removing this gross porosity. Submicron powder was held in air at 500 C for 3 hours to remove organics, including TEA and polyethylene contaminants from the milling operation. Upon granulation and cold pressing granule breakdown was complete and sintered bodies without gross porosity were attained.

The oxygen content of the sintered AlN compact, prepared from the submicron powder, was 4.95 wt% for the compact held for 30 min. at 1950 C. This indicated some preferential removal of oxygen during sintering.

Coarse Powder

The coarse powder which was prepared showed no densification during sintering, and actually experienced a reduction of density at higher temperatures. Figure 7 shows that weight losses accompanied this reduction of density. As can be seen from Fig. 10, very little change in the microstructure is observed except for some neck growth and some recrystallization leading to a number of crystals with a platelike morphology. No evidence of a non-sinterable microstructure was observed though.

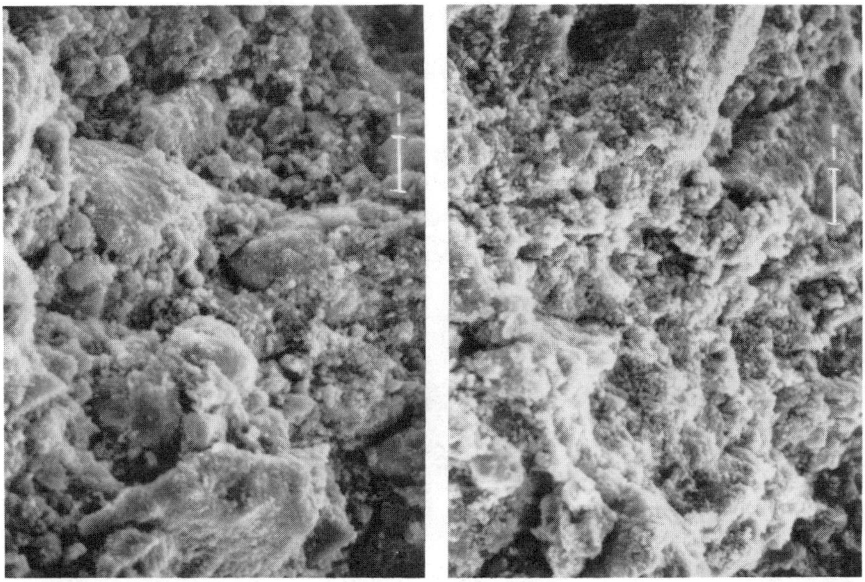

(a) 1025°C (b) 1820°C

(c) 1950°C (10 min.) (d) 1950°C (60 min.)

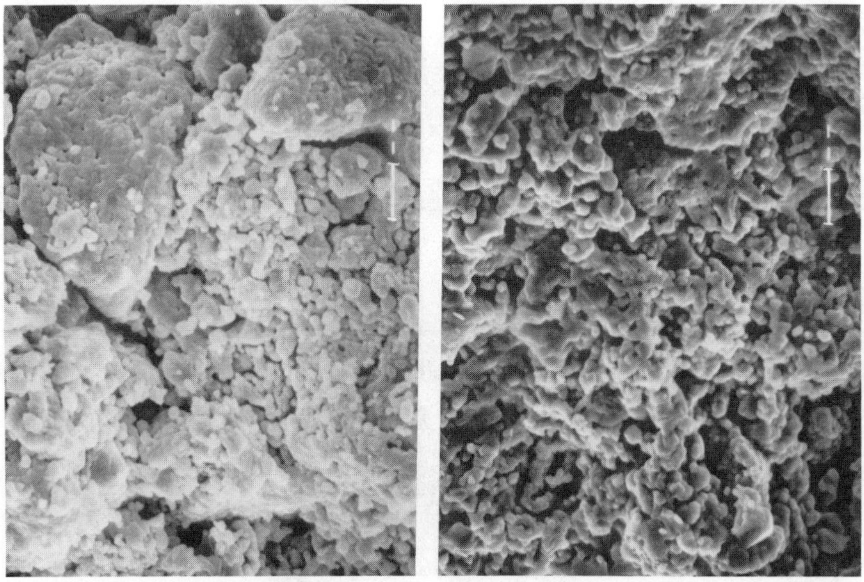

Figure 6 - Microstructural Development of As-received AℓN Powder (bar=10 um)

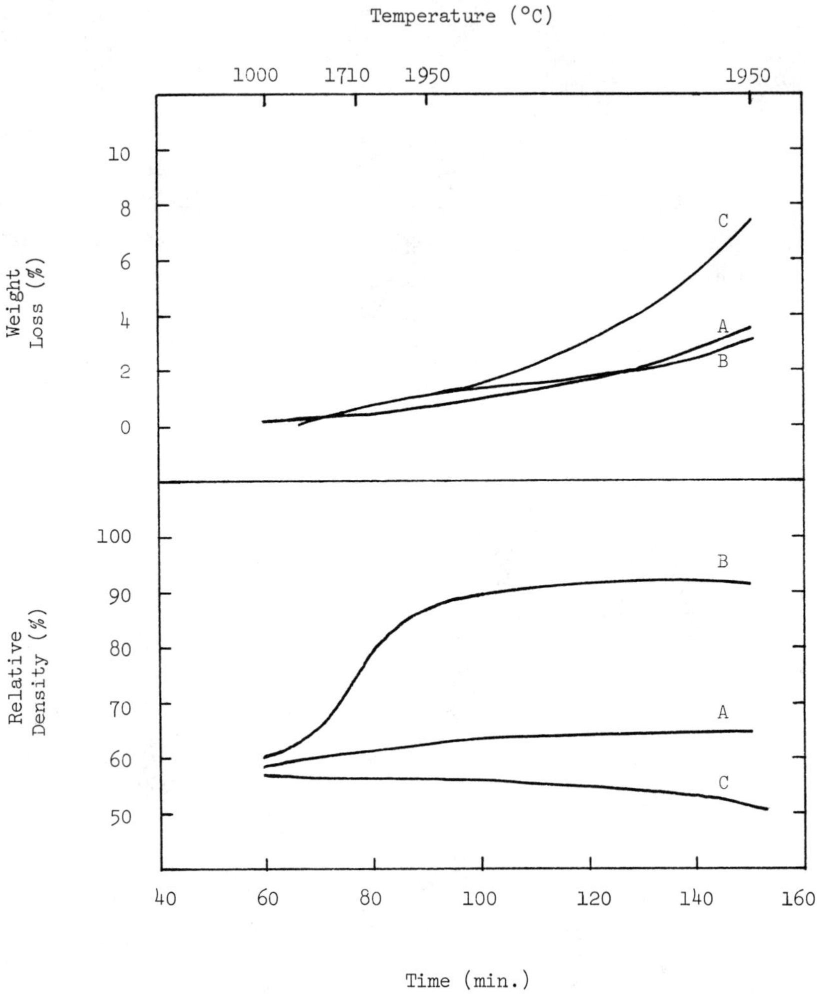

Figure 7 - Densification and weight loss of AℓN powders versus sintering schedule. (A-As Received, B-Submicron, C-Coarse)

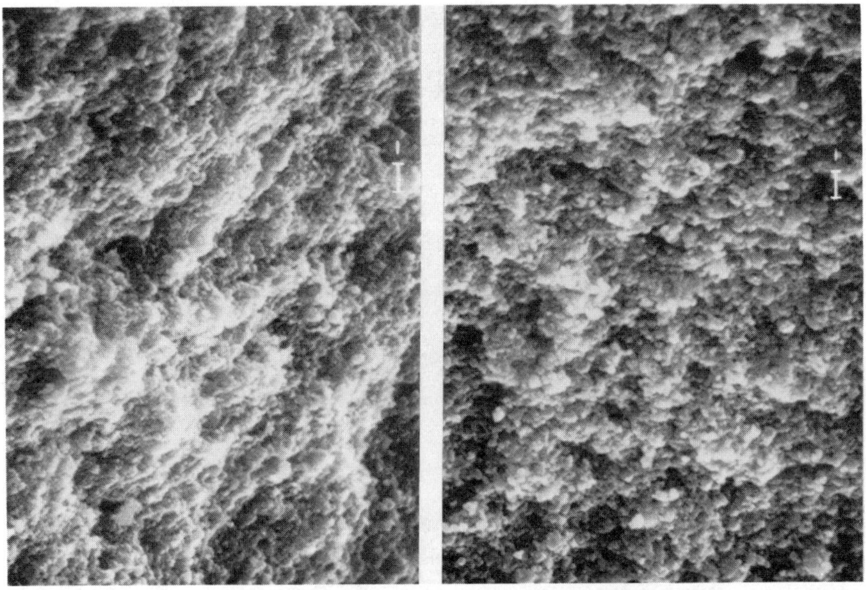

(a) 1025°C

(b) 1605°C

(c) 1820°C

(d) 1950°C (30 min.)

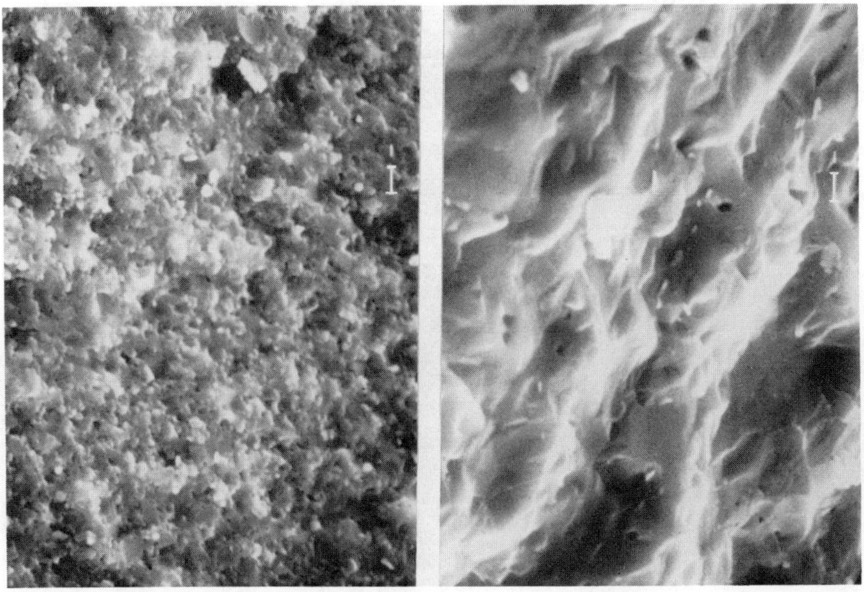

Figure 8 - Microstructural Development of Submicron Powder
(bar = 1 um)

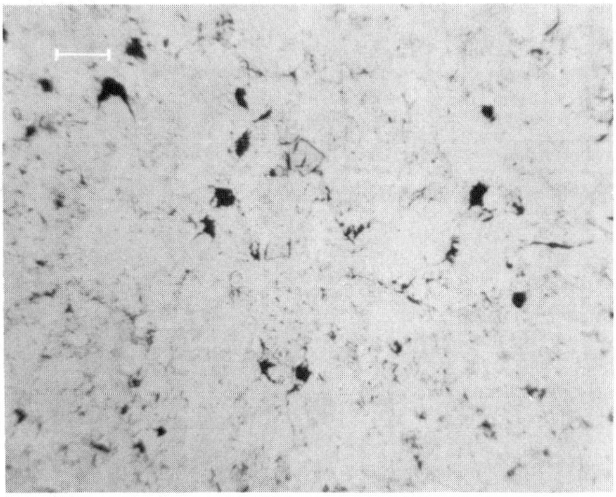

Figure 9 - Polished Section of sintered compact of submicron powder showing lack of granule break-down (Bar=100 um).

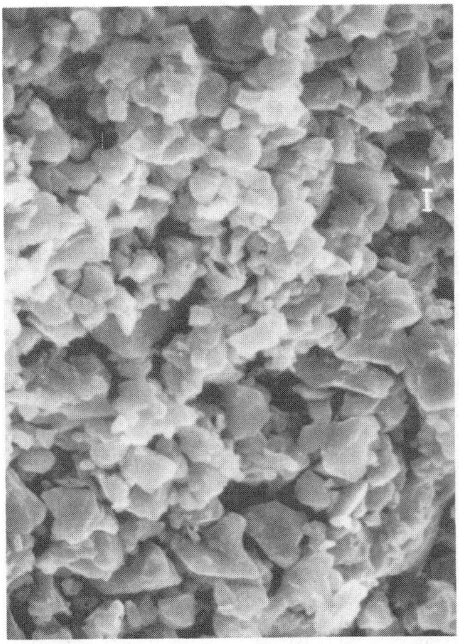

(a) 1025°C

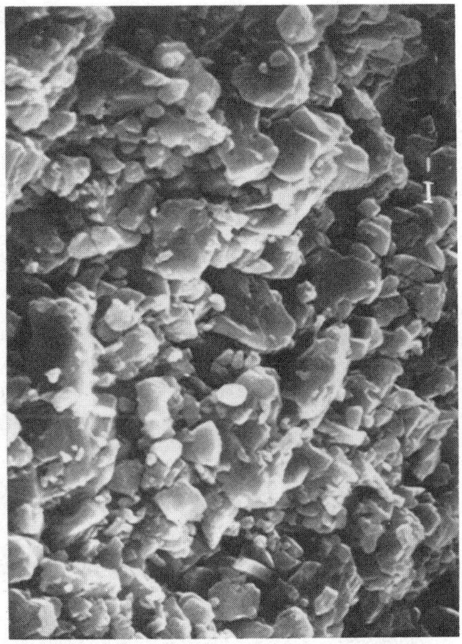

(b) 1820°C

(c) 1950°C (30 min.)

Figure 10 - Microstructural Development of Coarse Powder (bar-1 um)

A potentially important difference between the as-received powder and the two other powders was the oxygen content (2.35wt% vs. 5.71 and 5.90 for the processed powders). In an effort to determine whether the lower oxygen content was responsible for the development of the non-sinterable microstructure of the as-received powder, one additional experiment was conducted. Enough Al_2O_3 (0.05 um)** was mixed into the as-received powder to bring its oxygen content up to 5.71 wt%. This had no significant effect on the sintering behavior of the powder.

Conclusions

Submicron AlN can be sintered to high density provided the powder is composed of individual crystallites or very small aggregates. The presence of large aggregates can lead to the development of a non-sinterable microstructure. Because aggregates have such an important impact on the sintering behavior surface area measurements alone are not a sufficient indicator of the densification potential of AlN. This factor may explain the contradictory results which previous investigators have reported in sintering "submicron" AlN powders.

References

1. G. Long and L.M. Foster, "Aluminum Nitride, a Refractory for Aluminum to 2000°C", J. Am. Ceram. Soc., 42 (2) pp. 53-59 (1959).
2. C.F. Cooper, C.M. George, and L. Winter, "Aluminum Nitride Crucibles: Raw Materials Preparation, Characterization, and Fabrication", Special Ceramics, Vol. 4 Edited by P. Popper, British Ceramic Research Asso., London (1968).
3. S. Prochazka and C.F. Bobik, "Sintering of Aluminum Nitride," pp.321-331 in Sintering Processes, G.C. Kuczynski, ed.; Plenum Press, New York, N.Y. (1980).
4. C. Greskovich and J.H. Rosolowski, "Sintering of Covalent Solids", J. Am. Ceram. Soc., 59 (7-8) pp. 336-343 (1976).
5. M.A. Kuzenkova, et. al., "Sintering Behavior of Aluminum Nitride Synthesized in Low-Temperature Plasma," Poroshkovaya Metallurgiya, No. 4 (184), pp. 28-29, April, 1978.
6. M. Trontelj and D. Kolar, "Pressureless Sintering of Aluminum Nitride to High Density", J. Mater. Sci., 8, pp. 136-137 (1973).
7. D.N. Poluboyarinov, et. al., "Technology and Properties of High-Density Aluminum Nitride Materials", Ogneupory, No. 6, pp. 44-47 (1975).
8. W.H. Gritzen, "Alumina as a Ceramic Material," Chap. 14, American Ceramic Society, Columbus, Ohio (1970).
9. G. Slack and T.F. McNelly, "Growth of High Purity AℓN Crystals", J. Cryst. Growth, 34, pp. 263-79 (1976).

**High Purity Alpha Alumina, Type CR-30, Baikowski International Corp., NC

1. The authors wish to thank Forest Burns for his generous work to determine oxygen content by neutron activation analysis.
2. This work is based on the thesis of Glenn C. Culbertson in completing a M.S. in Engineering Materials, University of Maryland.

SINTERING BEHAVIOR OF Er_2O_3-STABILIZED HfO_2 POWDERS PRODUCED BY DIFFERENT PROCESSING METHODS*

G. W. Jordan, M. F. Berard, C. D. Wirkus†
Ames Laboratory
Iowa State University
Ames, Iowa 50011
USA

The sintering behavior of a broad range of erbium oxide-stabilized hafnium oxide compositions which have been processed in different ways is discussed. The processing methods include different ways of dewatering the precipitate (the hydroxide precursor for the oxide) and variations in subsequent heat treatment. Specifically, the dependence of sintered density on calcining temperature, sintering atmosphere, sintering temperature, and presintering treatment is reported. In addition, the morphologies of the precursor and the calcined oxide are discussed and are related to the precipitate dewatering method and to the sintered density.

---

* Supported by USDOE, Office of Basic Energy Sciences, Div. of Materials Sciences, under Contract No. W-7405-ENG-82.

† Now with North American Refractories Co., Curwensville, Pennsylvania, 16833.

The effects of various processing variables and methods upon the sintering behavior and properties of ceramic materials are poorly understood. The objective of ceramic processing investigations, in general, is to better understand these effects so that characteristics such as sintered density can be consistently produced. Most mechanical and physical properties of materials are degraded by the presence of porosity. Such porosity can render materials unsuitable for desired applications, particularly in high temperature environments. In addition, the interpretation of measurements of bulk properties of materials is much more difficult if specimens are porous.

The objective of this work was to determine the effects of variations in processing methods upon the powder morphology and the sintering behavior of a broad range of erbia-stabilized hafnia compositions. Processing methods investigated included different ways of dewatering the precipitate and variations in subsequent heat treatment. Specifically, the dependence of sintered density on calcining temperature, sintering atmosphere, sintering temperature, and presintering treatment was studied.

Hafnia is an appealing material for many high temperature applications because it is chemically inert and very refractory; its melting point is well over 2800°C. However, HfO_2 undergoes a destructive monoclinic/tetragonal polymorphic transformation (similar to that in ZrO_2) over the approximate temperature range 1500 to 1730°C. Therefore, in order to be useful for practical high temperature applications, the structure must be at least partially stabilized by doping with another component, such as Er_2O_3, so the polymorphic transformation does not cause disintegration of the specimen during thermal cycling.

The HfO_2-Er_2O_3 System

The phase diagram of the hafnia-erbia system proposed by Johnstone (1) is shown in Fig. 1. The high hafnia portion of this diagram agrees closely with another version of the hafnia-erbia system proposed by Spiridonov and Komissarova (2). A very broad range of compositions from <10 to >50 mol% Er_2O_3 are single phase fluorite structure solid solutions. This substantial range of solid solubility is the region where HfO_2 is fully stabilized by Er_2O_3. The breadth of this range implies that it may be possible to tailor properties of this phase by manipulating composition. Notice that the fluorite phase is exceptionally refractory and does not decompose below 2600°C unless the Er_2O_3 content exceeds 50 mol%.

The defect structure of the fluorite phase erbia-stabilized hafnia is consistent with the generation of an anion vacancy accompanying the incorporation of every two Er ions into the solution (1). According to this model, the theoretical density of a given composition is described by

$$\rho_{Theoretical} = \frac{6.641\ [x\ (MW_{Er_2O_3}) + (1-x)\ (MW_{HfO_2})]}{(x+1)a_o^3} . \qquad (1)$$

Here x = mole fraction Er_2O_3, $MW_{Er_2O_3}$ and MW_{HfO_2} are the molecular weights of Er_2O_3 (382.52) and HfO_2 (210.49) respectively, and a_o is the lattice parameter in angstroms. The theoretical density decreases from about 10.2 g cm^{-3} for 10 mol% erbia-stabilized hafnia (ErSH)to 9.2 g cm^{-3}

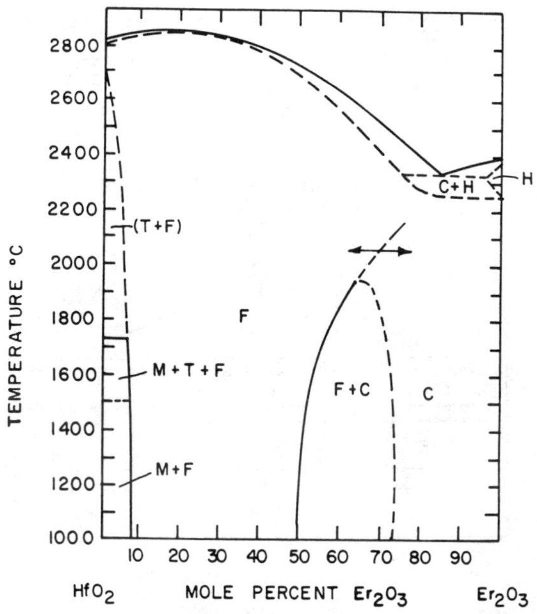

Figure 1 -- Proposed phase diagram for the hafnia-erbia system (1). Phases are: T, tetragonal structure; M, monoclinic structure; F, fluorite structure; C, rare earth C-type structure; and H, hexagonal structure.

for 50 mol% ErSH (in part, due to an increase in lattice parameter from 5.125 to 5.228 Å). In this paper, sintered bulk densities of specific compositions are referred to as a percentage of the theoretical density predicted by Eq. 1.

Experimental Procedure

A general flow diagram of the ceramic processing steps employed in this study to produce oxide specimens is shown in Fig. 2.

Precipitation

Stock solutions, prepared by dissolving erbium oxide in nitric acid and hafnium oxychloride in distilled water, were blended in proportions required to achieve the desired final compositions. Coprecipitation was effected by dripping this acidic mixture containing the cations into an excess amount of 2.5 N aqueous ammonia to assure that the excess nitric acid was neutralized and that precipitation was complete. It was necessary to add the acidic cation blended solution to the base (the so-called "reverse strike" method) because of the differing solubilities of the two cations. This acid-to-base method of precipitation caused the precipitation of both cations to occur instantaneously to produce an amorphous net-

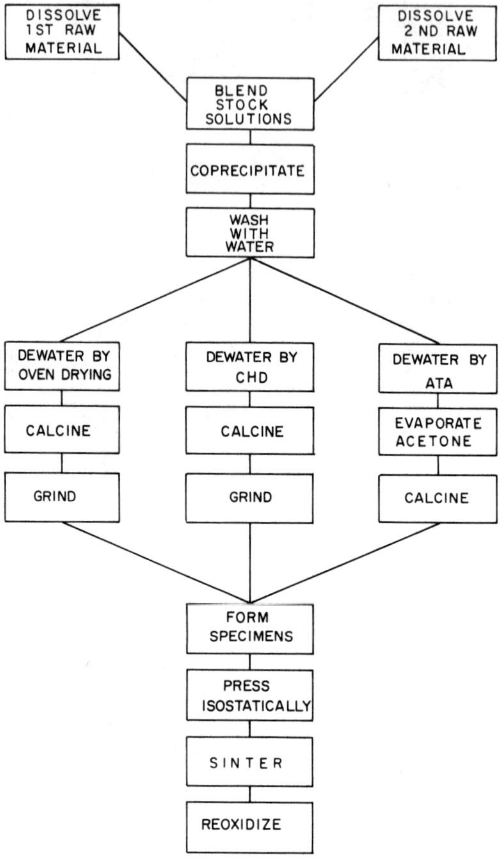

Figure 2 -- General ceramic processing flow diagram.

work which entrapped water and formed a gelatinous precipitate.* At least 150% of the theoretical amount of aqueous ammonia was necessary for the final pH to exceed about 9 so that both oxide components were quantitatively precipitated; lower final pH would have resulted in incomplete precipitation of the Er. The precipitate flocs were allowed to settle, the clear liquor was decanted, and the suspension was washed twice with distilled water to remove any ammonia salts present.

Dewatering of Precursors

The dewatering of reverse strike precipitates is a critical step in the production of powders which will exhibit good sintering behavior.

* It has been shown (3) that highly sinterable single component pure rare earth oxide powders, such as Y_2O_3, can be consistently produced by simply adding the base to the acidic cation solution, i.e. the "direct strike" precipitation method. Coprecipitation necessitates the reverse procedure to assure homogeneity on the micro scale.

Conventional dewatering, or oven drying, is accomplished by placing the water-washed precipitates in a circulating oven at 110°C and drying rapidly to constant weight. After calcination and grinding to pass a 60 mesh sieve, oxide powders produced from reverse strike hydroxides dewatered in this way exhibit inconsistent and generally poor sinterabilities; final sintered densities are usually <95% and often much <90% of theoretical density. For this reason, two alternative dewatering methods were explored in this study. Both of these alternative methods have been shown capable of producing highly sinterable single-component rare earth oxides from reverse-strike hydroxide precipitates.

The first alternative method of dewatering is the slow removal of water by controlled-humidity drying (CHD method) in a chamber in which both temperature and humidity are controlled. In a previous study, CHD conditions of 90°C and 95% R.H. were found to be the optimum conditions for producing highly sinterable single rare earth oxides from reverse-strike hydroxide precursors (4).

The second alternative dewatering method employs a series of organic washes (followed by filtrations) to strip the water out of the precipitate (5). The most frequently used sequence of washes involves first acetone, then toluene and finally acetone again. This so-called acetone-toluene-acetone (ATA) dewatering technique has proven most effective in producing sinterable oxide powders from a wide variety of reverse-strike hydroxide precursors. The final acetone wash was followed by stirring the powder in a pre-heated mortar and pestle to effect rapid evaporation of acetone in a fluidized bed of fine particles.

Specimen Preparation

Following dewatering by either CHD or ATA treatments, the precursors were calcined in air at various temperatures between 600 and 1200°C, usually for two hours. The oxide powders produced by the CHD dewatering method were then ground to pass a 60 mesh sieve since these powders were considerably more agglomerated than those produced by the ATA dewatering technique. The oxide powders were then formed without binder into small cylindrical specimens, first using uniaxial pressing in a double-acting die to ~42 MPa (6 ksi), followed by isostatic pressing to 207 MPa (30 ksi).

Specimens were sintered in vacuum or air to various temperatures between 1450 and 1880°C, sometimes with a prior presintering treatment at a lower temperature. During sintering, temperatures were measured with an automatic optical pyrometer corrected for sight glass absorption. Specimens were contained in covered erbia crucibles; the air firings were performed inside an alumina muffle tube inserted into a natural gas-air furnace. After vacuum sintering, all specimens were heated in air for two hours at 1100°C to correct for small deviations from stoichiometry caused by the mildly reducing conditions.

Specimen bulk densities were determined by standard water immersion procedures, involving an initial vacuum saturation step. For specimens of the size and density of those used in this study, this procedure has been shown to have a standard deviation of 0.013 gm/cm^3 (6).

Results and Discussion

Effect of Dewatering Method on Morphology

All the following micrographs were taken utilizing the scanning mode of a JEM-100 CX scanning transmission electron microscope.* Figure 3 compares the morphologies of an erbia-stabilized hafnia precursor and a pure rare earth oxide precursor when dewatered in various ways. Figure 3(A) shows the morphology of yttrium hydroxide precursor when it is dewatered by the ATA method; Figure 3(B) shows the morphology of the same precursor when it is dewatered by controlled humidity drying (90°C, 95% R.H.). Higher magnifications show clearly that, for these pure rare earth oxide precursors, the morphology consists of agglomerates of platey particles for both dewatering methods, although the plates are larger and thicker for the controlled humidity dewatered precursor. Both of these materials produced oxides which sintered to greater than 95% of theoretical density.

Figure 3(C) is the morphology of the hydroxide precursor of erbia-stabilized hafnia (40 mol% Er_2O_3) dewatered by the ATA method. The morphology is very similar to that of the ATA-dewatered pure rare earth oxide precursor described above. The erbia-stabilized hafnia powder produced from this ATA-dewatered precursor sintered fairly well (i.e., to nearly 95% of theoretical density). Figure 3(D) shows the shard-like morphology of the erbia-stabilized hafnia precursor when it has been dewatered under controlled humidity conditions (90°C, 95% R.H.). The oxide produced from this precursor sintered very poorly to only 70% of theoretical density. This shard-like morphology is also commonly observed for pure rare-earth hydroxide precursors which have been rapidly dewatered by conventional oven drying, a process which normally yields poorly sinterable oxide powders.

Figure 4 shows the morphology of the oxide powders after the dewatered precursors shown in Fig. 3 have been calcined. It is clear that the morphology of the yttrium oxide resulting from precursors dewatered by both ATA and CHD methods and the morphology of the erbia-stabilized hafnia (40 mol% Er_2O_3) resulting from a precursor dewatered by the ATA method are quite similar. In contrast, Fig. 4(D) shows that the poorly sinterable erbia-stabilized hafnia, whose precursor had been dewatered by controlled humidity drying, exhibits a shard-like morphology even after calcining.

Effect of Dewatering Method on Sintered Density

Figure 5 shows the sintered densities of various compositions of ErSH from hydroxide precursors dewatered by ATA and CHD treatments. The curves connect the mean values for three to six replications, and the bars delimit one standard deviation. The ATA dewatering method produced ErSH powders which sintered much better than those dewatered by controlled humidity drying at 90°C and 95% relative humidity at all compositions. This may be because the CHD conditions used, which have been shown to be optimum for precursors of pure rare earth oxides, are not the optimum for precursors of ErSH. Or it may be that precursor ageing during controlled humidity dewatering, which was so effective in producing well-developed platelike particles which calcined to yield highly sinterable rare earth oxides, simply cannot occur with the more gelatinous erbia-hafnia precursors.

---

* JEOL Ltd., Tokyo, Japan.

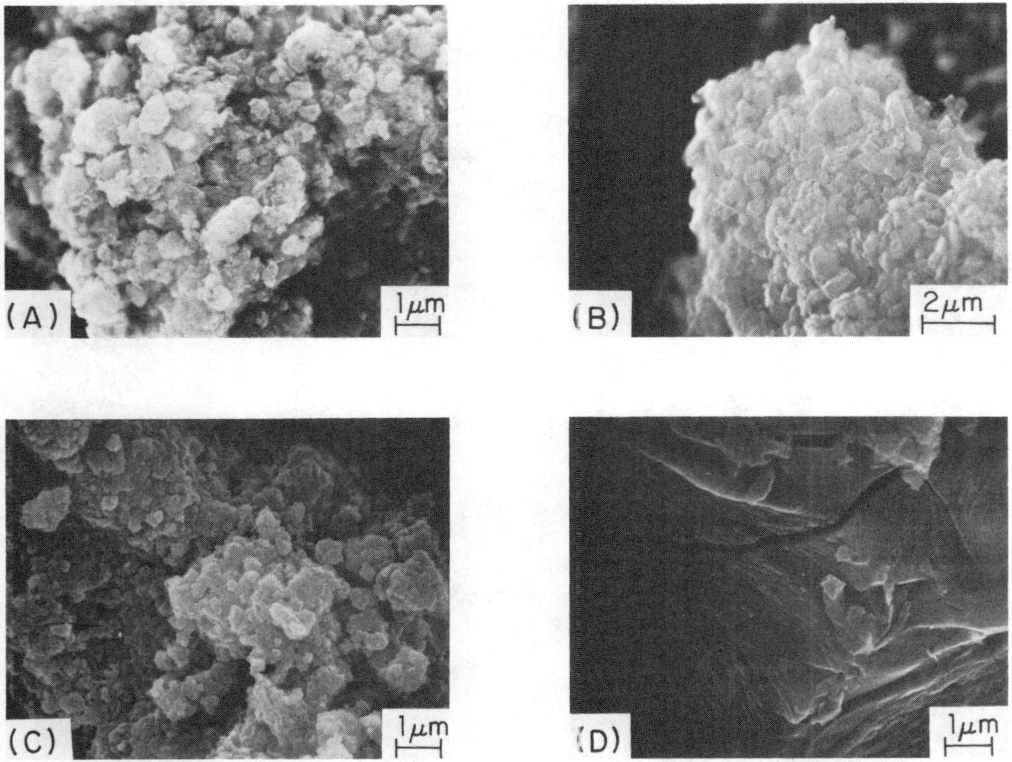

Figure 3 -- Comparison of morphologies of Y_2O_3 and ErSH precursors dewatered in different ways.
Top: Yttrium hydroxide precursor dewatered by the ATA method (A) and CHD method (B).
Bottom: ErSH hydroxide precursor dewatered by the ATA method (C) and CHD method (D).

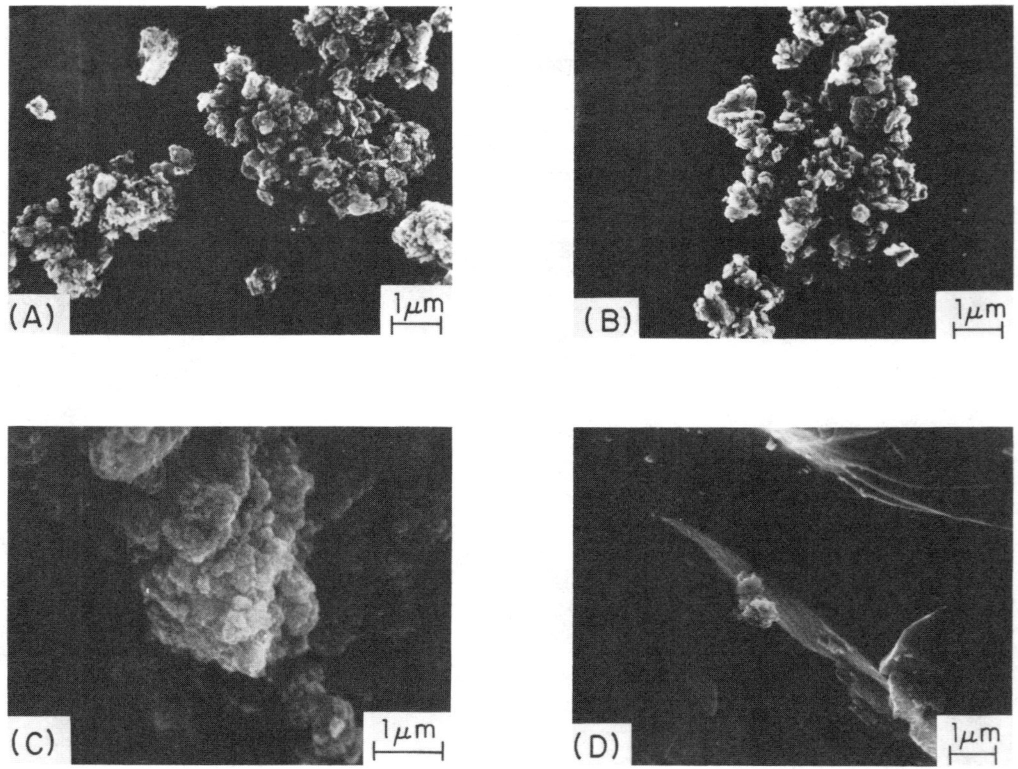

Figure 4 -- Comparison of morphologies of Y_2O_3 and ErSH powders following calcination when the precursor is dewatered in different ways. Top: Y_2O_3 from hydroxide precursor dewatered by the ATA method (A) and CHD method (B). Bottom: ErSH from hydroxide precursor dewatered by the ATA method (C) and CHD method (D).

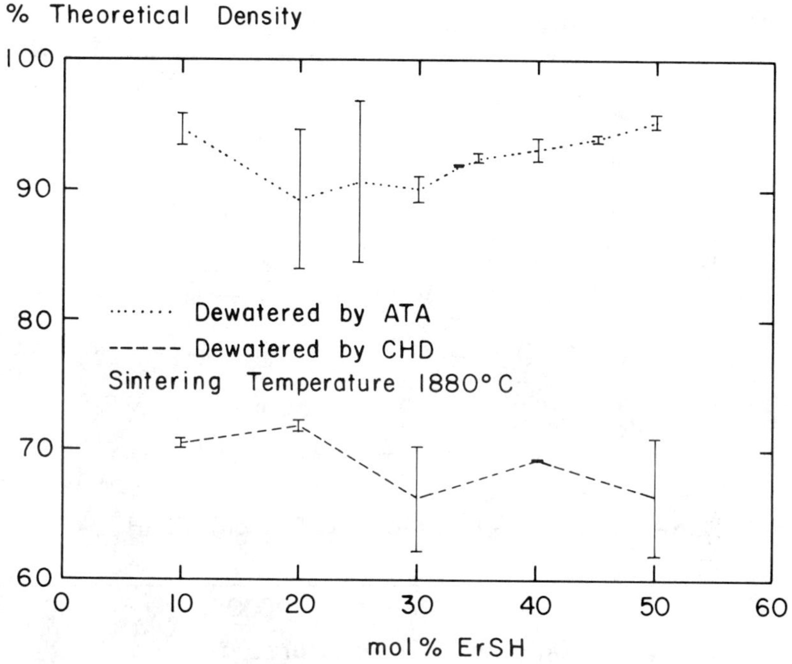

Figure 5 — Dependence of sintered density of erbia-stabilized hafnia on precursor dewatering method and composition.

No obvious differences in powder morphology were observed at different compositions. Nevertheless, the sinterability of the powders did show some dependence upon composition (Fig. 5) when dewatering was done by the ATA method (and when the ErSH specimens were vacuum sintered at a blackbody temperature of 1880°C). Consistent production of high density sintered specimens for the 20 and 25 mol% ErSH compositions could not be achieved with any combination of calcining and sintering conditions investigated. This poor sintering behavior is presently unexplained, although it may be related to the ordered substructures which are known in some anion deficient fluorite systems at comparable compositions (7).

Effect of Calcining Temperature on Sintered Density

The results of four experiments performed to determine the optimum calcining temperature of various erbia-stabilized hafnia compositions are shown in Fig. 6. In order to emphasize sinterability differences, a very low sintering temperature of 1450°C was used. The 10 mol% ErSH composition did exhibit an optimum calcining temperature. When sintered in air (the dotted curve), the effect was not great, but an optimum was found at about 750°C. When sintering was done in vacuum (the dot and dash curve), the optimum at 1100°C was much more pronounced. In contrast, the 20 and 25 mol% ErSH compositions, which have already been described as being difficult to sinter, did not exhibit an optimum calcining temperature. Instead, the density for both air and vacuum sintering treatments increased approximately monotonically as the calcining temperature was increased over the range 600 to 1200°C.

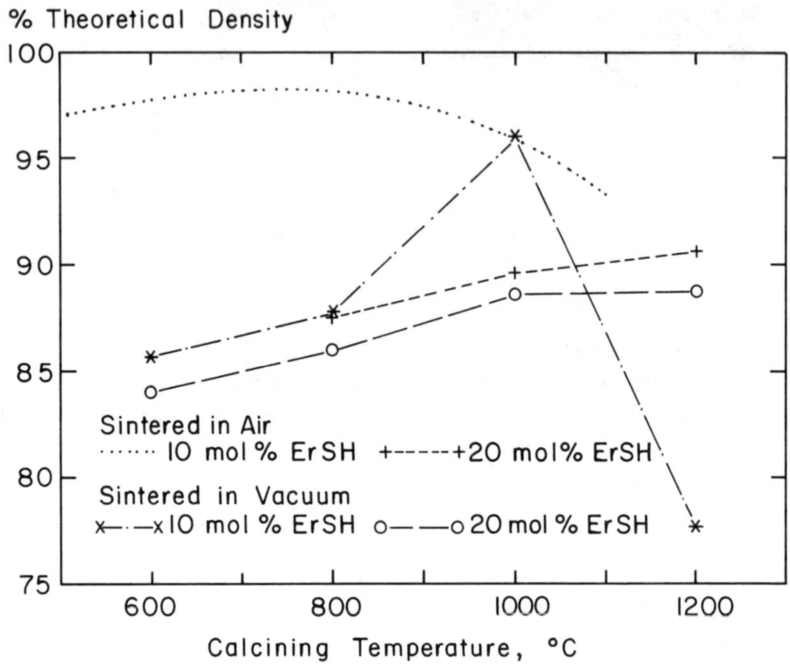

Figure 6 -- Dependence of sintered density of various erbia-stabilized hafnia compositions on calcining temperature. Precursors dewatered by ATA treatment. Sintering temperature was 1450°C.

Effect of Sintering Temperature and Atmosphere on Sintered Density

Figure 7 allows comparison of the sintered densities of various ErSH compositions obtained by sintering in vacuum for 2 hours at 1708 and 1880°C, as well some data from Tesch (6) for specimens sintered in vacuum for 2 hours at 1900°C followed by one-half hour at 2050°C. In general, the higher sintering temperatures produced increases in density, and suggest that vacuum sintering treatments should be carried out above 1880°C to produce densities exceeding 95% of theoretical, especially at the "problem" compositions of 20 and 25 mol% Er_2O_3.

Figure 7 also compares sintered densities produced by sintering in air or vacuum at slightly above 1700°C. With the exception of the 10 mol% composition, where there was no difference, air sintering produced consistently higher densities than vacuum sintering at the same temperature. A part of this improvement may be due to the slower heating rates inherent in a combustion furnace compared to a vacuum resistance furnace. This may allow for more complete pore removal at lower temperatures during heat up prior to the onset of rapid grain growth at the higher temperatures.

It has been suggested by Tesch (6) that a presintering treatment in air at 1400 to 1600°C improves the final density produced by vacuum sintering. This was investigated by presintering specimens in air for one hour at 1500° and 1600°C prior to a vacuum sintering treatment for two hours at

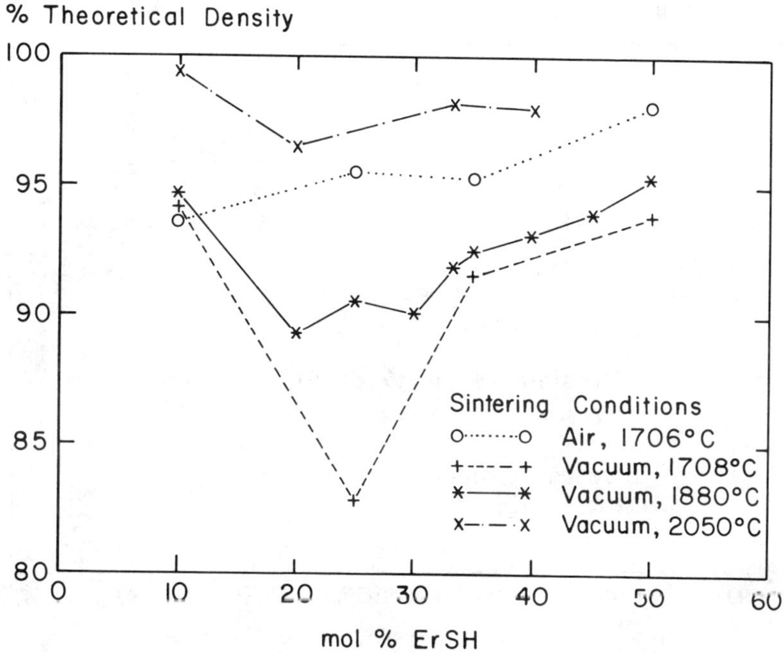

Figure 7 — Dependence of sintered density of erbia-stabilized hafnia on sintering atmosphere and vacuum sintering temperature. (Data at 2050°C from Reference 6). All powders derived from ATA treated precursors.

1880°C. Figure 8 shows the results of the final densities for the presintered specimens compared with those for a vacuum sintering treatment at the same temperature with no presintering. Except for the 10 mol% composition, significant improvement in final density was produced by using an air presintering treatment. Although not shown in Fig. 8, air presintering of powders derived from CHD treated precursors also somewhat improved the final densities produced by vacuum sintering, but these CHD-derived powders still remained poorly sinterable.

<u>Summary</u>

Highly sinterable erbia-stabilized hafnia (ErSH) powders can be produced from ATA dewatering of hydroxide precursors; however, controlled-humidity dewatering of these precursors produces poorly sinterable powders. ATA dewatering of ErSH precursors produces a platey morphology which is essentially retained when the precursors are calcined to the oxides. The same morphologies have been observed in ATA dewatered pure rare-earth hydroxide precursors and the resulting oxides. When ErSH hydroxide precursors are subjected to a controlled-humidity dewatering (CHD) treatment, a shard-like morphology is produced; whereas CHD treatment of pure rare-earth hydroxide precursors produces a well-developed platey morphology. There is a strong correlation observed between platey

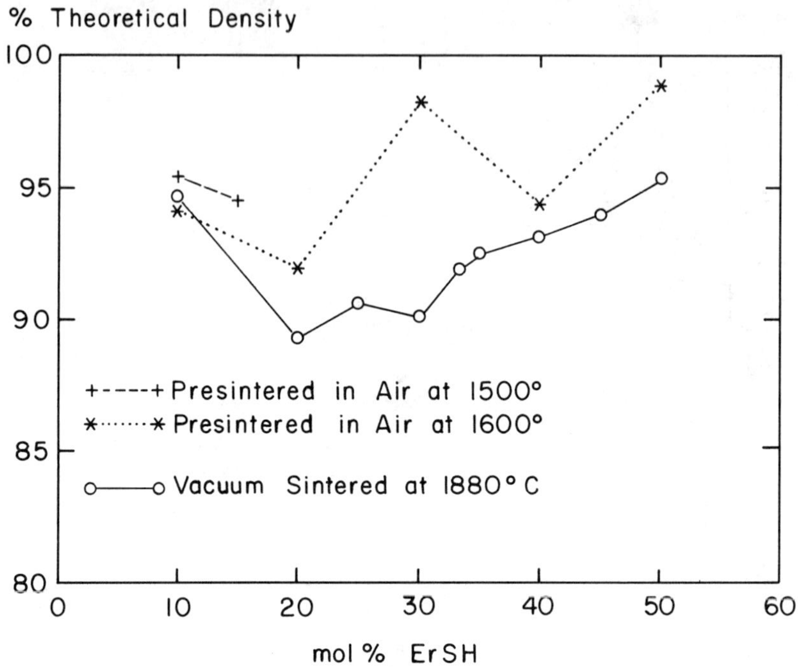

Figure 8 -- Dependence of sintered density of erbia-stabilized hafnia on presintering treatment. Precursors dewatered by ATA treatment.

morphology for precursors and calcined powders and high sintered density.

For a given set of processing conditions, powder sinterability was found to be composition dependent, with the 20-25 mol% ErSH compositions being particularly difficult to sinter to high density. The existence and value of an optimum calcining temperature was also found to be dependent on composition as well as on the sintering atmosphere.

When sintering was performed in vacuum, the density of most compositions was found to increase with increasing temperature between 1700 and 2050°C. When sintering was performed in air at 1700°C, a significant increase in density was obtained for most compositions over that produced by sintering in vacuum at the same temperature. Limited data showed that presintering in air at 1500 or 1600°C prior to vacuum sintering at 1880°C produced a marked increase in density compared to that produced by vacuum sintering without the air presintering step.

References

1. J. K. Johnstone, "The Erbia-Hafnia System," Ph.D. Thesis, Iowa State University, Ames, Iowa, 1970.

2. F. M. Spiridonov and L. N. Komissarova, "HfO_2-Er_2O_3 System," Russian Journal of Inorganic Chemistry, 15 (3) (1970) pp. 445-447.

3. Private Communication, Mark D. Rasmussen, Ames Laboratory, Iowa State University, Nov. 1980.

4. G. W. Jordan and M. F. Berard, "Production of Highly Sinterable Oxide Powders by Controlled Humidity Dewatering of Precursors," paper presented at American Ceramic Society 83rd Annual Meeting, Washington, DC, May 1981.

5. S. L. Dole, R. W. Schiedecker, L. E. Shiers, M. F. Berard, and O. Hunter, Jr., "Technique for Preparing Highly-Sinterable Oxide Powders," Materials Science and Engineering, 32 (1978) pp. 277-281.

6. R. J. Tesch, "Self-Diffusion of Hf in Er_2O_3-Doped Polycrystalline HfO_2," M.S. Thesis, Iowa State University, Ames, Iowa, 1981.

7. H. J. Rossell and H. G. Scott, "Ordering in Fluorite-Related Oxide Systems," J. de Physique, 38 (12) (1977) pp. C728-C731.

POWDER PROCESSING OF POTASSIUM ALUMINOSILICATES

Lawrence P. Cook

Center for Materials Science
National Bureau of Standards
Washington, DC 20234

Two processing alternatives have been investigated for overcoming the poor densification obtained during sintering of potassium aluminosilicates: 1--reaction sintering, employing $KAlSi_3O_8$ glass and $KAlO_2$ in the proper proportions to yield $KAlSi_2O_6$ or $KAlSiO_4$ and 2--use of submicron-sized single phase powders of $KAlSi_2O_6$ and $KAlSiO_4$. Method 1 works reasonably well for production of $KAlSi_2O_6$ approaching 90 percent of theoretical density, but not well for $KAlSiO_4$. Method 2 yields densities approaching 90 percent of theoretical density for both $KAlSi_2O_6$ and $KAlSiO_4$, provided powder particle sizes are below a certain threshold. Data describing the kinetics of these enhanced densifications at 1400 °C have been obtained.

Introduction

The potassium aluminosilicates comprise a structurally related group of compounds and solid solutions on the join $KAlO_2$-SiO_2 (1). These materials have in common the fact that the silicon and aluminum atoms are in tetrahedral coordination and, further, that all tetrahedra are linked to one another through mutual sharing of their oxygens, in this way building up a continuous three dimensional framework. Several of these materials have high melting points and two, $KAlSiO_4$ and $KAlSi_2O_6$, are inherently stable under conditions of high alkali vapor pressure and in contact with alkali-rich slag. They appear, from a purely chemical standpoint, to have considerable promise as potassium resistant refractories for MHD applicatio ns, provided suitable ceramics can be processed (2). Work has concentrated on developing the science of processing $KAlSiO_4$ and $KAlSi_2O_6$ to the point where dense polycrystalline ceramics for testing can be prepared. This has proved to be a challenging goal. Experience has shown (see below) that when micron-sized powders of potassium aluminosilicates are thermally sintered, little, if any densification results. However, substantially increased densification has been achieved using a reaction sintering method in which $KAlSi_3O_8$ glass was reacted with $KAlO_2$ to produce $KAlSi_2O_6$ or $KAlSiO_4$ (2). The partial success of this method is thought to be due to enhanced volume diffusion associated with presence of viscous $KAlSi_3O_8$ glass, which behaves essentially as a solid, but one with high atomic mobility, even at temperatures above its melting point. Unfortunately, ceramics processed by this method seem to have a substantial amount of residual porosity, mostly in the form of very large pores. Efforts have therefore concentrated on methods for improving the densification to 90 percent of theoretical or more. Two main avenues have been explored: one is essentially a follow-up of the reaction sintering method, and the other explores the use of very fine grained single phase powders.

Reaction Sintering Approach

As noted above, this method depends upon the fact that in the absence of water pressure $KAlSi_3O_8$ does not crystallize from its own highly viscous glass. Therefore one can, in principle, simultaneously synthesize and sinter potassium aluminosilicates according to the following reactions:

$$KAlO_2 + 2(KAlSi_3O_8)^{glass} \rightarrow 3KAlSi_2O_6 \quad (1)$$

$$2KAlO_2 + (KAlSi_3O_8)^{glass} \rightarrow 3KAlSiO_4 \quad (2)$$

To test these possibilities, potassium aluminate was prepared and handled according to methods outlined in ref. (1). $KAlSi_3O_8$ glass was prepared by melting K_2CO_3, Al_2O_3 and low-iron silica at 1575-1600 °C for several hours. The aluminate/glass mixture was ball-milled briefly to homogenize it and dried, and then a commercial cellulose binder with a small amount of ammonium stearate was added. Pellets of 6.35 mm dia. were pressed and fired at a preselected temperature according to methods in ref. (3), except that a prefiring temperature of 600 °C was employed.

Although $KAlSi_3O_8$ glass powders are extremely difficult to mill below a few microns, it appears to be necessary to accomplish this feat if satisfactory elimination of pores is to be achieved. Fig. 1 indicates how closely the porosity of the final $KAlSi_2O_6$ product depends upon the particle size of the $KAlSi_3O_8$ glass powder used in sintering reaction [1]. For example, estimated average particle sizes of 7, 3, and 1 μm produce products

Fig. 1.
Effect of particle size of $KAlSi_3O_8$ glass upon product of reaction sintering (SEM micrographs). (a)-(c) Powdered glass of average particle size 7, 3, and 1 μm; (d)-(f) $KAlSi_2O_6$ ceramics produced by sintering powders. (a)-(c) with $KAlO_2$ at 1400 °C for 1000 min. $KAlSi_2O_6$ ceramics prepared from these powders are designated LC-1, LC-2, and LC-3 respectively.

with estimated densities of 75, 85, and 90 percent of theoretical (these densities are estimates made taking into account the density determined by either geometrical or immersion methods and compensating for the amount of Al_2O_3 introduced during milling). Although parallel improvements are also noted for production of $KAlSiO_4$ by the analogous reaction scheme [2], the maximum achievable density is far less (see Fig. 2), and $KAlSiO_4$ ceramics with densities greater than 75 percent of theoretical have not been prepared by this method.

It was hoped that densities of $KAlSi_2O_6$ ceramics could be substantially increased by varying processing conditions (time and temperature). A shrinkage vs time curve at 950 °C for the three batches of $KAlSi_2O_6$ powder corresponding to Fig. 1 is given in Fig. 3. Increasing time beyond 17 h does not appreciably increase densification. Similar data for 1400 °C are shown in Fig. 4; from this diagram it can be seen that processing time is

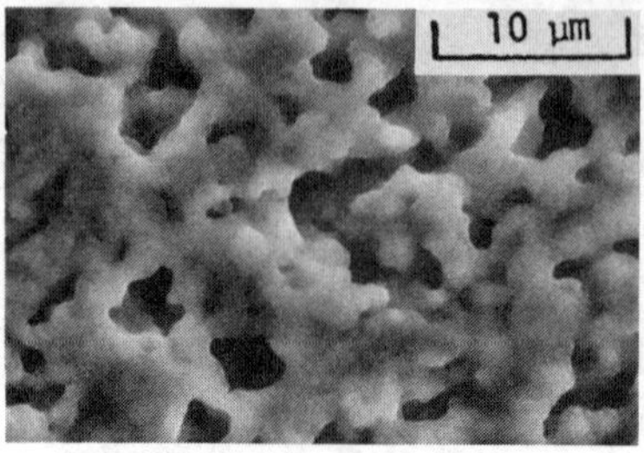

Fig. 2
$KAlSiO_4$ ceramic prepared by reaction using glass powder in Fig. 1-C, 1400 °C, 1000 min (SEM micrograph).

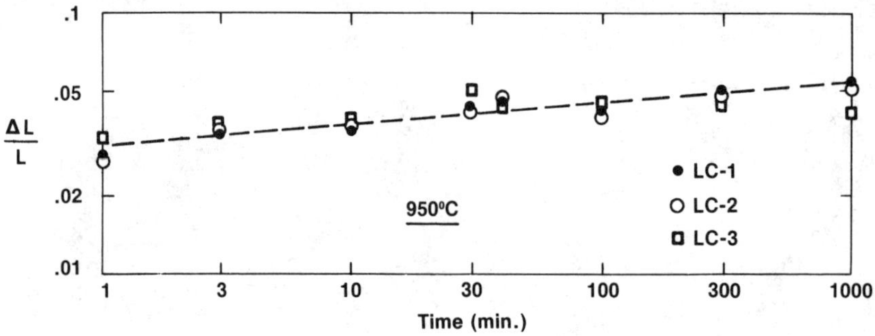

Fig. 3
Shrinkage vs time (log/log) plot for LC-1, LC-2, LC-3 ceramics sintered at 950 °C.

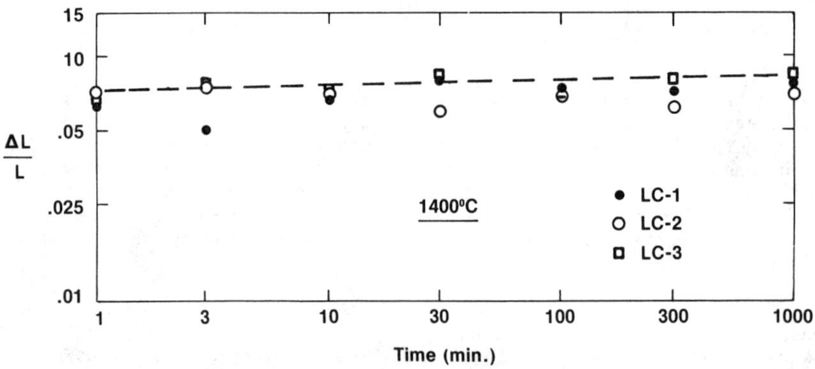

Fig. 4.
Same as Fig. 3, except data taken at 1400 °C.

again not a significant factor at this temperature. While increasing the sintering temperature from 950 to 1400 °C increases the net shrinkage reached in 17 h appreciably, densities still do not exceed 90 percent of theoretical. Examination of Fig. 5 confirms that essentially all of the densification occurs within a short period of time and that subsequent changes involve primarily increases of both pore size and grain size. Other experiments have shown that densification begins at a temperature between 900 and 950 °C.

A B

C D

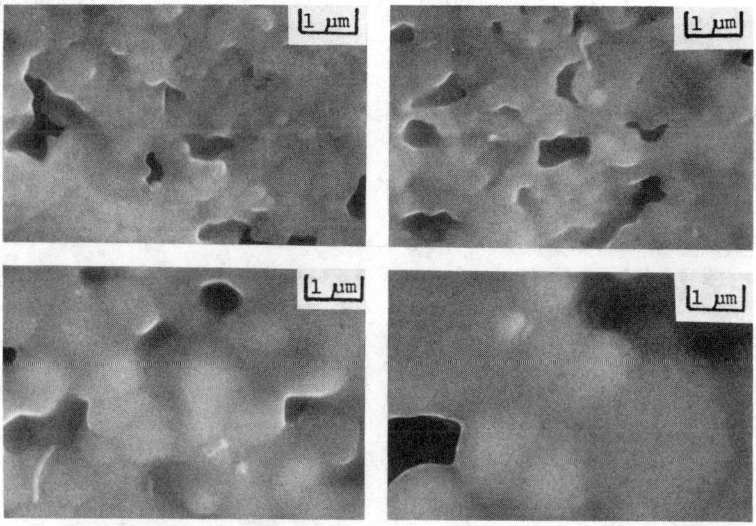

Fig. 5.
Effect of sintering time on microstructure of LC-3 ceramics. SEM micrographs from specimens sintered by reaction method at 1400 °C for times of 1, 10, 100, and 1000 min, respectively.

Use of Submicron-Sized Powders

It has been found experimentally that in certain situations, use of ultra-fine grained powders results in greatly improved densification (4). However, production of the necessary homogeneous fine-grained powders for this approach is a major technical problem in the case of potassium aluminosilicates. Three oxides are involved and the required low temperature phase equilibrium data are lacking if coprecipitation or gel methods are to be employed. Mechanical communition was, therefore, the only means readily available to the author for production of submicron sized powders. A ball mill with alumina grinding media was used for this purpose.

It was found that marked improvement in densification can indeed be achieved by extended ball milling of $KAlSi_2O_6$ and $KAlSiO_4$ powders. For comparison, Fig. 6 shows results for sintering at 1400 °C of powders of 3 μm average particle size (prior to extended ball milling). Densification has been negligible, although appreciable sintering has occurred. Note that the microstructural difference between the $KAlSi_2O_6$ and $KAlSiO_4$

ceramics is marked in spite of their proximity in chemical composition. Presumably this is due primarily to high symmetry of $KAlSi_2O_6$ (cubic) vs $KAlSiO_4$ (orthorhombic). If the powders in Fig. 6-A, B are extensively ball milled to produce the powders of particle size 0.5 µm shown in Figs. 7-A, B, appreciably more dense ceramics result from sintering at 1400 °C, as shown in Figs. 7-C, D. Relative densities for these ceramics approach 90 percent. Strengths are dramatically improved over those reported in (2); strengths of $KAlSi_2O_6$ test bars measured by the three point bend method averaged close to 100 MPa.

The kinetics of densification for submicron powders of $KAlSi_2O_6$ and $KAlSiO_4$ at 1400 °C were investigated using methods given in ref. (3). Results are shown in Fig. 8. Shrinkage appears to be proportional to $t^{0.2}$ for $KAlSi_2O_6$ and $t^{0.3}$ for $KAlSiO_4$. An end point in the densification process appears to have been reached at 12-15 percent shrinkage and 300 min firing time. Much of the densification occurs very quickly at 1400 °C. Fig. 9 is a micrograph of $KAlSi_2O_6$ powder sintered for one min at this temperature. Many of the fines have already sintered, resulting in extensive particle bridging.

Fig. 6.
SEM micrographs of single phase 1 µm powders of $KAlSi_2O_6$ (a) and $KAlSiO_4$ (b), and the ceramics produced after firing at 1400 °C, 1000 min (c and d respectively).

Fig. 7.
Same as Fig. 6, after extended ball milling of powders to 0.5 μm average size.

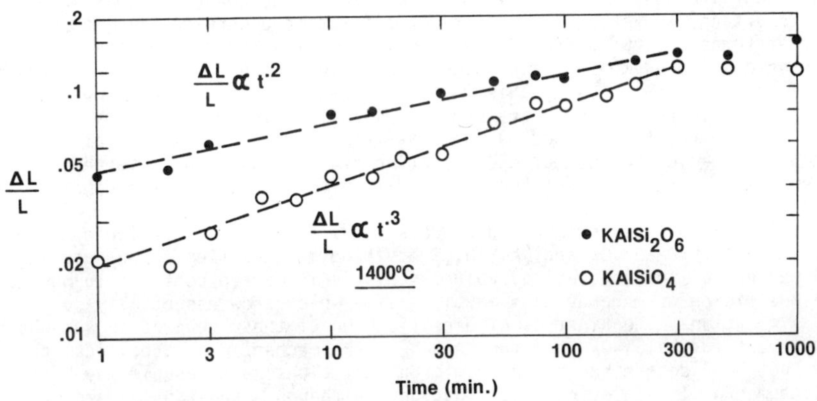

Fig. 8
Shrinkage vs time (log/log) plot for 0.5 μm powders produced by extended ball-milling.

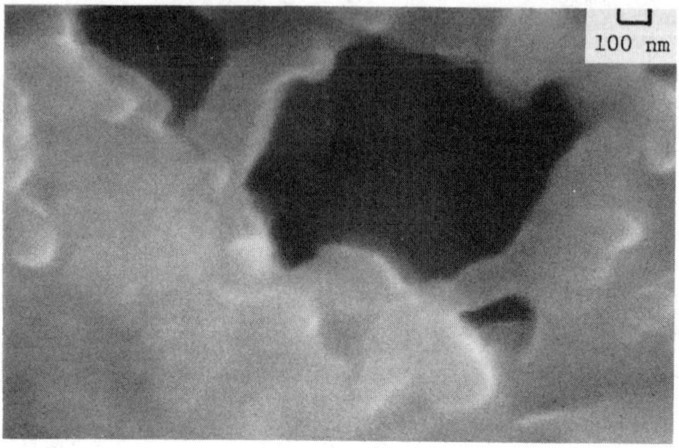

Fig. 9.
Initial stage of sintering in 0.5 μm $KAlSi_2O_6$ powder, sintered at 1400 °C, 1 min (SEM micrograph).

Conclusions

Use of reaction sintering according to the reactions outlined ($KAlSi_3O_8$ glass plus $KAlO_2$) results in improved densification of $KAlSi_2O_6$, to a degree depending upon the particle size of the powdered glass in the starting material. Much of the densification occurs quickly, at temperatures as low as 950 °C, perhaps as the glass transition point is exceeded. Achievement of densities greater than 90 percent of the theoretical value for $KAlSi_2O_6$ by this method may require use of glass powders in the submicron range, which are exceedingly difficult to produce by mechanical means. The reaction method may be unsatisfactory for production of $KAlSiO_4$ ceramics because of the formation of $KAlSi_2O_6$ as an intermediate step and resultant loss of the volume diffusion path needed for densification.

The use of submicron-sized powders leads to greatly improved densification for both $KAlSi_2O_6$ and $KAlSiO_4$, resulting in densities in approaching 90 percent of the theoretical value. This is in marked contrast to powders of one micron or more average particle size which show essentially no densification. The kinetics of densification of these powders is somewhat different for $KAlSi_2O_6$ than for $KAlSiO_4$. The mechanism of densification has not been determined but a submicron particle size threshold may be instrumental in achieving densification. Although a substantial amount of alumina may be introduced during the extended ball milling required to produce these fine powders, the effect of this on sintering and densification is expected to be negligible since $KAlSi_2O_6$-Al_2O_3 and $KAlSiO_4$-Al_2O_3 are stable joins (5). This has been confirmed by additional experiments which will be reported in a subsequent paper (6).

Acknowledgment

The author is indebted to Carl Robbins for use of powder preparation equipment.

References

1. L. P. Cook, R. S. Roth, H. S. Parker and T. Negas, "The System K_2O-Al_2O_3-SiO_2. Part 1. Phases on the $KAlSiO_4$-$KAlO_2$ Join," Amer. Mineral. 62 (11-12) (1977) pp. 1180-1190.

2. L. P. Cook, "Development of Potassium Aluminosilicate Ceramics for MHD Application", pp. 15.3.1-15.3.6 in Proceedings of 19th Symposium on Engineering Aspects of MHD; Univ. Tenn. Space Inst., Tullahoma, TN, 1981.

3. L. P. Cook, "SEM, TEM, and STEM Investigation of Sintering in Anorthite", Bull. Amer. Ceram. Soc. 60 (10) (1981) pp. 851-852. (See also this volume).

4. M. Hoch and K. M. Nair, "Ultrafine Powders of Oxide and Non-Oxide Ceramic Materials and Their Sinterability", pp. 33-40 in Processing of Crystalline Ceramics, Hayne Palmour III, R. F. Davis and T. M. Hare, eds., Plenum Press, New York, NY, 1978.

5. Ernest M. Levin, Carl R. Robbins and Howard F. McMurdie, Phase Diagrams for Ceramists, p. 156; American Ceramic Society, Inc., Columbus, OH, 1964.

6. L. P. Cook, "Critical Minimum Partice Size Hypothesis for Densification of (K, Al)-Silicate Powders", paper to be presented at American Ceramic Society Annual Meeting, Cincinatti, OH, May, 1982.

THE INFLUENCE OF PARTICLE SIZE ON THE SINTERING

KINETICS OF ULTRAPURE SODIUM CHLORIDE

R. J. Thompson and Z. A. Munir

Division of Materials Science and Engineering
College of Engineering
University of California
Davis, California 95616

Sodium chloride particles with size fractions covering the range < 38 to 180 μm were sintered in air in the temperature range 973–1048 K. At any given time and temperature, the degree of densification was found to be a decreasing function of particle size. Calculated activation energies from these densification results showed a corresponding dependence on particle size, i.e., higher energies were associated with the sintering of larger size fractions. Surface area reduction measurements showed a similar trend with respect to particle size. These results are discussed in terms of simultaneous contributions from densifying (bulk diffusion) and non-densifying (evaporation-condensation) processes in the sintering of sodium chloride.

Introduction

The sintering of sodium chloride has been studied by several investigators (1-5). The first definitive study of the nature of the mass-transport mechanism in the sintering of NaCl was reported by Kingery and Berg (1). On the basis of neck growth measurements between spheres of 60-70 μm diameter at temperatures in the range 700-1023 K these authors concluded that the dominant mass-transport mechanism leading to the formation and growth of the observed necks was evaporation-condensation. It was observed that the center-to-center distance of the experimental spheres remained constant and thus mechanisms associated with densifications were not contributing. Moser and Whitman (2) did a similar study and also concluded that the mechanism responsible for the neck growth between contacting NaCl spheres was evaporation-condensation. The latter authors' experiments were conducted in an argon atmosphere utilizing spheres with average diameters ranging from 127-227 μm. In contrast to these two classical experiments, the three most recent investigations (3-5) have demonstrated that densification does accompany the sintering of NaCl, and thus evaporation-condensation is not the only operative mass-transport mechanism. Morgan, Hall, and Yust (3) showed that compacts of a submicron-sized powder densified up to 21.6% when held for only one minute at 1023 K. Ammar and Budworth (4) investigated the effect of particle size on the sintering of NaCl and concluded that densification is possible when the particle size was less than 150 μm. Compacts made of particles with larger sizes showed no densification. In the most recent work on the sintering of sodium chloride, Tomlinson and Astle (5) reported that densification does occur if the compacts were made of particles smaller than 130 μm.

Despite these experimental observations, Swinkels and Ashby (6), in their recent comprehensive analysis of sintering diagrams concluded that NaCl is expected to sinter by an evaporation-condensation mechanism in agreement with the early observations (1,2).

In this paper we present the results of an investigation of the role of particle size in the sintering of NaCl. Six particle size fractions in the range of < 38 to 180 μm were utilized over the temperature range 973-1048 K.

Experimental Materials and Methods

Ultra pure sodium chloride* with the following levels of impurities was used in this work: Al, Cd, Cu, Pb, Co, Ni, Zn, Fe, Mn, Tl, and Mg at ≤ 0.05 ppm; Sr, Ca, and Li at 0.1 ppm; Rb at 1 ppm; and Ba, K, and Cs at 5 ppm.

The powder was milled for 72 hrs using one inch diameter resin-coated stainless steel milling balls. The milled product was then dried at 130° C for 60 minutes, cooled and sieved into fractions with the nominal sizes < 38, 38-63, 63-90, and 90-125 μm. To obtain sufficient amounts of powder in the ranges 125-150 and 150-180 μm, it was necessary to sieve the powder without prior milling. For each size fraction 48 compacts were made through the application of a 10^3 N load on a 0.6502 cm diameter piston. The densities of these compacts ranged from 68.18 to 70.02 of theoretical with an average value of 69.22 and a standard deviation of 0.65.

---

* Purchased from Alfa Products, Ventron Corp.

Sintering was carried out in air inside a Lindberg "Hevi-Duty" 7.62 cm (3 inch) diameter tube furnace at 973, 998, 1023, and 1048 K. Temperatures were controlled to ± 3° C for each of the above settings. Sintering was carried out to 200, 340, 600, 1000, 1600, and 2500 minute intervals. Dimensional and weight measurements were made before and after each sintering run, and surface area measurements were made on groups of three compacts sintered for selected temperatures and times from the 38-63, 63-90, and 90-125 μm particle size fractions. This latter determination was made using a Quantasorb Sorption (BET) System.

Particle image analyses were made on the size fractions listed above. These analyses gave results which markedly differed from those based on the sieve analysis data. The weighted average particle sizes calculated from these image analyses for each of the sieve sizes of < 38, 38-63, 63-90, 90-125, 125-150, and 150-180 μm are 11.9, 15.2, 16.0, 21.2, 18.9, and 17.7 μm, respectively. These results indicate that agglomeration of smaller particles gives rise to a larger nominal sieve size distribution. The problem of agglomeration is especially acute in the two size fractions which were not milled, the 125-150 and 150-180 μm ranges. Figures 1(a) through 1(c) are scanning electron microscope (SEM) images of particles in the nominal size fractions 38-63, 90-125, and 150-180 μm, respectively. It is obvious from these SEM micrographs that numerous fine particles are agglomerated onto the surfaces of much larger ones. In some cases e.g., the 150-180 μm particle size fraction, the fine particles are at least an order of magnitude smaller than the larger, cubic-shaped particles. This non-spherical geometry of the particles was evident in all of the size fractions studied. A general observation can be stated with respect to the particle size disparity within any fraction as examined by the SEM: those fractions which were collected without prior milling had a larger percentage of much finer particles (see for example Figure 1(c) in comparison to Figure 1(b)). The severity of agglomeration would suggest that the nominal (sieve) distribution is a better measure of the effective size distribution of the powder, and thus all reference to particle sizes will be made in terms of the fractions stated above. Observations of the microstructures of the various powder fractions justify this approach.

Results

The influence of particle size on the densification of NaCl is shown in Figures 2(a) and 2(b) for the milled samples sintered at 973 and 1048 K, respectively. The results of the unmilled samples (125-150 and 150-180 μm) showed more of a scatter and generally less uniformity in behavior in contrast to the milled samples. Densification results obtained with these two size fractions are shown in Figures 3(a) and 3(b) for the two stated temperatures.

The temperature dependence of the densification of milled and unmilled size fractions is shown in Figures 4(a) and 4(b), respectively, for the data obtained after 340 minutes of sintering. Similar behaviors were observed for the data for the 600 and 1000 min sintering. These figures show that as the particle size increases, the absolute value of the slope of $\ell n\ \Delta\rho/\rho_0$ vs $1/T$ (where ρ is the density and ρ_0 is the initial density) also increases.

Surface area measurements revealed a dependence on particle size as shown in Figure 5 for the 38-63 and 90-125 μm size fractions. The reduction in surface area increases, as expected, with increasing time and is higher at any given time for the smaller particle size. The temperature

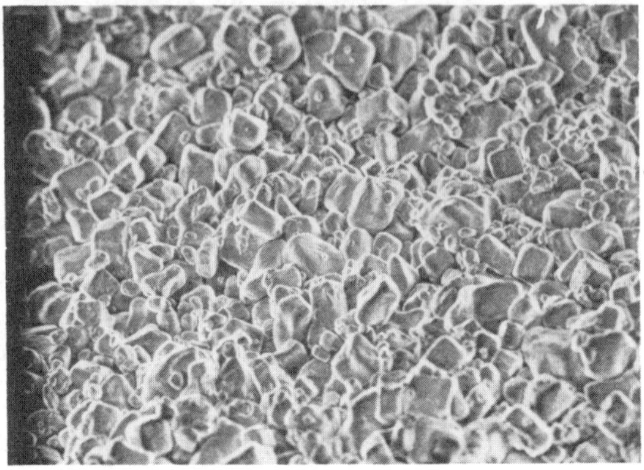

(a) 38–63 μm

(b) 90–125 μm

(c) 150–180 μm

Figure 1 – Unsintered NaCl powder

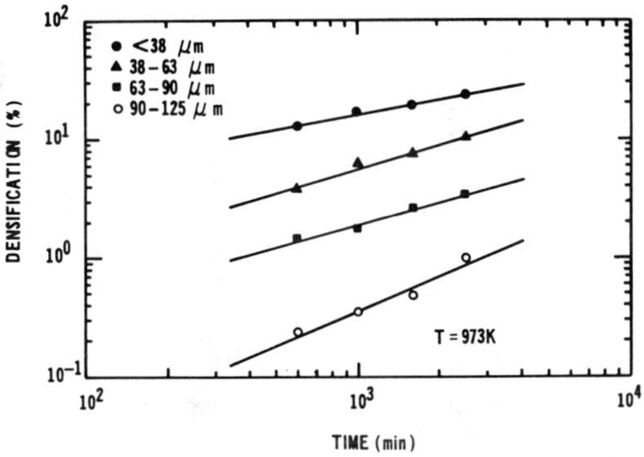

(a) T = 973 K

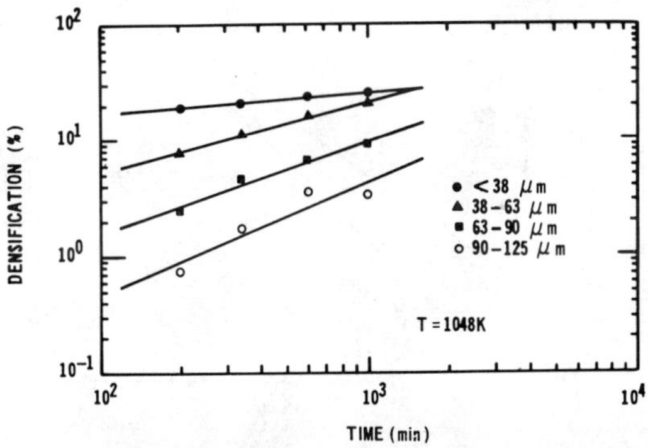

(b) T = 1048 K

Figure 2 – Densification vs time for milled NaCl powders in the size range < 38 to 125 μm

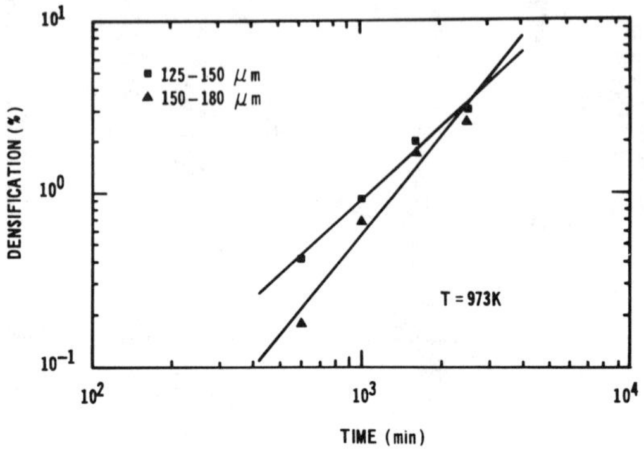

(a) T = 973 K

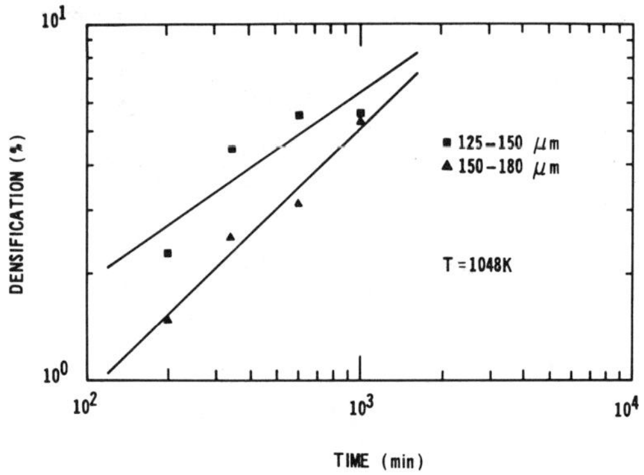

(b) T = 1048 K

Figure 3 – Densification vs time for unmilled NaCl powder in the size range 125 to 180 μm

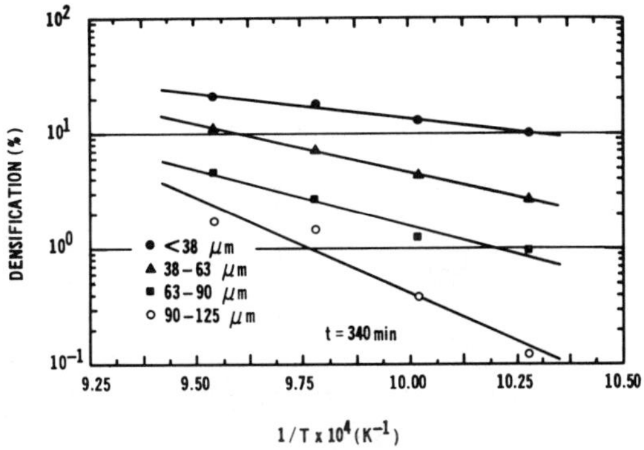

(a) t = 340 min; < 38 to 125 μm

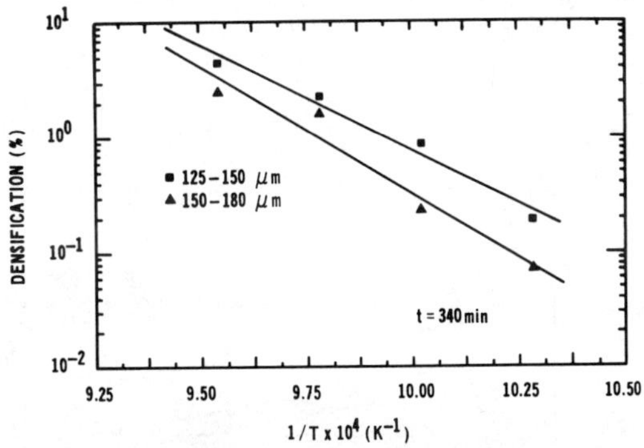

(b) t = 340 min; 125 to 180 μm

Figure 4 - The temperature dependence of the densification of NaCl powders

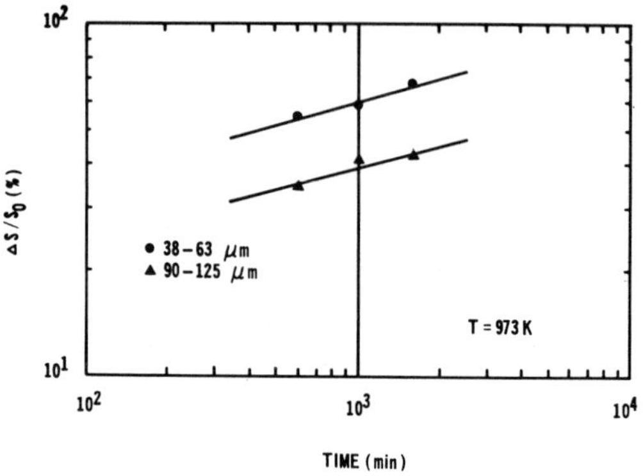

Figure 5 – Time dependence of surface area reduction of two powder fractions

dependence of the surface area reduction is shown in Figure 6 for three size fractions as evaluated from compacts sintered for 340 min. For any given temperature, surface area reduction increases with decreasing particle size.

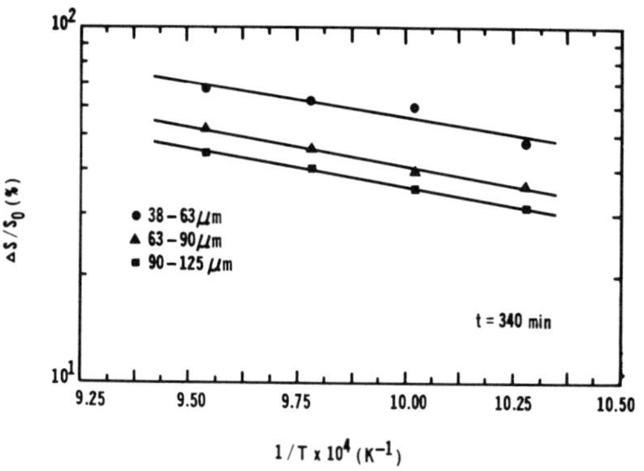

Figure 6 – Temperature dependence of surface area reduction of powders in the range 38 to 125 μm

The morphological changes resulting from sintering at 1023 K for 1000 min are shown by the SEM micrograph sequences of Figures 7, 8, and 9. Figures 7(a) and 7(b) are micrographs of unsintered and sintered compacts for the 38–63 μm size fraction, respectively. Figures 8 and 9 depict similar behavior for the 90–125 and 150–180 μm size fractions, respectively.

Discussion

That the sintering of NaCl exhibits densification is evident from the results given in the previous section. Maximum densification of about 20 was measured for the smaller particle size fractions (Figure 2(b)). However, as stated earlier, the degree of densification depended on the particle size. Figure 10 shows this trend where the data are plotted as ℓn (particle size) vs. ℓn (densification) for four sintering times at 973 K for the four smallest size fractions. Similar trends are obtained when the data generated at the higher temperatures are treated in this fashion. Within the range of the nominal sizes of the present powder, a threshold value for lack of densification is not observed. The work of Ammar and Budworth (4), and Tomlinson and Astle (5) showed that densification occurred when the particle size was less than 150 μm and 130 μm, respectively.

Since the slope of $\ell n\ \Delta\rho/\rho_0$ vs 1/T is directly proportional to the activation energy of sintering, an evaluation of this parameter offers a useful indication of the effect of particle size on the sintering kinetics. Values of the slopes were calculated for each size fraction at three experimental times, and are listed in Table 1. The increase in the absolute value of the slope is indicative of an increase in the activation energy for sintering if mechanism dominence is unaltered. In order to calculate activation energies, knowledge of the sintering mechanism exponent, N, in the relationship $(\Delta\rho/\rho_0)^N = Bt$ must be assumed. In the above relationship B is a materials and temperature related constant and t is time. We have calculated the activation energies by two methods. In the first method we assumed values of N associated with sintering mechanisms which give rise to densification. The second method utilizes average experimental values of N. The results of the calculation by the last method are given in Table 1 with the corresponding activation energies. In both methods, the activation energy increases as the particle size increases for the milled powders. The unmilled powders show a definite inconsistency and this may be the consequence of a wider particle size distribution, as stated earlier. Although the use of the average N values would implicity assume a change in mechanism, use of a constant value would result in a qualitatively similar trend in the activation energies. In all cases the calculated average mechanism exponent increases as the particle size decreases.

In the recent analysis of Swinkels and Ashby (6) it was concluded that only two mechanisms are likely to be operative for NaCl:evaporation-condensation and grain boundary diffusion. The latter mechanism corresponds classically to an N value of 6. As seen from Table 1 the experimental values for the milled powder fractions ranged from 2.2 to 4.9. The fact that densification did occur for all four size fractions implies that even for the largest particle size, evaporation-condensation is not exclusively operative. However, indications of the existence of evaporation-condensation are clear when one examines the SEM micrographs. Figure 11 shows pronounced facetting which is characteristic of evaporation processes (7). Facetting was observed more commonly in samples sintered for longer periods or at higher temperatures.

(a) unsintered (t = 0)

(b) t = 1000 min

Figure 7 - Microstructural changes of a 38-63 μm sized NaCl powder sintered at 1023 K

(a) unsintered (t = 0)

(b) t = 1000 min

Figure 8 - Microstructural changes of a 90–125 μm sized $NaCl$ powder sintered at 1023 K

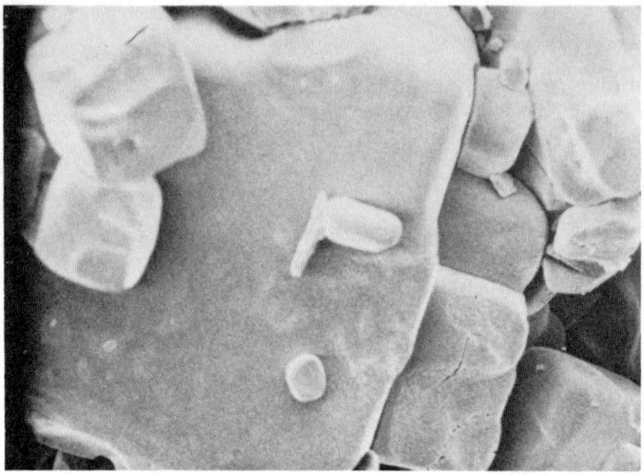

(a) unsintered (t = 0)

(b) t = 1000 min

Figure 9 - Microstructural changes of a 150–180 μm sized NaCl powder sintered at 1023 K

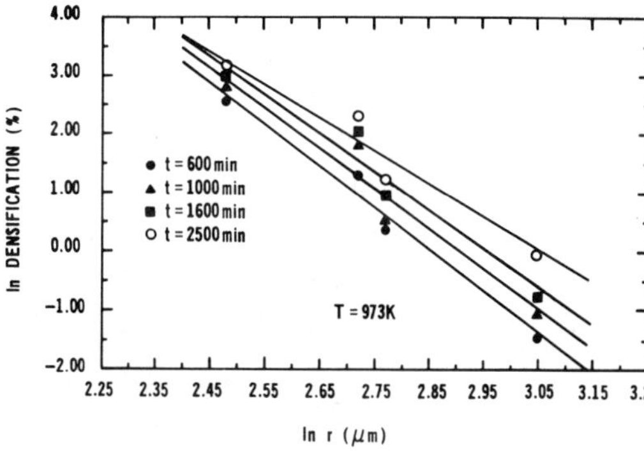

Figure 10 – The dependence of densification on the average particle size for NaCl sintered at 973 K for various times

The present observations contradict the conclusions of Swinkels and Ashby (6) regarding the probable sintering mechanism for NaCl. For the conditions of time and temperature of this study, the sintering diagram predicts evaporation-condensation as the dominant mechanism. It is likely that the discrepancy between the present results and the predictions (6) is a consequence of the tenuous nature of the available diffusion data (6).

The activation energy calculated by Tomlinson and Astle (5) for NaCl compacts with particle sizes in the range 44–50 μm was reported to be 231 ± 5 $kJ.mol^{-1}$. In comparison, the corresponding value calculated in this work for the compacts with the size fraction closest to the above (38–60 μm) ranged from 203 to 251 $kJ.mol^{-1}$, Table I. Tomlinson and Astle (5) interpreted their observations of shrinkage in terms of a volume diffusion mechanism. They cited a reasonable agreement between their calculated activation energy (231 ± 5 $kJ.mol^{-1}$) and that reported for the diffusion of Cl^- in sodium chloride (8), 260 $kJ.mol^{-1}$. The values used by Swinkels and Ashby (6) for the lattice, grain boundary, and surface diffusion activtion energies are 217, 155 and 217 $kJ.mol^{-1}$, respectively. The value estimated by these authors for evaporation is 182 $kJ.mol^{-1}$. The values calculated from the present results for the smaller particle sizes are closer to the reported volume diffusion data than to the boundary diffusion data. A comparison with the evaporation data is not possible at this point since these calculations are based on densification and not neck growth observations.

A more appropriate base for such a comparison, however, can be obtained from the results of surface area reduction experiments. The results plotted in Figure 5 show a decrease in the surface area with increasing time at 973 K for the 38–63 and 90–125 μm size fractions. In accordance with the analysis presented by German and Munir (9), the dependence of surface reduction on time is

$$(\Delta S/S_0)^{\gamma} = Kt \qquad (1)$$

Table I

Calculated values for the activation energy Q using an averaged experimental N value for each particle size range

Time (min)	Particle Size (μm)	Slope ($x10^{-3}$)	N (Ave)	Q (kJ/mole)
340	< 38	-10.067	4.9	206
	38–63	-19.426	3.1	251
	63–90	-22.277	2.7	251
	90–125	-37.706	2.2	339
	125–150	-42.227	2.4	427
	150–180	-51.177	1.7	362
600	< 38	- 8.759	4.9	180
	38–63	-19.642	3.1	254
	63–90	-20.543	2.7	231
	90–125	-35.375	2.2	318
	125–150	-35.032	2.4	354
	150–180	-38.651	1.7	273
1000	< 38	- 6.167	4.9	126
	38–63	-15.729	3.1	203
	63–90	-22.339	2.7	252
	90–125	-28.843	2.2	259
	125–150	-24.782	2.4	250
	150–180	-27.185	1.7	192

where $\Delta S = S_0 - S$, S is the surface area, S_0 is the initial surface area, K is a constant, t is time, and the exponent γ is related to the sintering mechanism N by $\gamma = N/m$, where m is a coordination and mechanism dependent constant. Details of the derivation of Eq. (1) and the origin of γ are reported elsewhere (10,11). From the two lines in Figure 5, we calculate γ values of 1.97 and 2.13 for the 38–63 and 90–125 μm size fractions, respectively. Theoretically calculated γ values are based on, as stated above, knowledge of the coordination number and the operative mechanism. In this study the unsintered compacts had average fractional densities of 0.69 which correspond to a coordination number, $N_c = 10$. For this N_c value, γ is calculated to be 1.4, 2.7, and 3.2 for evaporation-condensation, volume diffusion, and grain boundary diffusion mechanisms, respectively. Since application of the surface area reduction model (10) is valid up to a $\Delta S/S_0 \simeq 50\%$, the analyis of the 38–63 μm size fraction by this model may not be entirely valid. Nevertheless, the

Figure 11 - Pronounced facetting in NaCl powders sintered at 1048 K for 600 min, 38-63 μm size fraction

difference in the γ values calculated for this and the other size fractions of Figure 5, is not significant. The γ values calculated from the present results fall between those corresponding to the evaporation-condensation and volume diffusion mechanisms. Since the theoretically calculated γ has been shown to have a range of values, the existence of evaporation-condensation and volume diffusion as simultaneous mechanisms can only be inferred from these calculations.

The activation energies calculated from the temperature dependence of the surface area reduction data provide a more convincing argument, however. Figure 6 shows such temperature dependences for the size fractions 38-63, 63-90, and 90-125 μm after 340 min of sintering. Calculation of the activation energies (13) was made using the experimentally calculated γ values (1.97 and 2.13) and the slopes of the lines of Figure 6. The calculations gave a range of 168-203 $kJ.mol^{-1}$ for the activation energy. If we exclude the data which correspond to $\Delta S/S_0 \simeq 50\%$, i.e., the data of the 38-63 μm size fraction, then this energy range is reduced to 199-203 $kJ.mol^{-1}$, or approximately 200 $kJ.mol^{-1}$ for both the 63-90 and 90-125 μm size fractions. The corresponding values of Q for these size fractions obtained from densification data are 251 and 339 $kJ.mol^{-1}$, respectively. Since surface area reduction represents contributions from both densifying and non-densifying processes, the difference between the activation energies is indicative of the relative contributions of evaporation-condensation and bulk (volume or boundary) diffusion. For the 63-90 μm fraction, the activation energy obtained from densification data is 25 higher than that calculated from surface area reduction experiments. In contrast, the activation energy based on shrinkage of the 90-125 μm is 70 higher than the corresponding value obtained from surface area measurements. The relatively small difference between the activation energies for the smaller (63-90 μm) fraction suggests that the contributions of the non-densifying process are less significant. Analogous reasoning leads to the conclusion that the contributions of the non-densifying process are substantial in the sintering of the 90-125 μm particle size fraction.

The considerations presented above lead to the general conclusion that the relative significance of the non-densifying process (presumed to be evaporation-condensation in this case) increases with increasing particle size. This further leads to the concept of an upper particle size limit above which no densification would be expected to take place. This expectation is in agreement with the conclusions of Ammar and Budworth (4) and Tomlinson and Astle (5) who reported 150 and 130 μm, respectively, as upper particle size limits for densification in NaCl powders.

Acknowledgements

This work was partially supported by the Department of Energy.

References

1. W. D. Kingery and M. Berg, "Study of the Initial Stages of Sintering Solids by Viscous Flow, Evaporation-Condensation, and Self-Diffusion," Journal of Applied Physics, 26 (1955) pp. 1205-1212.

2. J. B. Moser and D. H. Whitmore, "Kinetics of Sintering of Sodium Chloride in the Presence of an Inert Gas," Journal of Applied Physics, 31 (1960) pp. 488-493.

3. C. S. Morgan, L. L. Hall, and C. S. Yost, "Sintering of Sodium Chloride," Journal of American Ceramic Society, 46 (1963) pp. 559-560.

4. A. A. Ammar and D. W. Bundworth, "Sintering of Sodium Chloride," Proceedings British Ceramic Society, 3 (1965) pp. 185-189.

5. W. J. Tomlinson and G. Astle, "Sintering Kinetics of Pure and $CdCl_2$-Doped NaCl in Air," Journal of Materials Science, 11 (1976) pp. 2162-2165.

6. F. B. Swinkels and M. F. Ashby, "A Second Report on Sintering Diagrams," Acta Metallurgica, 29 (1981) pp. 259-281.

7. J. P. Hirth and G. M. Pound, "Condensation and Evaporation," Prog. Materials Science, 11 (1963) pp. 107-148.

8. W. D. Kingery, Introduction to Ceramics, p. 229; John Wiley, New York, N.Y., 1960.

9. R. M. German and Z. A. Munir, "A Kinetic Model for the Reduction in Surface Area During Initial Stage Sintering," pp. 249-257, in Materials Science Research, 10: Sintering and Catalysis, G. C. Kuczynski, ed.; Plenum Press, New York, N.Y., 1975.

10. R. M. German and Z. A. Munir, "Morphology Relations During Surface-Transport Controlled Sintering," Metallurgical Transactions, 6B (1975) pp. 289-294.

11. R. M. German and Z. A. Munir, "Surface Area Reduction During Isothermal Sintering," Journal of American Ceramic Society, 59 (1976) pp. 379-383.

12. Z. A. Munir, P. K. Higgins, and R. M. German, "The Variation of the Mechanism Dependent Exponent with Coordination Number and Neck Size in the Sintering Kinetics of Powders," pp. 26-34, in Sintering: New Developments, Materials Science Monographs, No. 4, M. M. Ristic, ed.; Elsevier Publishing Co., Amsterdam, 1979.

13. P. K. Higgins and Z. A. Munir, "The Influence of a Surface Oxide Layer on the Sintering Process of Lead," Powder Metallurgy, 21 (1978) pp. 188-194.

SEM and TEM Investigation of Sintering in Anorthite

Lawrence P. Cook

Center for Materials Science
National Bureau of Standards
Washington, DC 20234

By comparison with other tectosilicates, powdered crystalline anorthite [$CaAl_2Si_2O_8$] shows a strong tendency for densification during sintering. Experiments have been conducted to determine the conditions under which this behavior is reproducible and to determine the mechanism or mechanisms responsible. Results indicate that while minor changes in grain size distribution have a large effect upon the kinetics of sintering and densification, relatively coarse powders ultimately yield as dense a product as powders with a much higher percentage of fines. A two-stage hypothesis for densification at 1400 °C is proposed: (1) an early rapid densification associated with grain boundary diffusion driven by minimization of highly anisotropic surface energy; (2) volume diffusion, primarily along closely spaced parallel twin planes.

Introduction

The need for new materials--especially inexpensive, durable ones--is continually created as raw materials for established products become in short supply, or as new applications appear (1). In recent years the cost of energy required for processing has become an important consideration. Also, particularly in ceramics, the focus has also been on improving processing to the point where properties can be expected to reproducibly reach the theoretical levels which make these materials so desirable for applications demanding high temperature strength and corrosion resistance. Thus, basic understanding of the mechanisms operative during sintering and densification remains an important goal.

In this context, study of the sintering behavior of crystalline anorthite [$CaAl_2Si_2O_8$] is of interest. It was observed in this laboratory (see Fig. 1) that anorthite, in contrast to structurally related tectosilicates such as leucite [$KAlSi_2O_6$] and $KAlSiO_4$ (2), shows a marked tendency for sintering and densification. Thus from a purely scientific standpoint, further investigation seemed desirable. There are some practical implications as well, stemming from the low cost of the raw materials needed for anorthite ceramics, the potential of anorthite as a refractory in slagging applications of moderate temperature, and the suggestion that the relative ease with which dense, pure anorthite ceramics can be processed might lead to increased applications in whitewares and electronics.

The present study thus had essentially two aims: (1) to determine under what conditions the sintering behavior of anorthite is reproducible and (2) to determine, if possible, the mechanisms of densification.

Powder Processing

Starting materials used to prepare anorthite were crystalline $CaCO_3$, SiO_2 (quartz), and α-Al_2O_3. The $CaCO_3$ was of analytical reagent grade and contained less than 0.03 wt.% alkalies. This material, when dried at 160 °C and calcined at 1200 °C, typically showed a weight loss corresponding to stoichiometric $CaCO_3$. Two grades of silica were used. The purest grade was optical quality, colorless Brazilian quartz, pulverized to $\sim$ 10 μm average grain size, for which spectroscopic analysis showed less than 0.01 wt.% Fe and less than 0.02 wt.% Al. A less pure variety used was a commercial 1.5 micron-sized particulate quartz for which spectroscopic analysis showed 0.05 wt.% Fe and 0.1 wt.% Al. The alumina used was 0.3 μm "Linde A"[1] for which spectroscopic analysis indicated less than 0.01 wt.% Na or Fe. All materials were dried at 160 °C overnight and weighed together on a high-precision balance in a glove box in the proper proportions to maintain anorthite stoichiometry. Following this, they were ground together under acetone in a mortar and pestle and dried.

Three batches of anorthite powder for experimentation were prepared from the above materials according to Table I. The prime differences are in: (a) type of silica used; (b) type of precalcination homogenization; and (c) length of post-calcination ball-milling. For the ball milling step, alumina balls were used, which inevitably lead to slight contamination;

---

[1]Certain commercial equipment, instruments, or materials are identified in this paper in order to adequately specify the experimental procedure. Such identification does not imply recommendation or endorsement by the National Bureau of Standards, nor does it imply that the materials or equipment identified are necessarily the best available for the purpose.

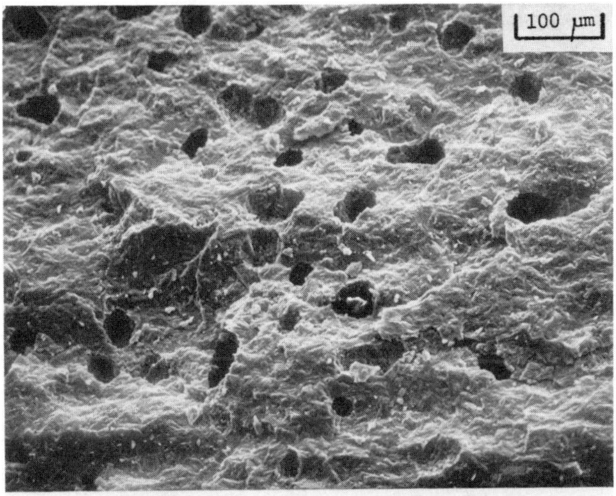

A

B

Fig. 1.
"Self densification" of anorthite, illustrated by uncompacted single-phase powder sintered without pressing for 12 h at 1500 °C. (a) Low magnification SEM micrograph of fracture surface. Note large pores surrounded by pore-free matrix. (b) Higher magnification micrograph showing structure on free surface of grains in pore.

efforts were made to minimize this by using large amounts of sample. For the milling times employed, it is estimated that perhaps as much as 2 wt.% Al_2O_3 may have been added to batch An-2 during milling, whereas for batches An-1 and An-3, contamination was probably at the 1 wt.% level or less, based upon contamination vs milling time curves constructed for the ball mill employed. These small amounts of alumina should not affect the sintering behavior--no melts are formed between alumina and anorthite until temperatures approaching 1550 °C are reached (3). During the 1300 °C calcination step in Table I, virtually complete reaction to produce anorthite occurred, as judged from x-ray diffraction patterns.

Table I. Preparation of Anorthite Powders

Powder Batch	Starting Materials	First Homogenization	Calcination	Second Homogenization	Est. Avg. Particle Size
AN-1	Brazilian Quartz + AR Grade $CaCO_3$ + 0.3 μm Alumina	Ball-Milled 1/2 h	1300 °C 12 h	Ball-Milled 1/2 h	2 μm
AN-2	Brazilian Quartz + AR Grade $CaCO_3$ + 0.3 μm Alumina	Mechanically Mixed	1300 °C 12 h	Ball-Milled 1 h	1 μm
AN-3	1.5 μm Quartz (0.01 wt.% Fe) + AR Grade $CaCO_3$ + 0.3 μm Alumina	Mechanically Mixed	1300 °C 12 h	Ball-Milled 1/2 h	3 μm

The most important differences between powders An-1, An-2, and An-3 are the differences in estimated average grain size (2, 1 and 3 μm, respectively). Micrographs illustrating grain size and grain size distribution are shown in Fig. 2. Also of possible significance is the precalcination ball milling of An-1.

Powders prepared according to Table I were examined by TEM/STEM methods to determine if other differences existed. In general, powders appear to be well-crystallized with no discernible evidence of second phase materials (Fig. 3A). The electron diffraction patterns (Fig. 3B) are sharp, with no evidence for extra reflections, indicating that the material has the primitive unit cell at room temperature (4). One feature worth noting, however, is the abundance of polysynthetic twinning (Fig. 4). At least two twin laws are operative in Fig. 4. Such twins occur frequently in the material examined, with a density of several per hundred nanometers. The number of these twins which represent growth twins present at high temperature is not known (see discussion below).

Sintering Experiments

Pellets of small mass (~ 100 mg) were pressed from the powders described above using a 6.35 mm dia. die at pressures of 34.5 MPa. A polyvinyl alcohol binder was used with a small amount of ammonium stearate added as a lubricant.

A  B

C

Fig. 2.
SEM micrographs of powders AN-1, AN-2, and AN-3, respectively.

By using a small die of this type, internal pressure variations were minimized during pressing of the pellets.

One of the first possibilities considered as an explanation of the high tendency for densification exhibited by anorthite was the existence of an intergranular melt phase, presumably due to non-stoichiometry. To test this possibility, pellets of batch An-2, prepared as described above, were sintered at 1300 °C and 1500 °C for similar lengths of time. As examination of Fig. 5 shows, the amount of any melt present due to nonstoichiometry should increase manyfold from 1300 °C to 1500 °C; the presence of this melt should become obvious at 1500 °C. Fig. 6 gives SEM micrographs of specimens sintered at these two temperatures. As can be seen, extensive sintering and grain growth occur at 1300 °C as well as 1500 °C. The 1500 °C specimen has larger grains and fewer pores, but no evidence of large amounts of an intergranular melt. Thus, it appears that existence of an equilibrium melt phase is ruled out. However, presence of a thin intergranular layer is still a possibility, and this will be commented on below.

To conduct kinetic studies of anorthite sintering, pressed pellets were first prefired at 1100 °C for about five minutes. In general, no measurable shrinkage was recorded during this step, although a slight expansion was measured upon cooling. Next, the pellets were individually placed in small boats of Pt foil (covered for the longer experiments) and inserted in a rapid recovery silicon carbide resistance furnace. Specimens were allowed one minute to reach thermal equilibrium, timed, removed, cooled, and measured to 0.03 mm with a micrometer. Four measurements were taken on each pellet and the results were averaged. From these data, the shrinkage vs time curves shown in Figs. 7 and 8 were constructed. Figure 7 shows shrinkage vs time

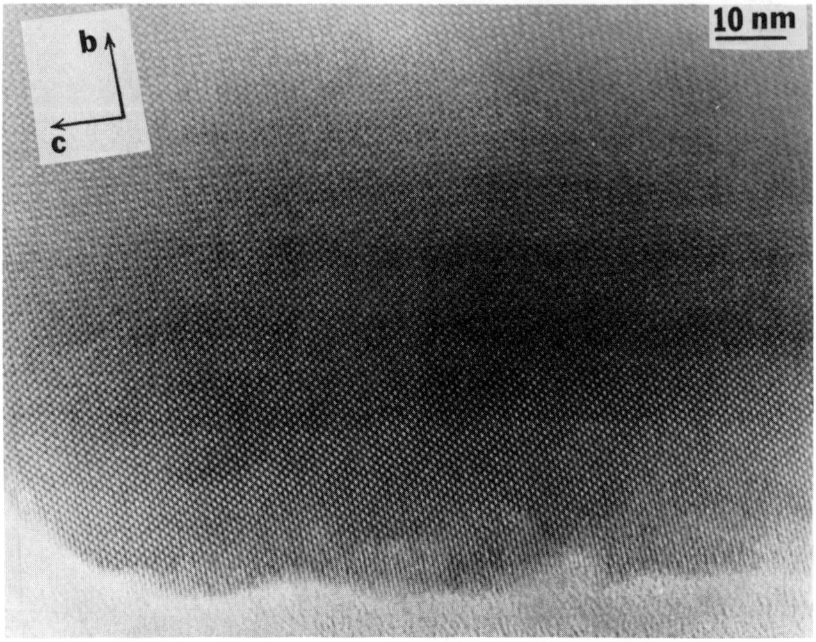

A

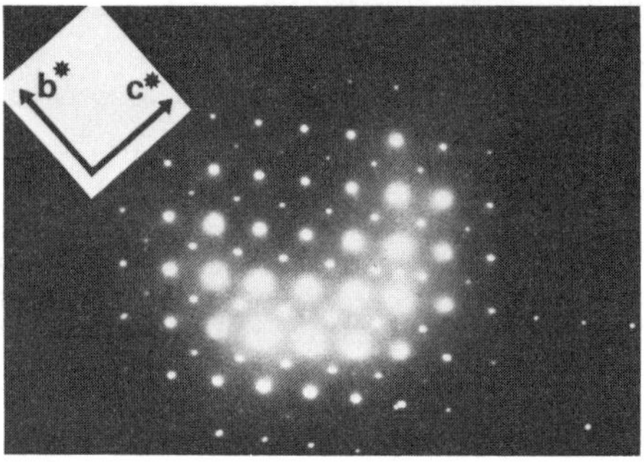

B

Fig. 3.
TEM investigation of powders. (a) Lattice image of anorthite particle from powder AN-1. (b) Electron diffraction pattern from crystal in (a).

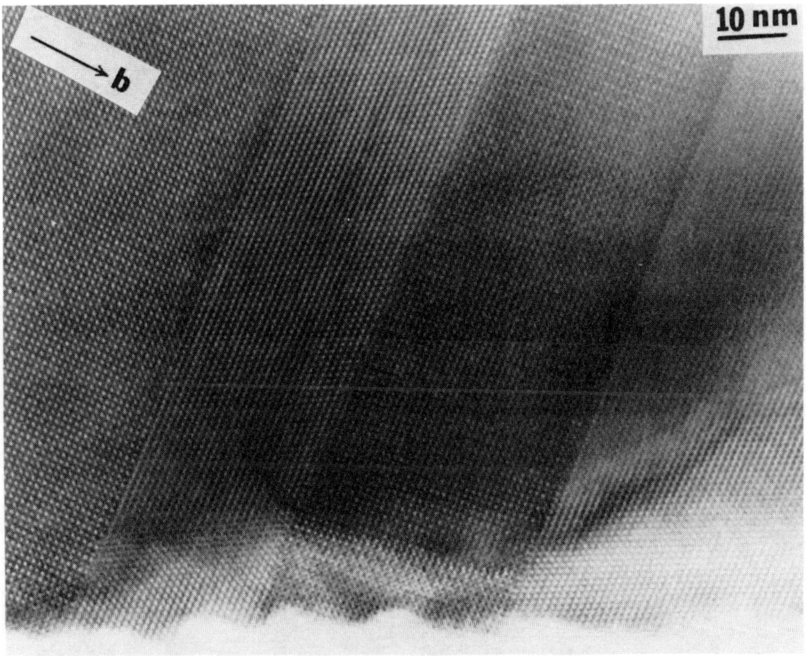

Fig. 4.
Lattice image of anorthite particle from powder AN-1
showing multiple twinning.

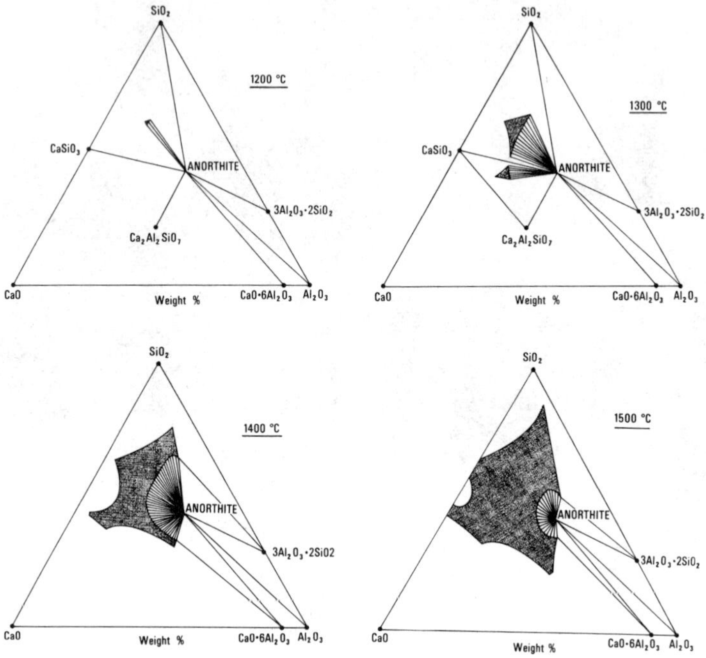

Fig. 5.
Isothermal sections for the system CaO-Al_2O_3-SiO_2, after ref. (3). Shading shows area of equilibrium melt phase.

A 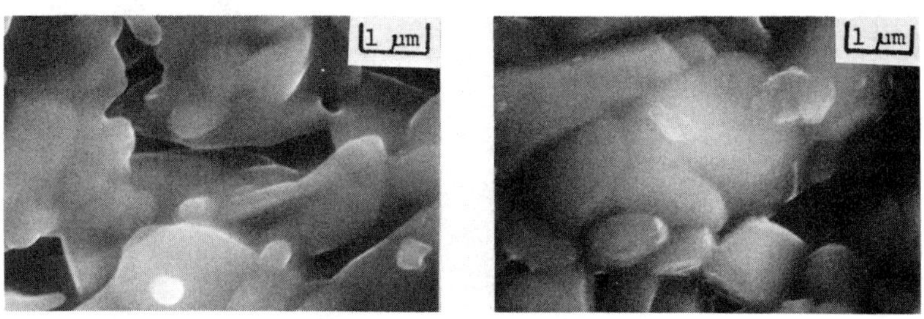 B

Fig. 6.
Effect of temperature on sintering of powder AN-2. (a) SEM micrograph of specimen sintered 1300 °C, 1.5 h. (b) Same, sintered for 1500 °C, 1 h.

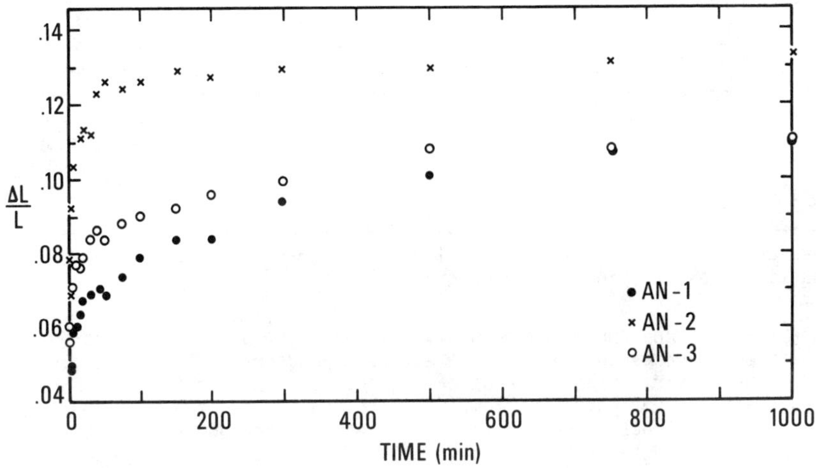

Fig. 7.
Shrinkage vs sintering time at 1400 °C for three anorthite powders (linear scales).

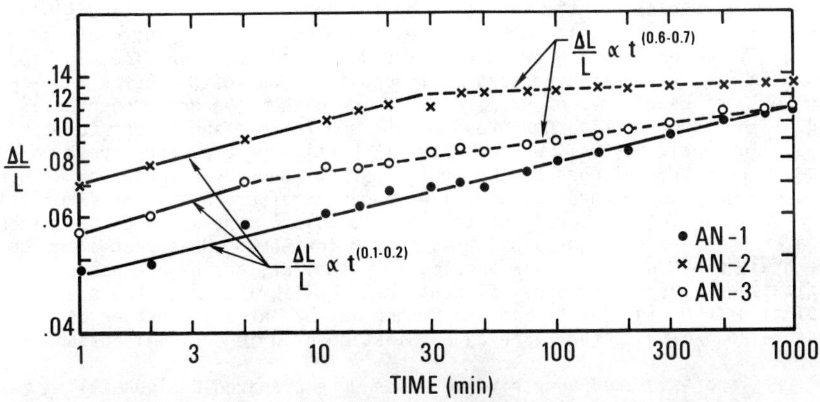

Fig. 8.
Shrinkage vs sintering time at 1400 °C for three anorthite powders (log/log plot).

on a linear scale, and it is apparent that the bulk of the densification occurs within the first few minutes for powder An-2, with little if any increase thereafter; whereas for An-1 and An-3, behavior appears to be asymptotic and shrinkages are much lower. Table II summarizes densities measured (using an immersion method) after 17 h; densities are similar for all three powders and are clearly in excess of 90 percent of theoretical, even allowing for alumina contamination. Figure 8 shows the same data plotted on a log/log basis. While data for An-1 plots on an essentially straight line,

Table II. Densities of Sintered Anorthite Powders*

Powder	Sintering Conditions	Density, % Theoretical
AN-1	1400 °C, 17 h	92
AN-2	1400 °C, 17 h	96
AN-3	1400 °C, 17 h	94

*Measured by adjusting liquid density to the point where the ceramic particle was indefinitely suspended.

data for An-2 and An-3 require two lines; however, one of these for each powder may have essentially the same time dependence as An-1; i.e., proportionality to $t^{(0.1-0.2)}$. The other segments of data for An-2 and An-3 are essentially proportional to $t^{(0.6-0.7)}$. A series of SEM micrographs showing microstructural evolution of pellets of each powder with time is given in Figs. 9 through 11. In general, the evolution is marked by the following changes: an increase in pore size as the fine fraction rapidly sinters; continued increase in grain size to a point where pore size begins to decrease. This point was apparently reached at between 30 and 100 min of sintering time for all three powders. This feature is most pronounced in the case of An-3 (Fig. 11).

Additional changes occur during continued sintering which have to do with the morphology of the anorthite grains and which are not as visible in Figs. 9-11. In contrast to the changes noted in pore size and grain size, changes in morphology appear to be related to the breaks in slope in Fig. 8. Shown in Fig. 12 are micrographs from opposite sides of the break in slope for the shrinkage curve of An-2, Fig. 8. Note that the grain boundaries in Fig. 12-A (time shorter than break point) are rounded and appear to lack crystallographic control, whereas Fig. 12-B (time greater than break point) shows marked crystallographic control of grain boundary orientation. This is a change which occurs somewhat independently of grain size as examination of Fig. 13 suggests. Specimens in Figs. 13A, 13-B were sintered for times greater than the slope breakpoint in Fig. 8 for An-3. Note that although the grain size in Fig. 13A is considerably smaller, grains are euhedral, indicating a large degree of orientational equilibrium among the grains. Evolution from Fig. 13A to Fig. 13-B then may be largely a matter of overall grain size growth, rather than an elimination of orientational instabilities.

Foils of sintered anorthite ceramics were prepared by ion milling to electron transparency suitable for TEM/STEM examination. However, due to the action of the ion beam on the sample, which produces an amorphous layer as the sample becomes thin, results to date have been inconclusive on the existence of an intergranular phase. Fig. 14 illustrates the problem.

Discussion

The data in Fig. 8, specifically the breaks in slopes, might under certain conditions be explained as due to a bimodal grain size distribution (5). Examination of Fig. 2 does indeed suggest an increased tendency for bimodality of An-2 and An-3 relative to An-1. However, the author would like to suggest that the breaks in slope for An-2 and An-3 are not the direct result

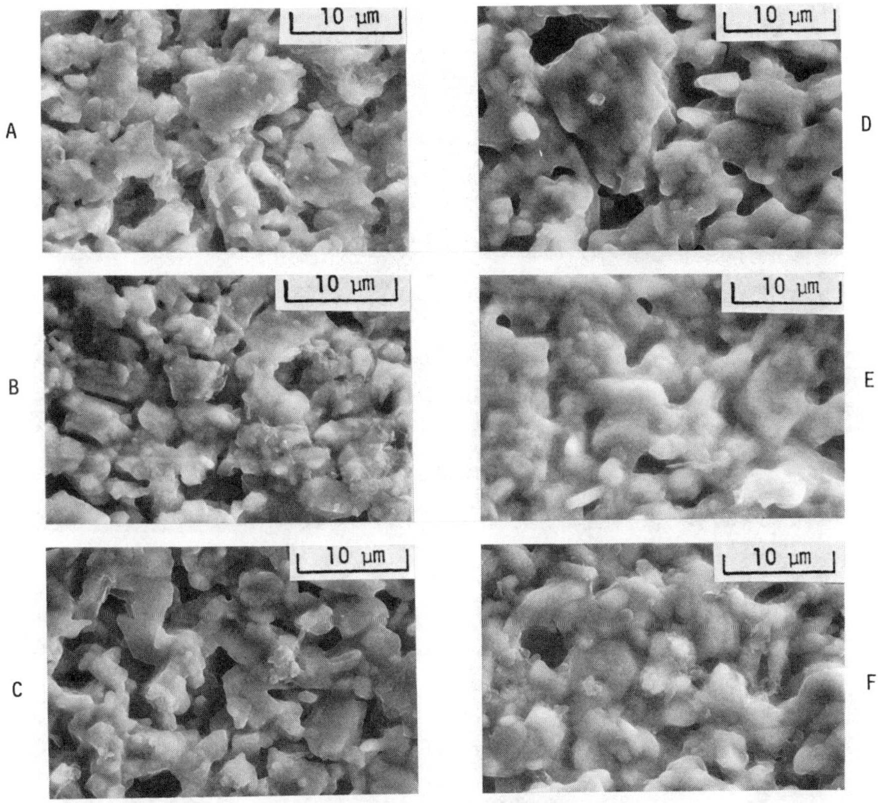

Fig. 9.
Microstructural evolution of powder AN-1. (a)-(f) are SEM micrographs of thermally etched surfaces for sintering times of 5, 10, 30, 100, 300, and 1000 min, respectively, at 1400 °C.

Fig. 10.
Microstructural evolution of powder AN-2. (a)-(f) are SEM micrographs of thermally etched surfaces for sintering times of 5, 10, 30, 100, 300, and 1000 min, respectively, at 1400 °C.

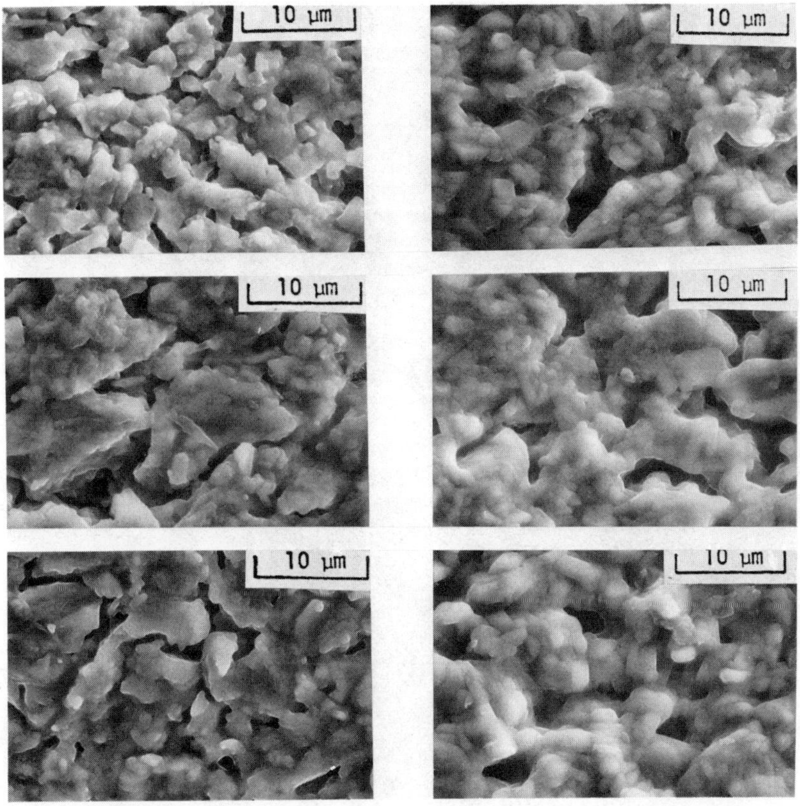

A D

B E

F

Fig. 11.
Microstructural evolution of powder AN-3. (a)-(f) are SEM micrographs of thermally etched surfaces for sintering times of 5, 10, 30, 100, 300, and 1000 min, respectively, at 1400 °C.

A

B

Fig. 12.
SEM micrographs of AN-2 specimens on opposite sides of break in slope in Fig. 8. (a) 1400 °C, 5 min, (b) 1400 °C, 1000 min.

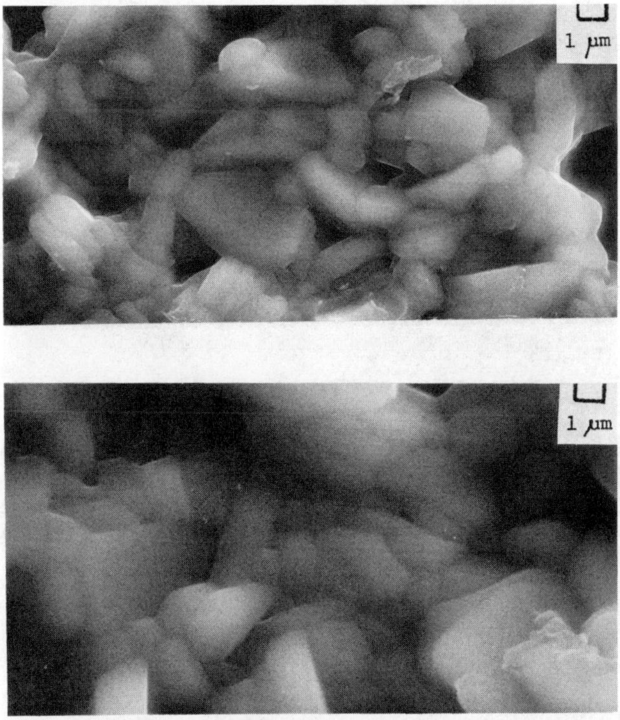

Fig. 13
SEM micrographs of AN-3 specimens corresponding to times longer than the break in slope in Fig. 8.
(a) 1400 °C, 100 min, (b) 1400 °C, 1000 min.

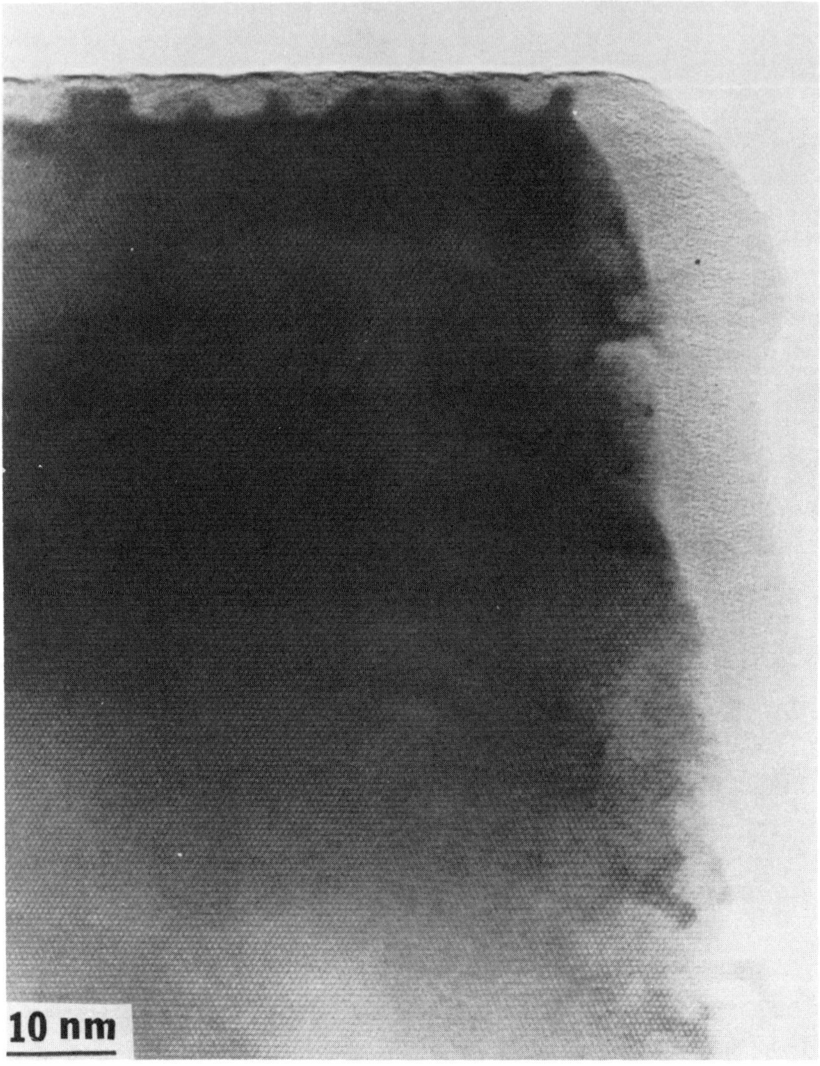

Fig. 14
TEM micrograph of ion-thinned anorthite ceramic (AN-2 sintered 1400 °C, 1000 min). Note lattice fringes grading into amorphous area at edge. The latter probably represents an artifact produced by ion thinning; it makes the confirmation of intergranular amorphous phases difficult.

of the bimodality, but, rather, are the result of fundamentally different sintering mechanisms which operate because of the bimodality. Alternatively stated: rather than a single process which operates at different rates on different grain sizes, the author suggests two different sintering mechanisms, each dominant for one of the widely disparate grain sizes, which both lead to densification.

The hypothesis for anorthite densification proposed here is as follows: because of low symmetry (triclinic primitive at room temperature, going to body centered above 250 °C) and the so called "aluminum avoidance" rule (6), the interfacial energies associated with intergranular intergrowths are highly anisotropic. Thus there may be relatively few mutual orientations which are both crystallographically and energetically favorable. Therefore, in a fine grained anorthite compact, where many grains are placed into random orientation with each other, surface energies are relatively high and a large driving force for grain boundary diffusion exists, resulting primarily in the destruction of unfavorable orientations and the creation of new, favorable interfacial orientations. This need not result in appreciable grain coarsening, except by virtue of destruction of the smaller, unstable grains (see Fig. 12A). Once an equilibrium intergranular configuration occurs (e.g., Fig. 13A), then the primary driving force for rapid densification and sintering is spent, and a slower densification mechanism becomes dominant. This latter is postulated as a volume diffusion mechanism, perhaps related to the presence of abundant polysynthetic growth twin planes on a scale of a few hundred nanometers (see Fig. 4; also it may be suggested that the stepped appearance of the prominant grain in Fig. 1-B is related to the effect of such twin planes upon recrystallization). Further densification occurs up to the point where twin planes no longer exit in the pores and diffusion paths are effectively blocked. A diagrammatic representation is given in Fig. 15. According to this model,

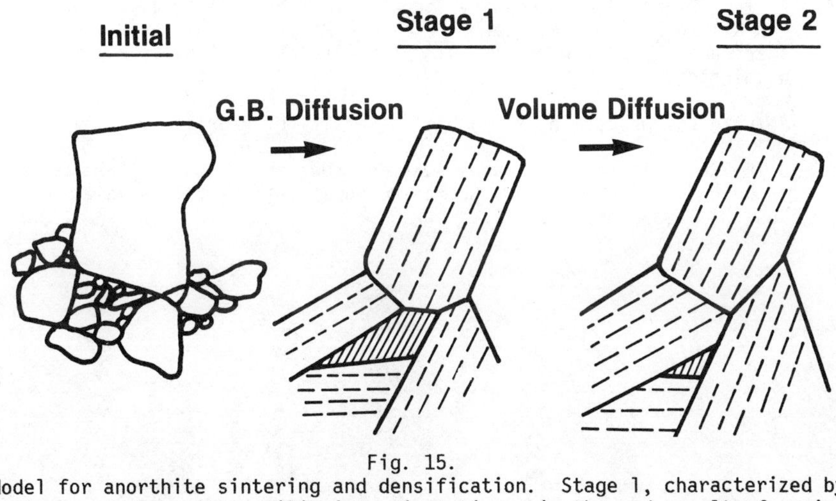

Fig. 15.
Model for anorthite sintering and densification. Stage 1, characterized by grain boundaries with equilibrium orientations, is the end result of grain boundary diffusion, which operates on initial randomly oriented powder. Stage 2, in which pores have been further eliminated until diffusion into the pores along twin planes is no longer possible, is the product of lattice diffusion, which becomes dominant after grain boundary diffusion ceases.

the reason why An-1 never reaches orientational equilibrium (volume diffusion dominant) is that interfacial reorientation involving large grains in the absence of abundant fines (see Fig. 2) is a much slower process.

While this hypothesis effectively explains the rather unique sintering behavior of anorthite, it rests upon two assumptions which must be further investigated: existence of a high grain boundary diffusion constant for certain mutual orientations and existence of a relatively high volume diffusion constant associated with twinning.

Acknowledgments

I am grateful to Dr. Stane Pejovnik of the Josef Stefan Institute, Ljubjana, Yugoslavia, who kindled my interest in sintering of anorthite during a visit in 1980. Dr. Arne Olsen, Dr. Bernard Hockey, and Mr. Gene Danko assisted with the TEM part of the investigation.

References

1. J. B. Wachtman, "National Materials Policy: Critical Materials and Opportunities," Bull. Amer. Ceram. Soc., 61 (2) (1982) pp. 214-220.

2. L. P. Cook, "Powder Processing of Potassium Aluminosilicates," Bull. Amer. Ceram. Soc., 60 (10) (1981) p. 849; (see also this volume).

3. Ernest M. Levin, Carl R. Robbins, and Howard F. McMurdie, Phase Diagrams for Ceramists, p. 219; American Ceramic Society, Inc., Columbus, OH, 1964.

4. Paul H. Ribbe, "The Chemistry Structure and Nomenclature of Feldspars," pp. R-1 to R-72 in Feldspar Mineralogy, Mineralogical Society of America Short Course Notes, vol. 2; Mineralogical Society of America, Washington, DC, 1975.

5. Private Communication, M. Sachs, University of Florida, Oct., 1981.

6. J. R. Goldsmith and F. Laves, "Cation Order in Anorthite ($CaAl_2Si_2O_8$) as Revealed by Gallium and Germanium Substitutions," Z. Kristallogr. 106 (3) (1955) pp. 213-226.

MODIFICATION OF THE SINTERING KINETICS OF PALLADIUM BY A SURFACE OXIDE

P. K. Higgins and Z. A. Munir

Division of Materials Science and Engineering
University of California
Davis, California 95616

The presence of a surface oxide layer on palladium powders gives rise to a densification-free sintering process controlled by a surface diffusion mechanism with an activation energy of 59.2 $kJ.mol^{-1}$. At temperatures above the oxide dissociation temperature in air, 1062 K, palladium powders sinter with an activation energy of 120.4 $kJ.mol^{-1}$, indicative of a grain boundary diffusion. At $T < 1062$ K and under a 10^{-8} atm vacuum, sintered powders showed densification but with an activation energy too high to be grain boundary diffusion-related. These results are interpreted in terms of the role of the surface oxide and are compared with previously published results on the sintering of Pd.

Introduction

The presence of a surface oxide layer on metallic powders has long been recognized as contributing to the kinetics of sintering. Earlier investigations attributed the unsinterability (in reality, lack of densification) of low melting metals to the presence of stable surface oxide layers (1,2). Others, however, have suggested that the presence of surface oxides contribute to enhancement of sintering. It has been suggested that metal powders whose surface oxides are rendered unstable just prior to their sintering are more "active" and hence sinterable (3). In fact, suggestions have been made that the oxide layers themselves are the reason for the sintering enhancement in copper, iron, and nickel (4). A more recent investigation has demonstrated that the presence of the oxide layer shifts the dominance of the sintering process from bulk to surface mass transport in the case of lead (5).

An analytical assessment of the role of surface oxide layers in the sintering of metals has been recently provided (6) and systematically applied to a large number of metals (7). Predictions of this model were found to be consistent with available experimental observations. Three of these observations concern the sintering of palladium powders (8-10). Palladium has an oxide which is stable in air up to 1062 K (11). Thus sintering in air above this dissociation temperature, T_d, proceeds unhindered by a surface layer. However, unlike the case of lead (5), palladium oxide can be made thermodynamically unstable by heating it in an ordinary vacuum (10^{-8} atm) at a much lower temperature than T_d. Anderson (8), using submicron Pd powders (average particle size 600 Å), demonstrated that sintering in a 10^{-8} atm vacuum was accompanied by densification even at temperatures as low as 573 K. He concluded that under these conditions palladium sinters through grain boundary diffusion. Using powders with a much larger size range (60 - 525 μm), Anderson (8) sintered Pd at 1783 K in air, and on the basis of calculated diffusion coefficients he concluded that the dominant mass transport process in sintering is volume diffusion. In a more recent investigation by Imm and Saeger (9), the dependence of densification on various experimental conditions was studied. The dependence of densification on temperature showed a maximum in the range 1073-1273 K. In the most recent of the investigations on palladium, German and Ham (10) concluded that grain boundary diffusion is the rate controlling mechanism for the sintering of < 44 μm sized sponge powder in the range 1017-1223 K under a flowing argon atmosphere. In this paper we describe the results of investigations of the sintering of palladium powders conducted in air at temperatures higher than the dissociation temperature, T_d, in air at temperatures lower than T_d, and in vacuum at $T < T_d$.

Experimental Materials and Methods

Palladium powders* with a 99.9+ purity were utilized in this work. Powder impurities and their levels are shown in Table I. Particle size distribution analysis showed a size spectrum between 0.5 and 6 μm with the largest proportion being between 1 and 3 μm. Approximately one gram samples were pressed in a graphite die at a pressure of about 10^8 Pa to give cylindrical compacts of approximately 0.45 cm high and 0.64 cm

---

* Purchased from the Western Gold and Platinum Co., Belmont, CA.

Table I - Impurity Content of the Palladium Powder

Element	Concentration (ppm)	Element	Concentration (ppm)
Ag	10	Mn	10
Al	60	Ni	60
Au	30	Pb	30
Ca	10	Pt	40
Cu	150	Rh	50
Fe	80	Si	60
Mg	5	Ti	40

diameter. Typically these compacts had a relative density of 56% and a specific surface area of 1.770 m^2/gm. This latter value compares with 2.616 m^2/gm for the as-received loose powder.

Densification studies were made in a sintering dilatometer utilizing a linear variable differential transformer. In this apparatus studies can be made under vacuum, air, or other atmospheres to temperatures in excess of 1300 K. Sample temperatures were measured by a Pt/Pt-10% Rh thermocouple. Specific surface area measurements were made in a Quantachrome Corp., BET apparatus. For total surface areas greater than 0.1 m^2 nitrogen is used as the adsorbate gas with helium as the carrier gas, and when the total area is below this value, krypton is used as the adsorbate gas.

Five groups of compacts were sintered at different conditions. The first group was sintered in air at 773, 823, 873, and 898 K. These temperatures correspond to 0.42, 0.45, 0.48, and 0.49 of the absolute melting point of Pd (1823 K) and 0.73, 0.78, 0.82, and 0.85 of the oxide dissociation temperature, T_d (1062 K). Under these conditions the oxide is thermodynamically stable. The compacts were sintered for 30, 60, 90, and 120 min at each temperature. A second group of pellets was sintered in air at 1073, 1098, 1123, and 1148 K. These temperatures correspond to 0.59, 0.60, 0.62, and 0.63 of the melting point of Pd and 1.01, 1.03, 1.06, and 1.08 of T_d. Under these conditions the oxide is thermodynamically unstable. A third group of pellets was sintered in a vacuum of 10^{-8} atm at the same temperatures as those for the first group. At this oxygen partial pressure (~ 2 x 10^{-9} atm) the dissoiation pressure of palladium oxide is 569 K. Thus for this series of experiments the conditions of temperature represents 1.36, 1.45, 1.53, and 1.58 of the dissociation temperature of the oxide. Pellets in this group were sintered in the dilatometer for a period of 120 min. The fourth group of pellets was sintered in air in the dilatometer at 923, 948, 973, 998, 1023, and 1048 K which correspond to 0.51, 0.52, 0.53, 0.55, 0.56 and 0.57 of the melting point of Pd and 0.87, 0.89, 0.92, 0.94, 0.96, and 0.99 of T_d, respectively. A final group of powder compacts was sintered in an electrobalance. Samples were heated in air at a constant heating rate up to 773, 873, 973, and 1073 K. Once at temperature the system was evacuated and weight change measurements were made as a function of time at a constant temperature. One specimen from each of the five groups described above was sectioned for microscopic observation.

Results

Palladium pellets sintered in the electrobalance under the condition described above showed weight changes which were functions of temperature and oxygen partial pressure as depicted in Figure 1. Pellets sintered in air below 898 K oxidized and had weight gains between 0.8 and 0.9 . Those sintered above T_d (1062 K) lost weight which was constant except for the value corresponding to the temperature 1073 K. Specimens sintered in air above 898 K but below T_d experienced weight loss with a slight temperature dependence. Samples sintered in vacuum (10^{-8} atm) between 773 and 898 K experienced weight loss as indicated in Figure 1.

Linear shrinkage and densification results for 120 min sintering, shown in Figures 2 and 3 follow patterns closely related to the weight change behavior (Figure 1). Pellets sintered in air at low temperatures experienced a weight gain and a volume expansion. Surface area measurements for these pellets gave results shown in Figure 4 plotted as a function of sintering time. In accordance with the analysis of German and Munir (12), the relationship between surface area reduction and time is expressed as

$$\frac{\Delta S}{S_0} = K\ (Bt)^{m/N} \qquad (1)$$

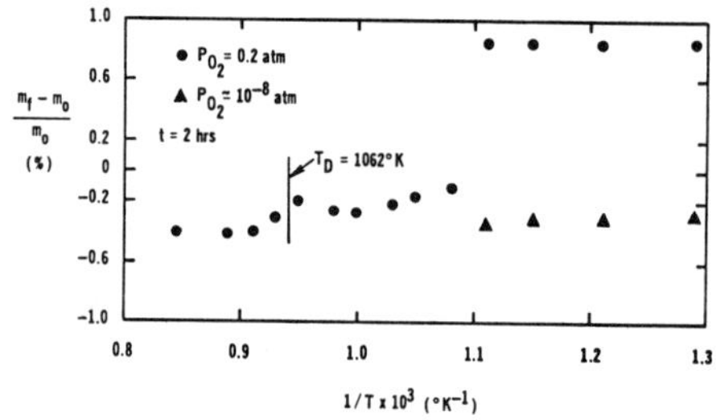

Figure 1 - Weight change of palladium pellets as a function of sintering temperature

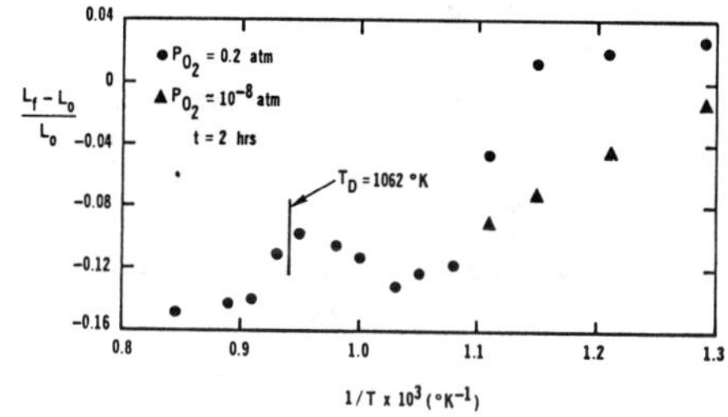

Figure 2 - Shrinkage of palladium pellets as a function of sintering temperature

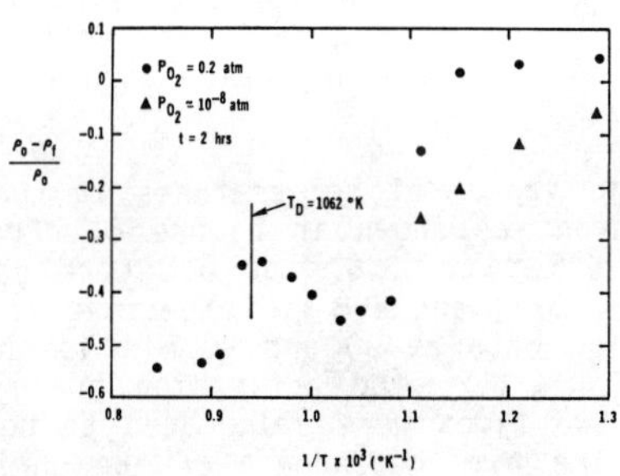

Figure 3 - Densification of palladium pellets as a function of sintering temperature

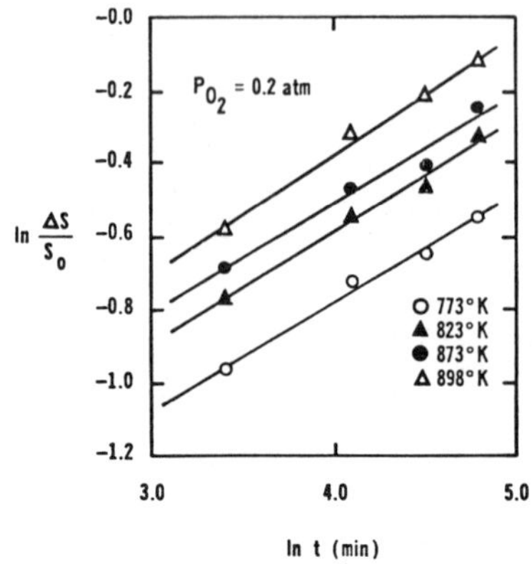

Figure 4 - Time dependence of the surface area reduction of palladium pellets sintered in air

where ΔS is the change in surface area, S_0 is the initial surface area value, K and m are constants whose values have been calculated for the various sintering mechanisms (12), B is a constant containing materials parameters and temperature, N is the mechanism related constant, and t is time. From the results presented in Figure 4, values of N were calculated as 6.8, 6.6, 6.7, and 6.2 for the data at 773, 823, 873, and 898 K, respectively. The temperature dependence of these surface area reduction data is shown in Figure 5 for t values of 45 and 105 min. Using an average N value of 6.6 obtained from the results of Figure 4, the activation energies for sintering for the two specified times are calculated to be 57.7 and 60.6 $kJ.mol^{-1}$, respectively.

Specimens sintered in vacuum at temperatures in the range 773–898 K showed considerable skrinkage as shown in Figure 6. From the slopes of these lines N values of 6.5, 6.4, 5.6, and 6.3 were calculated for the data gathered at 773, 823, 873, and 898 K, respectively. The temperature dependence of the skrinkage data at 30 and 90 min is shown in Figure 7. With an average value of N = 6.2, the activation energy values for the sintering data at these two times were calculated to be 265.0 and 269.2 $kJ.mol^{-1}$, respectively. The time dependence of the shrinkage of pellets sintered in air at temperatures higher than T_d is shown in Figure 8 for temperatures ranging from 1073 to 1148 K. Again, N values were found to be 6.1, 6.1, 6.1, and 6.0 for the temperatures 1073, 1098, 1123, 1148 K, respectively. Activation energies, calculated at t = 30 and 90 min, were determined from the slopes of the data of Figure 9 and found to be 120.4 $kJ.mol^{-1}$ for both times.

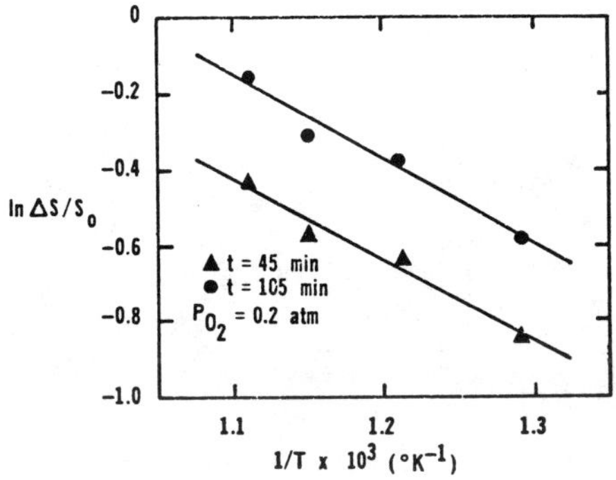

Figure 5 – Temperature dependence of the surface area reduction of sintered Pd pellets

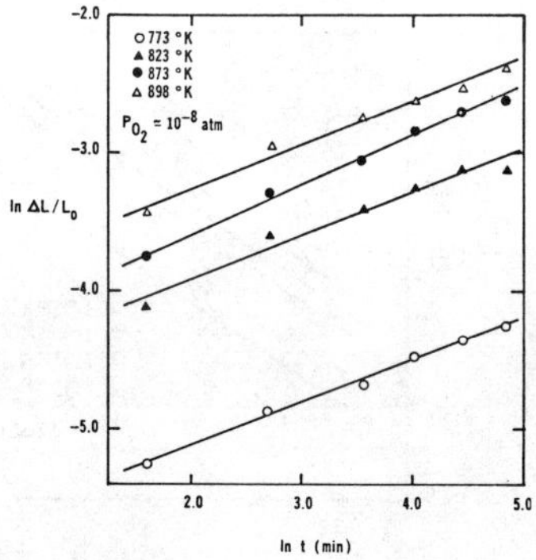

Figure 6 – Time dependence of the shrinkage of Pd pellets sintered under vacuum

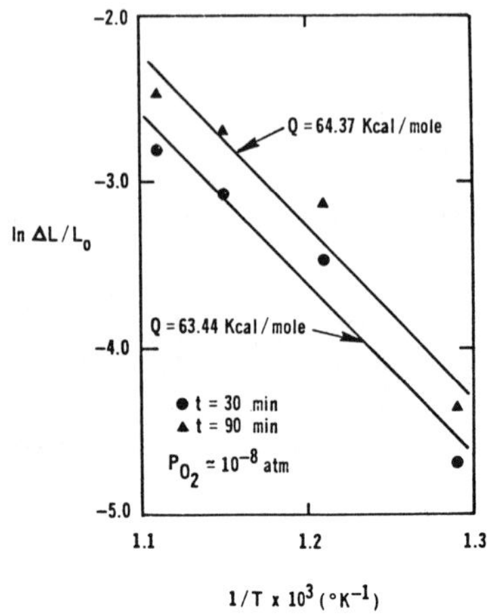

Figure 7 – Temperature dependence of the shrinkage of palladium powders sintered under vacuum

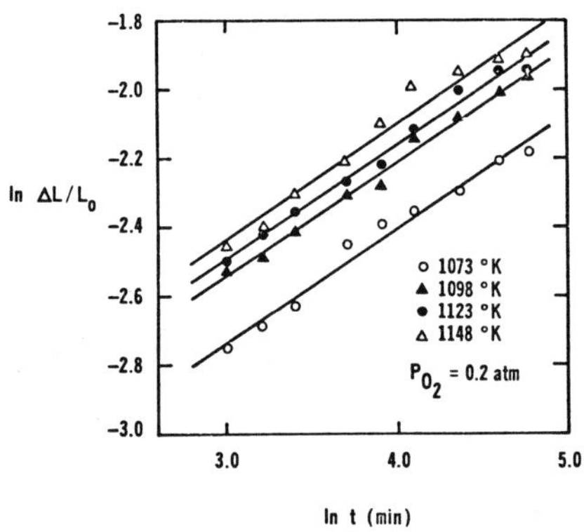

Figure 8 – Variation of the shrinkage with sintering time for air sintered Pd powders

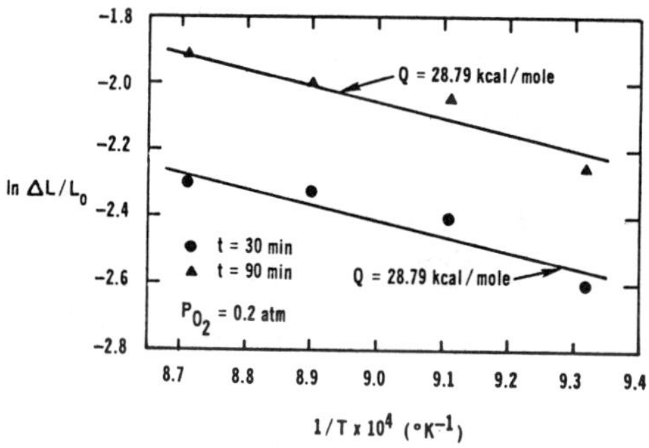

Figure 9 - The influence of temperature on the shrinkage of air sintered palladium powders

Discussion

Observed weight loss characteristics and their dependence on temperature are replotted in Figure 10 for the range 923–1048 K for pellets sintered in air for 120 min. As can be seen in this figure, the weight loss increases with temperature up to a maximum value between 973 and 998 K. As the temperature increases beyond this region the weight loss decreases steadily up to the dissociation temperature of the oxide. These observations indicate that the initial (as-received) palladium powder possessed an oxide layer which had been thermally equilibrated at a temperature in the range 973–998 K. Shrinkage data, Figure 11, show a similar anomaly with temperature. Maximum shrinkage is obtained at approximately the same temperature as that corresponding to maximum weight loss, Figure 10. With the concept of an initial oxide which was equilibrated at 973–998 K, the above observations are consistent with the general conclusion that the presence of the surface layer is responsible for lack of densification, i.e., higher weight losses correlate with higher densification.

This is clearly demonstrated by the results of the low temperature sintering in air. Sintering data obtained in air between 773–898 K (a range in which the oxide is stable) showed little or no densification. However, as pointed out earlier, evidence for sintering is provided by the observed surface area reduction, Figure 4. Additionally scanning electron microscope observations show neck formation between powders sintered in this temperature range. Figures 12(a) through (c) show micrographs of the initial powder, a powder sintered at 823 K for 120 min, and a powder sintered at 898 K for 120 min, respectively. The activation energy for sintering in air in this temperature range is 58.9 $kJ.mol^{-1}$. Lack of densification indicates that this energy is associated with a surface

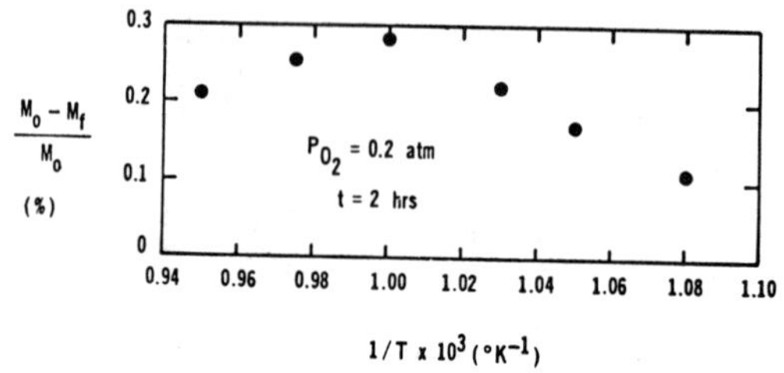

Figure 10 – Weight change of Pd pellets after a two-hour sintering

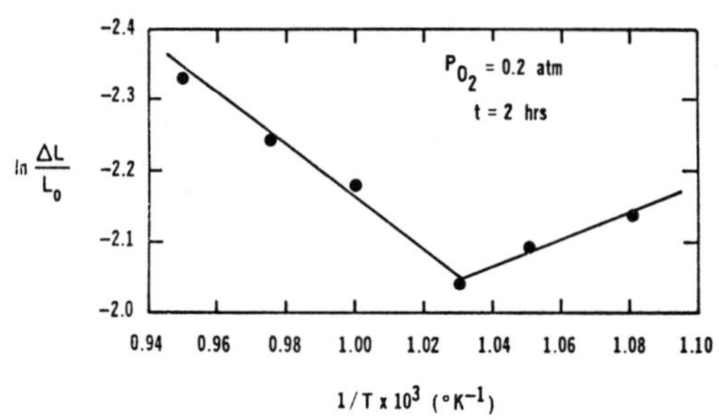

Figure 11 – The temperature dependence of shrinkage of Pd powders sintered in air for 120 min

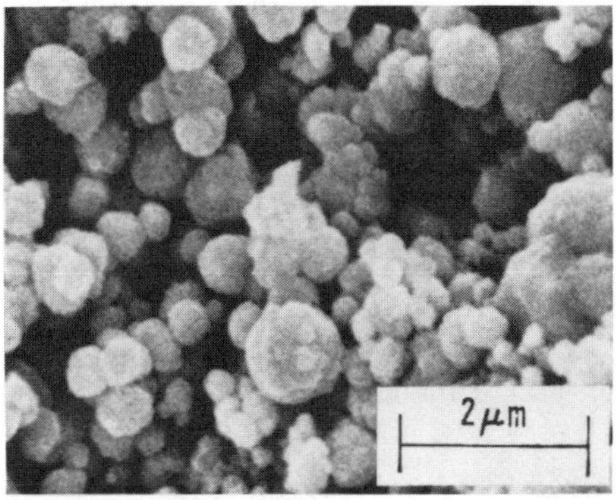

(a) SEM micrograph of unsintered palladium powder

(b) Air sintered palladium powder, 823 K for 120 min.

(c) Air sintered Pd powder, 898 K for 120 min

Figure 12

transport mechanism (13), and since the vapor pressure of Pd in this range is exceedingly low (14) (~ 10^{-14} atm at 900 K) it is concluded that the dominant sintering mechanism is surface diffusion. Unfortunately there are no reported data for the surface diffusivity of Pd, however, from a published value (15) for the activation energy for volume diffusion (265.8 $kJ.mol^{-1}$) an activation energy for surface diffusion is estimated to be approximately one fourth of this value i.e., 66.5 $kJ.mol^{-1}$. Agreement between this value and that obtained from the surface area reduction data is quite reasonable. Implied in the suggestion that the dominant sintering mechanism for oxide coated powders is surface diffusion, is the existence of discontinuities in the surface layer caused by the compaction process. The activation energy for diffusion of Pd through PdO is not available but is estimated (7) to be significantly higher than the experimentally determined value for sintering, and thus neck formation by diffusion through the oxide layer is energetically unlikely.

In comparison to the above observations, pellets sintered in vacuum over the same temperature range (773–898 K) show considerable densification as indicated by the results of Figure 6. Additionally, evidence for extensive neck formation and densification is readily observable from SEM micrographs, Figures 13(a) and (b), especially when compared to corresponding ones for air-sintered pellets (Figures 12(b) and (c)). The activation energy for the sintering of Pd in vacuum at these low temperatures is 265.8 $kJ.mol^{-1}$ which is more than 50% higher than the value reported by Anderson (8) (166.8 $kJ.mol^{-1}$) for the sintering of submicron powder in vacuum between 573 and 823 K. Anderson (8) concluded that sintering under these conditions is controlled by a grain boundary diffusion process with the estimated activation energy of the latter being 132.9 $kJ.mol^{-1}$. The value obtained in this study, however, is closer to the volume diffusion activation energy (267.1 $kJ.mol^{-1}$) than it is to the grain boundary diffusion value. Although values of the mechanism exponent, N, cannot be relied upon alone to identify the dominant process (16), they can give general indications, however. The average N value for this set of observations is 6.2 with 6.0 being the nominal value for grain boundary diffusion and 5.0 is the value for volume diffusion. Since, as will be discussed below, the high temperature sintering data of oxide-free Pd showed a grain-boundary controlled process, and in view of relative magnitudes of the volume and grain boundary activation energies, it is unlikely that volume diffusion is the dominant mechanism at low temperatures. Of course this conclusion is based on the assumption that the powders are oxide-free. It is reasonable to assume, therefore, that in this study the reduction of the oxide layer caused by heating in vacuum at low temperatures (< 900 K) is relatively slow such that sintering is taking place in the presence of the oxide during the initial stages. Johnson (17) has shown that the existence of surface diffusion as an initial or simultaneous mechanism has the effect of increasing the apparent activation energy. In this regard the difference between the present observations and those of Anderson (8) may be attributable to differences in the reduction kinetics which in turn are related to the powder characteristics. The powder utilized in the work referred to above is about two orders of magnitude smaller and the compacts made therefrom are about a factor of two more porous than the corresponding values for the palladium of the present study. Table II gives a comparison between the results of this study and those reported in the literature under the stated conditions.

Palladium sintered above the oxide dissociation temperaure, T_d, in air showed considerable densification, Figure 8. Under these conditions pellets densified up to 89% of the theoretical density, well into the second stage of sintering. The activation energy calculated from the

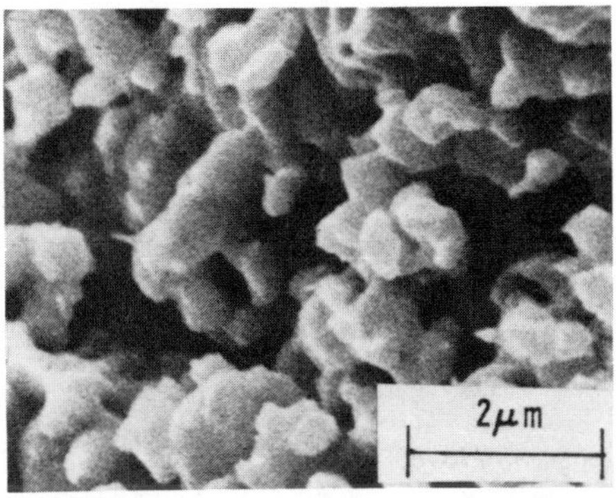

(a) SEM micrograph of vacuum sintered Pd powder,
823 K for 120 min

(b) Vacuum sintered Pd powder, 898 K for 120 min

Figure 13

Table II - Summary of Results of the Sintering of Palladium

Particle Size, μm	Sintering Atmosphere	Temperature Range, K	Activation Energy kJ.mol-1	Suggested Mechanism[a]	Reference
1-3	air	1073-1148	120.4 ± 2.0	GBD	this work
	air	773-898	59.2 ± 1.6	SD	this work
	10^{-8} atm	773-898	267.1 ± 2.1	GBD	this work
0.06	10^{-8} atm	573-823	166.8	GBD	8
60-525	air	1783	-	VD[b]	8
1[c]	10^{-6} atm	873-1273	150-280	GBD,VD[d]	9
< 44	Ar	1047-1223	210 ± 70	GBD	10

a GBD: grain boundary diffusion; SD: surface diffusion; VD: volume diffusion.

b This was concluded on the basis of agreement between a calculated and reported diffusion coefficient at the indicated temperature.

c This value represents the diameter of the fine particles forming the agglomerates.

d The authors (ref. 9) suggest a GBD for the low temperature end of the range and VD for the high temperature end.

present results is 120.4 $kJ.mol^{-1}$ which compares favorably with 132.9 $kJ.mol^{-1}$, a value estimated from the published volume diffusion activation energy (15). As indicated in the "Results" section the mechanism exponent N calculated for this region is 6.1. German and Ham (10), who sintered palladium sponge powder (< 44 μm) between 1017 and 1223 K in flowing argon obtained an activation energy of 210 ± 70 $kJ.mol^{-1}$ and concluded that the sintering is grain-boundary diffusion controlled during the initial stages. Imm and Saeger (9) reported a range of 150 to 280 $kJ.mol^{-1}$ for the activation energy for sintering in the 873-1273 K temperature range. These authors suggested that the activation energy is temperature dependent, being the result of grain boundary diffusion at the lower temperatures and of volume diffusion at the higher temperatures. The anomalous dependence of densification on temperature observed by these authors are likely to be the consequence of grain growth as has been demonstrated by the work of German and Ham (10). This anomalous behavior was observed in samples sintered in vacuum and under an argon atmosphere. However, those sintered in vacuum reached higher density maxima than corresponding samples sintered in argon. The concept of gas entrapment and subsequent pore swelling was advanced by Imm and Saeger (9) as a possible explanation for this observation.

Summary

The sintering in air of 1-3 μm sized Pd powders at temperatures above the dissociation temperature of PdO (1062 K) proceeds through a grain boundary diffusion process with an activation energy of 120.4 $kJ.mol^{-1}$. Although reporting values with larger uncertainties, other investigators have concluded a similar rate controlling process. The sintering of these powders at T < 1062 K in air showed no densification but a surface area reduction whose temperature dependence is indicative of a surface diffusion mechanism. An activation energy of 59.2 $kJ.mol^{-1}$ was calculated for sintering under these conditions. Palladium powders sintered at T < 1062 K in vacuum showed shrinkage which is interpreted in terms of a grain boundary diffusion process modified by surface diffusion. Under these conditions the reported concept of an "active" surface created by the reduction of the oxide layer is not substantiated. Shrinkage and weight loss measurements showed an inverse correlation. Thus in agreement with previous observations the presence of a surface oxide shifts the dominance of sintering mechanism from bulk to surface transport.

Acknowledgements

This work was supported primarily by a grant from the Lawrence Livermore National Laboratory. The authors are grateful to Dr. Ralph Condit for his continued interest in and contributions to this research. Partial support came from the Department of Energy under Contract No. DE-AM03-76SF00031.

References

1. R. C. Smith, "Sintering: It's Nature and Cause," J. Chem. Soc. of London, 123 (1923) pp. 2088-2094.

2. R. F. Smart and E. C. Ellwood, "Sintering of Tin Powder," Nature, 181 (1958) pp. 833-834.

3. W. Dienst and O. Werner, "Investigations of the Fundamentals of the Sintering Process of Metal Powders with Special Consideration of the Influence of Surface Oxide Layers," Z. Metalkun., 51 (1960) pp. 45-52.

4. P. Ramakrishnan and G. S. Tendolkar, "Influence of Thin Oxide Films on the Mechanical Properties of Sintered Metal Powder Compacts," Powder Met., 7 (1964) pp. 34-39.

5. P. K. Higgins and Z. A. Munir, "Influence of Surface Oxide Layer on Sintering Process of Leads," Powder Met., 21 (1978) pp. 188-194.

6. Z. A. Munir, "Analytical Treatment of the Role of Surface Oxide Layers in the Sintering of Metals," J. Mater. Sci., 14 (1979) pp. 2733-2740.

7. P. K. Higgins and Z. A. Munir, "Application of the Surface Oxide Model in the Sintering of Metallic Particles," Powder Met. Int. (1981) in press.

8. H. U. Anderson, "Initial Sintering of Palladium Powder Compacts," Sci. Sinter., 6 (1974) pp. 45-58.

9. R. Imm and K. E. Saeger, "Sintering Behavior of Palladium Powder," Metall. (Berlin), 29 (1975) pp. 687–691.

10. R. M. German and V. Ham, "Palladium Sintering: Shrinkage Kinetics for a Sponge Powder," Metall. Trans. A, 10A (1979) pp. 1593–1596.

11. J. P. Coughlin, "Heats and Free Energies of Formation of Inorganic Oxides," W. S. Bur. Mines Bull. 542, Washington, D.C., 1954.

12. R. M. German and Z. A. Munir, "A Kinetic Model for the Reduction in Surface Area During Initial Stage Sintering," pp. 249–257, in Materials Science Research Vol. 10: Sintering and Catalysis, G. C. Kuczynski, ed.; Plenum Press, New York, N.Y., 1975.

13. R. M. German and Z. A. Munir, "Morphology Relations During Surface-Transport Controlled Sintering," Metall. Trans. B, 6B (1975) pp. 289–294.

14. R. E. Honig, "Vapor Pressure Data for the Elements," pp. 475–494, in The Characterization of High-Temperature Vapors, J. L. Margrave, ed.; J. Wiley, New York, N.Y., 1967.

15. N. L. Peterson, "Isotope Effect in Self-Diffusion in Palladium," Phys. Rev., 136 (1964) pp. A568–A574.

16. Z. A. Munir, P. K. Higgins, and R. M. German, "The Variation of the Mechanism Dependent Exponent with Coordination Number and Neck Size in the Sintering Kinetics of Powders," pp. 26–34, Vol. 4, in Sintering-New Developments, Materials Science Monographs, M. M. Ristic, ed.; Elsevier, Amsterdam, 1979.

17. D. L. Johnson, "Sintering Kinetics for Combined Volume Grain Boundary and Surface Diffusion," Phys. Sinter., 1 (1970) pp. B1–B22.

ACTIVATED SINTERING OF STEEL POWDERS

A. Crowson and J. W. Burlingame

US Army Armament Research and Development Command
Dover, New Jersey 07801

Activation of the sintering of steel powder compacts was studied using a momentary iron-phosphorus liquid phase. Stereological properties of the evolving microstructure were used to ascertain the densification mechanism. Volume diffusion was observed to be the transport mechanism based on geometric parameters, but with a calculated diffusion coefficient two orders of magnitude larger than steel compacts without a liquid phase. On the other hand, viscous flow was observed to be the transport mechanism based on surface area change. Therefore, from the experimental data it was concluded that volume diffusion, enhanced by grain boundary sliding, was the mechanism for densification in the presence of a momentary liquid phase. Exaggerated grain growth during the latter stage of sintering was observed to be the limiting factor in achieving full densification.

Introduction

There have been two general techniques developed for enhancing the densification of metal powders during sintering: (1) the addition of low melting point metallic elements or alloys that form a liquid phase during sintering and (2) the addition of organic or inorganic compounds that modify the surface of the powder particles before sintering starts. Such techniques have been employed successfully for densifying tungsten-, nickel-, cobalt- and chromium-base alloys, as well as tool steels (1-5). However, their application to the densification of low alloy steel powders has been limited to the addition of phosphorus to iron powder (6). In this study, the technique of using a low melting point alloy to form a transient liquid phase was employed. Liquid phase medias investigated were iron-rich eutectic alloys of iron-phosphorus and iron-silicon. To quantify the effect of these alloys on sintering, densification and microstructural properties of the alloys were determined as a function of sintering time and temperature. Using appropriate sintering models, measured metric and topological properties were then utilized to ascertain the predominate transport processes and densification mechanisms.

Experimental Procedure

A commercially available atomized steel powder was used in this investigation. Representative properties of the powder are shown in Table I. The powder was a low-alloy steel powder with minor amounts of alloying and impurity constituents. The powder was unimodal in particle size distribution and had an apparent density of 3.00 g/cm^3, which is typical of an atomized steel powder.

Individual powder particles, evaluated by scanning electron microscopy (SEM) and standard optical metallography, are shown in Figure 1. It was observed in the SEM examination that the particles were highly irregular in shape; which aided in compactibility. It was observed, using optical metallography, that the microstructure of the particles had a generally uniform grain size of ∿ 16μm (ASTM No. 9).

In the initial phase of the investigation, iron-phosphorus and iron-silicon eutectic alloys were employed for the liquid phase additions. These alloys were made by vacuum melting reagent grade iron-phosphide, or 98.5 percent purity silicon, and the atomized steel powder to achieve eutectic compositions of 90Fe-10P and 81Fe-19Si. The eutectic alloys were then pulverized using a mortar and pestal.

For the sintering studies, the powdered eutectic alloys were mixed with the atomized steel powder to achieve one-half and one percent concentrations, respectively, of the secondary alloying element (phosphorus or silicon). These mixtures were compacted into 3.175 cm x 1.27 cm x 1.27 cm rectangular bars to 80 percent of theoretical density. The compacted preforms were then sintered in a vacuum at a temperature in the range of 1300 to 1450°C for 30 minutes. Density changes after sintering were determined by the Archimedean method (ASTM Method B328) and converted into a densification parameter ω,

$$\omega = \frac{\rho_s - \rho_p}{\rho_t - \rho_p} \qquad (1)$$

Table I. Representative Powder Properties of Atomized Steel Powder*

Sieve Analysis**		Chemical Analysis	
Screen Size	Weight %	Element	%
+ 60	0	Iron	99.
-60 + 80	Trace	Hydrogen Loss	0.16
-80 + 100	2.0	Carbon	0.015
-100 + 150	17.0	Manganese	0.20
-150 + 200	28.0	Insolubles	0.17
-200 + 325	31.0	Sulfur	0.015
-325	22.0	Phosphorus	0.01

* **Ancorsteel** 1000 Atomized Steel Powder, Hoeganaes Corporation, Riverton, NJ

** ASTM Method B214

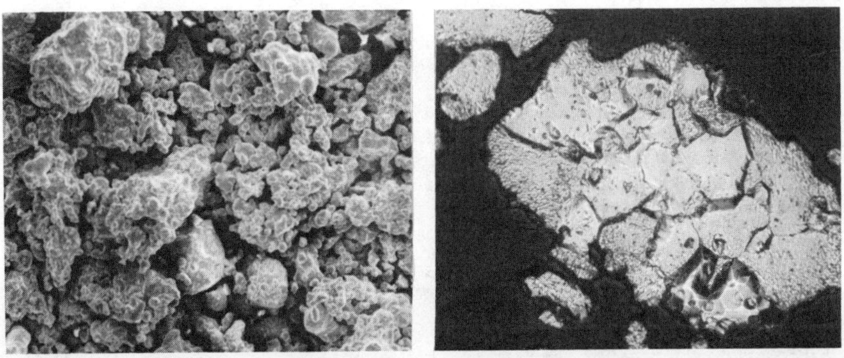

200μm Nital Etch 20μm

Figure 1. Particle shape and microstructure of atomized steel powder.

where ρ_s is the density after sintering, ρ_p is the density after pressing and ρ_t is the theoretical density.

In the next phase of the investigation (quantitative microscopy), only the iron-phosphorus eutectic was used as the liquid phase addition. The amount of the eutectic added was adjusted to achieve a 0.5 percent concentration of phosphorus in the powder mixture. The powders were compacted into rectangular bars of the same dimensions as before, but to only 70 percent of theoretical density. Sintering of these bars was conducted in a vacuum and/or dissociated ammonia at 1350°C for periods ranging from one minute up to 20 hours. Density and volume changes were used to calculate the densification parameter and the relative shrinkage.

Sample preparation, prior to the determination of the metric and topological properties, required obtaining agreement between the Archimedean density and the density calculated from a quantitative microscopic analysis of a polished sample. If the pore volume calculated from these two measurements differed by more than 0.025 (7), the sample was repolished and re-etched until the agreement was within the allowable limits.

Determination of the metric properties was accomplished by using the techniques and principles of quantitative reflection microscopy (8). The metric properties of surface area, mean surface curvature, mean pore intercept, mean grain intercept and length of triple line were obtained from measurable two dimensionable quantities using an automated image analyser. The measurable quantities included volume of pore phase, volume of solid phase, surface area of pore phase, surface area of solid phase, perimeter of pore phase, topological count, and triple point count. A statistical average from 50 fields of view was made using three specimens for each determination.

Determination of the topological properties was made using the technique of serial sectioning (9). Fifteen serial sections were obtained on the average, approximately 13×10^{-3} cm apart. Reconstruction of the photographs to evaluate the genus and number of separate parts was then achieved using the technique developed by Kronsbein, Buteau and DeHoff (10). Topological properties were determined over the pore volume range of 0.25 to 0.05; corresponding, respectively, to densities of 75 to 95 percent of theoretical.

Results and Discussion

Liquid Phase Addition

The effectiveness of employing eutectic alloy additions for densifying low alloy steel powder compacts is shown in Figure 2. Both the iron-phosphorus and the iron-silicon eutectics had a substantial influence on densification. This became more pronounced with increasing sintering temperature. Densification parameters from 0.8 up to 0.95 were achieved at 1450°C after 30 minutes of isothermal sintering in a vacuum. A densification parameter of only 0.3 was obtained for steel compacts without a liquid phase addition under similar sintering conditions (reference line designated S). An increase in the concentration of the eutectic additions also resulted in a slight increase in the densification factors. This was especially true for the iron-silicon eutectic.

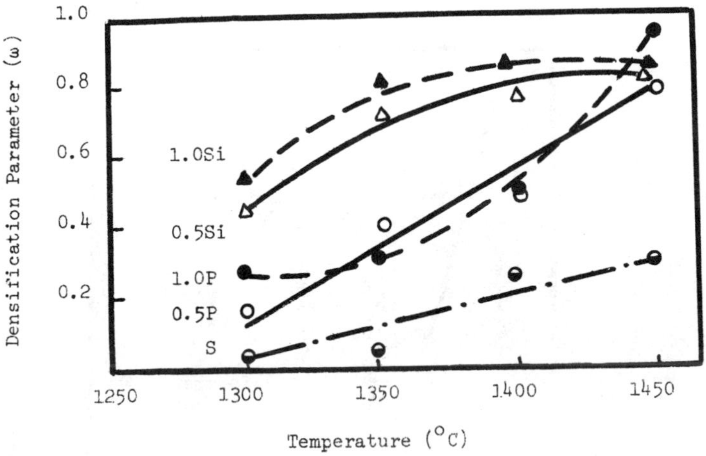

Figure 2. Dependence of densification parameter, ω, on sintering temperature for atomized steel powder (S) and 0.5 and 1.0 percent phosphorous and silicon additions (initial fractional density 0.80--sintered 30 minutes).

Stereology Investigation

To ascertain the transport process occurring during the sintering of a steel powder using a momentary liquid phase addition, a recent phenomenological approach developed by Rhines and DeHoff (11) was employed. Their approach uses a sophisticated mathematical procedure based on quantitatively characterizing the geometric parameters of the evolving microstructure. These parameters are divided into two categories: (1) those which depend upon the dimensions of the system--the metric properties and (2) those which characterize the shape of the system--the topological properties.

The results of the stereology investigation are shown in Figures 3 to 6. The topological properties, genus or total connectivity of surface-per-unit-volume (Gv), and number of separate or disconnected segments of surface-per-unit-volume of material (Nv), as well as the metric properties, mean surface curvature ($\overline{H}$), mean pore intercept ($\overline{\lambda}p$), mean grain intercept ($\overline{\lambda}_g$) and pore-solid interfacial area (S_V), are shown as a function of pore volume fraction (V_V). Pore volume fraction was used as the measure of densification as a function of time at a 1350°C sintering temperature. The temperature of 1350°C was selected as it represents an upper limit to practical sintering temperatures.

From the topological properties (Figure 3), only the intermediate and third stages of sintering could be identified. The genus, which gives a measure of the morphology of the network of pores, was seen to decrease in a linear fashion throughout most of the intermediate stage of sintering corresponding to a general decrease in porosity. It was concluded that the deviation from linearity (at about 0.15 pore volume fraction) during the latter stages of sintering was an indication that channel closure and

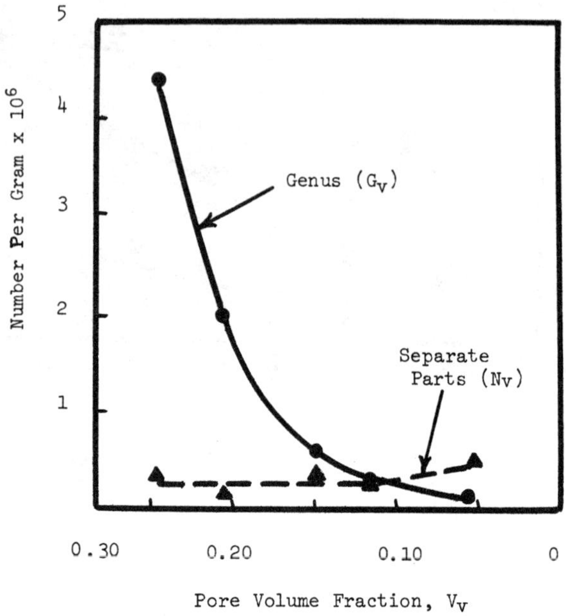

Figure 3. Variation of genus and number of separate parts with pore volume fraction for atomized steel powder with a 0.5 percent phosphorus addition.

isolation of the pore network had started. The number of separate parts, on the other hand, was seen to remain constant until a pore volume fraction of approximately 0.10. At smaller volume fractions, the number of separate parts increased indicating isolation of individual pores and pore networks. The transition from intermediate to third stage sintering was, therefore, concluded to be in the range 0.15 to 0.10 pore volume fraction.

The forces responsible for sintering during the intermediate stage varied depending on the densification parameter. During the early part of the intermediate stage, 0.30 to 0.20 pore volume, contraction of the powder compact was controlled by mechanisms of densification (volume and grain boundry diffusion and/or viscous flow). Surface curvature mechanisms (evaporation and condensation) were at a minimum, as noted in Figure 4. Operative forces involved in pore growth and grain growth, likewise, were negligible (Figure 5). As the pore volume decreased from 0.20 to 0.10 pore volume, surface smoothing mechanisms competed with the densification mechanisms. Thus, the pore-solid interfacial area decreased monoatomically with pore volume (Figure 6), in a similar manner to that reported by Rhines and DeHoff for the sintering of copper aggregates (11).

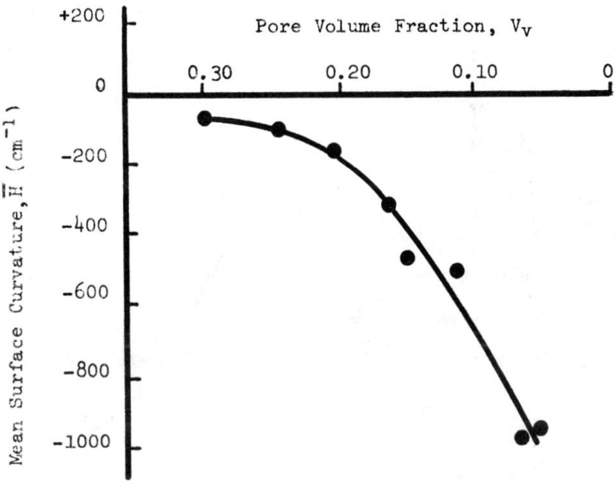

Figure 4. Variation of mean surface curvature with pore volume for atomized steel powder with a 0.5 percent phosphorus addition.

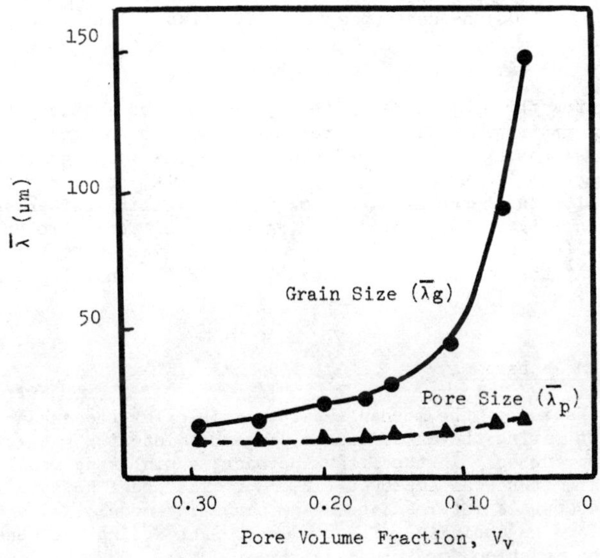

Figure 5. Variation of pore size ($\overline{\lambda}p$) and grain size ($\overline{\lambda}g$) with pore volume for atomized steel powder with a 0.5 percent phosphorus addition.

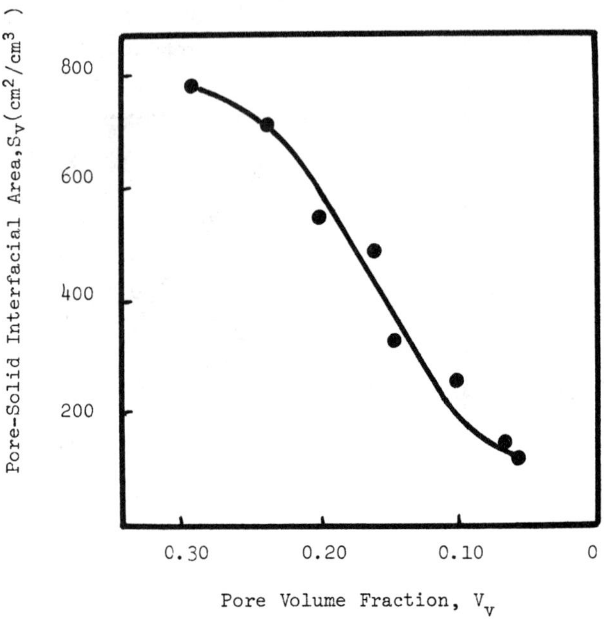

Figure 6. Variation of surface pore area with pore volume for atomized steel powder with a 0.5 percent phosphorus addition.

During the third stage of sintering, the forces involved with pore rounding and grain growth became operative. As the density of the compact increased (decreasing pore volume), Ostwald ripening and grain growth became more predominate. Eventually, grain growth, and the forces associated with this phenomena, became the controlling criterion for densification (Figure 5). Physically, grains grew to a size where the dislocations (pores) and the sinks (grain boundaries) were at such distances that the diffusion processes involved in densification became negligible, Figure 7 (V_v = 0.05).

Densification Mechanisms

Determination of the mechanisms responsible for the enhanced densification during the intermediate stage of sintering was established using two approaches. In the first approach, a sintering model based on geometric parameters was employed (12, 13). The model derived is similar to Coble and Gupta's intermediate stage sintering model (14) and is able to account for the shrinkage of a powder compact during the densification process. The equation for this model is:

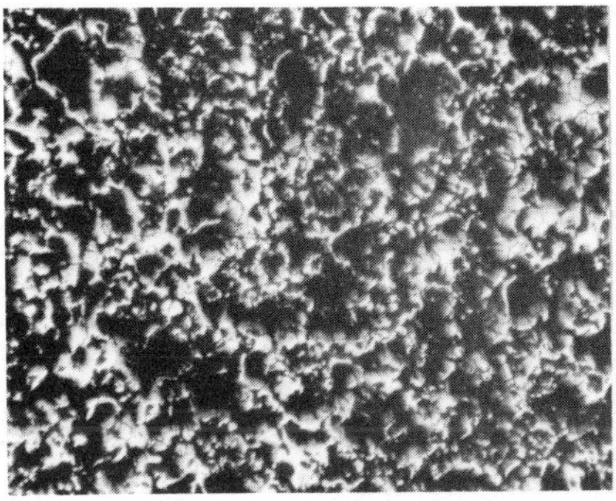

V_V = 0.25 200μm

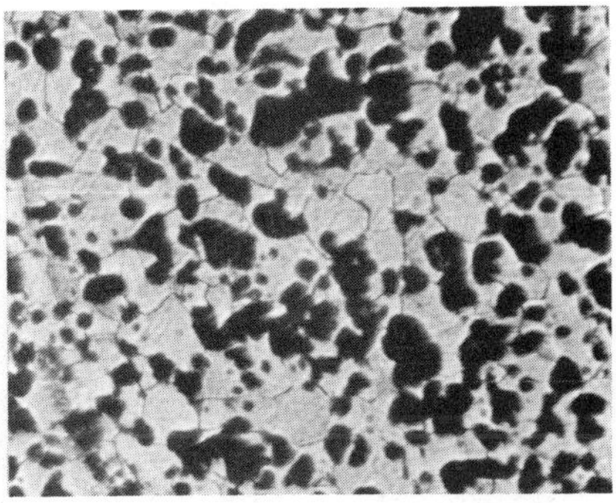

V_V = 0.15 200μm

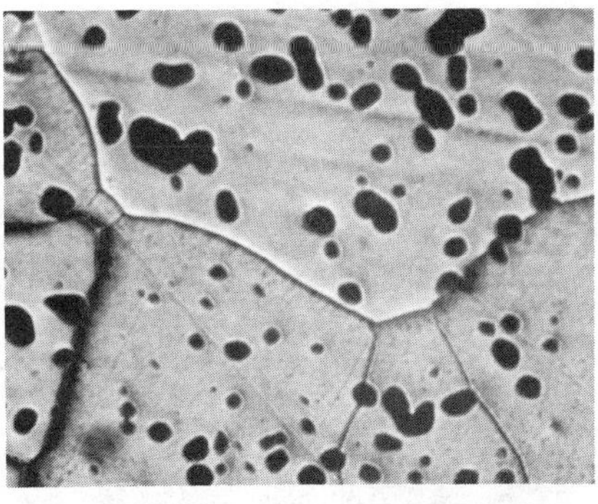

V_V = 0.05 200μm

Figure 7. Grain growth as a function of sintering time for atomized steel powder with an 0.5 percent phosphorus addition (sintering temperature 1350°C).

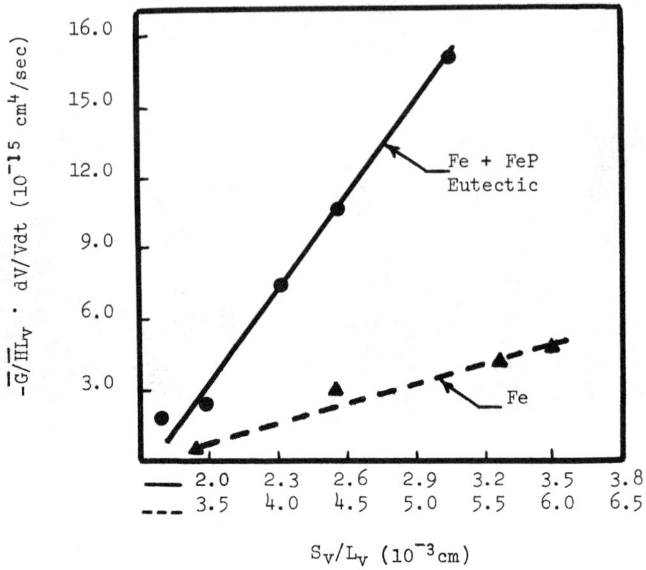

Figure 8. Determination of diffusion coefficients from measurable geometric parameters.

$$\frac{\overline{G}}{\overline{H}L_V} \cdot \frac{dV}{Vdt} = \frac{32\gamma\Omega D_V}{kT} \cdot \frac{S_V}{L_V} + \frac{32\gamma\Omega b D_b}{kT} \qquad (2)$$

where $\overline{G}$ = mean grain diameter, $\overline{H}$ = mean pore surface curvature, L_V = grain boundary-pore intersection length-per-unit-volume, $(1/V)\ (dV/dt)$ = rate of fractional volume shrinkage, V = volume fo compact, γ = surface tension, Ω = atomic volume, D_V = volume diffusion coefficient, S_V = pore surface area-per-unit-volume, D_b = grain boundary diffusion coefficient, b = grain boundary width, and kT has its usual meaining. In the second approach, a kinetic expression, based on surface area reduction was used in assessing the bulk-transport mechanisms. The equation, as developed by German and Munir (15, 16, 17), is:

$$\left(\frac{\Delta S}{S_o}\right)^{\delta} = Kt \qquad (3)$$

where ΔS = change in surface area, S_o = initial surface area, t = isothermal sintering time, K = constant and δ is a constant related to the sintering mechanism.

The sintering isotherms plotted from equation (2) at 1350°C, Figure 8, established volume diffusion as the predominate transport mechanism for steel powders with, or without, a momentary liquid phase. From the slopes of the lines, the apparent volume diffusion coefficients were calculated to be $3.86 \times 10^{-6}\ cm^2/sec$ for a steel powder with a momentary liquid phase and $5.97 \times 10^{-8}\ cm^2/sec$ for the same steel powder without a liquid phase

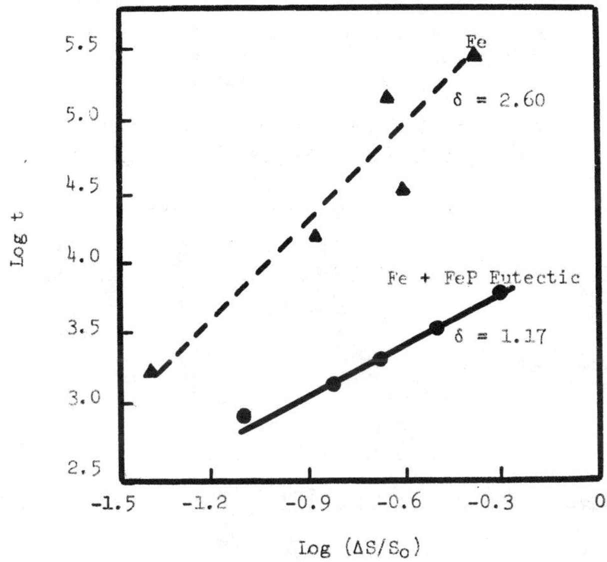

Figure 9. Determination of sintering mechanisms based on reduction of pore-solid interfacial area.

addition. From this difference in values, it was concluded that additional mechanism(s), besides volume diffusion, occurred when a liquid phase addition was used.

The surface area reduction analysis, Figure 9, indicated an exponent value of $\delta = 2.60$ for a steel powder without a liquid phase and $\delta = 1.17$ when a liquid phase was employed. The $\delta = 2.60$ value was typical for volume diffusion and verified the corresponding metric parameter analysis. A value of 1.17, however, for the liquid phase addition was more typical for a transport mechanism of viscous flow. Therefore, it was concluded that the enhanced densification noted when a momentary liquid phase was employed was the result of volume diffusion and grain boundary sliding. The role of the liquid phase, in this instance, was to wet the grain boundaries and thereby provide an extremely favorable situation for particle rearrangement. Full densification was prevented by the increasing grain size which reduced material transport by diffusion. A similar effect has been observed in the sintering of ceramic powders (18).

Calculation of the activation energies associated with the sintering of steel compacts with and without a momentary liquid phase addition was done to verify the diffusion mechanisms. The calculation was based on shrinkage rate as a function of sintering temperature (19). The values obtained are shown in Table II along with their respective diffusion coefficients. A steel powder compact without the liquid phase addition had an activation energy of 87kcal/mole, corresponding to a value representative of volume diffusion. The liquid phase addition, however, substantially lowered the activation energy (25kcal/mole) for densification. Its relatively low value was the result of the grain boundary slip phenomena accompanying volume diffusion. Activation energies and diffusion coefficients for other iron systems that have been reported are also shown in Table II for comparison.

Table II. Activation Energies and Diffusion Coefficients in Iron and Steel Systems

Materials	Diffusion Coefficient (cm^2/sec)	Activation Energy (kcal/mole)
Sintered Steel Powder	5.97×10^{-8}	87
Sintered Steel Powder +0.5% Phosphorus	3.86×10^{-6}	25
Iron Wires (Price, Hall, Greenough)[20]	4.52×10^{-9}	--
High Purity Iron (Buffington, Hirano, Cohen)[21]	6.61×10^{-9}	64.5
Gamma Iron (Birchenall and Mehl)[22]	---	74.2
Iron Wires (Shewmon)[23]	3.87×10^{-7} (1535°C, M.P.)	--

Conclusions

1. A momentary liquid phase can be used effectively to enhance the densification in steel powders.

2. In the presence of a momentary liquid phase, volume diffusion and viscous flow (grain boundary sliding) are the predominate mechanisms for densification.

3. During the third stage of sintering, exaggerated grain growth isolates the porosity and prevents full densification.

Acknowledgments

This work was performed in the Metallic Materials Branch, Materials and Manufacturing Technology Division of the Fire Control and Small Caliber Weapon Systems Laboratory as part of DA Project 1T161102AH61.

References

1. J. C. McLennan and C. J. Smithells, J. Sci. Instruments, 12 (1935) p. 159.

2. G. H. S. Price, C. J. Smithells, and S. V. Williams, J. Inst. Metals, 62 (1938) pp. 239 - 254.

3. V. N. DiGiambattista and W. E. Reilly, Mod. Dev. Pow. Met, 7 (1974) pp.153 - 173.

4. T. Levin and R. P. Hervey, Metal Progress, 115 (1979) pp. 31 - 34.

5. R. E. Matt, et. al., Patent 3 744 933.

6. P. Lindskog, Powder Met., 16 (32) (1973) pp. 374 - 86.

7. R. T. DeHoff and E. H. Aigeltinger, Advanced Experimental Techniques in Powder Metallurgy, J. S. Hirschhorn and K. H. Roll, ed., p. 96, Plenum Publishing Corporation, NY, 1970.

8. ibid, pp. 102 - 106.

9. ibid, pp. 107 - 131.

10. J. Kronsbein, L. T. Buteau, and R. T. DeHoff, AIME Trans., 233 (1965) pp. 1961 - 1969.

11. F. N. Rhines and R. T. DeHoff, Second European Sym. on Pow. Met., (1968) pp. 1 - 21.

12. D. L. Johnson, J. Appl. Phys., 40 (1969) pp. 192 - 200.

13. D. L. Johnson, J. Am. Cer. Soc., 53 (1970) pp. 574 - 577.

14. R. L. Coble and T. K. Gupta, Sintering and Related Phenomena, G. C. Kuczynski, N. A. Hooton and C. F. Gibson, eds., pp. 423 - 41, Gordon and Breach, NY, 1967.

15. R. M. German and Z. A. Munir, Sintering and Catalysis, G. C. Kuczynski, ed., pp. 249 - 257, Plenum Publishing Corporation, NY, 1975.

16. R. M. German and Z. A. Munir, Sintering and Catalysis, G. C. Kuczynski, ed., pp. 259 - 268, Plenum Publishing Corporation, NY, 1975.

17. R. M. German and Z. A. Munir, J. Am. Cer. Soc., 59 (1976) pp. 379 - 383.

18. P. Reynen, Bull. Soc. Franc. Ceram., 97 (1972) p. 3.

19. R. M. German, Met. Trans., 7A (1976) pp. 1879 - 1885.

20. A. T. Price, H. A. Hall, and A. P. Greenhough, Acta Met., 12 (1964) pp. 49 - 58.

21. F. S. Buffington, K. Hirano, and M. Cohen, Acta Met., 9 (1961) pp. 434 - 439.

22. C. E. Birchenall and R. F. Mehl, Trans. Amer. Inst. Min. Engrs., 188 (1950) pp. 144 - 149.

23. P. G. Shewmon, Diffusion in Solids, McGraw-Hill, NY, 1963, p. 66.

Properties

THE ROLES OF RAW MATERIAL SELECTION, POWDER PREPARATION, AND SINTERING PROCESSES ON THE PHYSICAL PROPERTIES OF BETA"-ALUMINA CERAMIC ELECTROLYTES

B. J. McEntire, G. R. Miller, and R. S. Gordon

Ceramatec, Inc.
Salt Lake City, Utah 84115
USA

Critical to the successful production of beta"-alumina ceramic electrolytes is the selection of appropriate raw materials, powder synthesis techniques, and sintering processes. Extensive investigations into these three aspects of beta"-alumina electrolyte fabrication have been conducted. A comparison of the physical properties of beta"-alumina electrolytes fabricated from different raw materials and manufacturing methods demonstrates that conductive, high performance and durable ceramics can be prepared from relatively low-cost, high purity raw materials using potentially economical fabrication processes.

Introduction

Over the past decade, much has been written concerning worldwide shortages of energy and the need for efficient use of fuel resources. This concern has spurred an interest in the development of novel energy conversion and storage devices. Kummer and Weber provided the stimulus for a new generation of electrochemical devices based on the use of beta or beta"-alumina (1). These two compounds are members of a family of polymorphic sodium aluminates possessing high sodium ion conductivity. This unique feature allows these ceramics to serve as solid electrolytes in various applications. In particular, sodium beta and beta"-alumina are considered to be the prime candidates for use in the sodium-sulfur battery (2). Of the two compounds, beta"-alumina possesses the highest sodium ion conductivity and for this reason it is generally preferred over beta-alumina. The high ionic conductivity of beta"-alumina results from its layered crystal structure. The rhombohedral unit cell consists of layers of four close-packed planes of oxygen and aluminum ions forming a spinel-like structure. These layers, often referred to as spinel blocks, are separated by loosely-packed planes in which sodium ions reside and move freely under the influence of an electric field. Beta"- alumina is believed to be unstable at elevated temperatures where sintering normally occurs but the material has been effectively stabilized by additions of small amounts of either Li_2O, MgO, or combinations of Li_2O and MgO. These stabilizing aids substitute in the spinel block for Al^{3+} ions and are compensated by a corresponding increase in the number of positive charge carriers in the two dimensional conduction planes.

Over the last several years, polycrystalline beta"-alumina electrolytes have been fabricated into cylindrical tubes by a number of techniques including isostatic pressing (3), electrophoretic deposition (4), extrusion (5,6), tape casting (7), and slip casting (8). Densification of these shapes has been accomplished by zone (9), continuous, and batch (10) sintering methods. The properties of these electrolytes are attractive. Mass densities over 3.18 g/cm^3 with ionic resistivities between 3 and 15 ohm-cm (300°C) and reasonably high strengths (170 to 320 MPa) are now routinely achieved. Despite this fact, electrolytes possessing these properties exhibit varying performance characteristics in sodium-sulfur electrochemical cells. Electrolytic degradation resulting in premature cell failure, cell resistance increases, and asymmetric cell polarization behavior has been frequently observed. While these problems are undoubtedly caused, in part, by improper handling of the electrolytes prior to cell assembly and factors unrelated to the electrolyte (e.g., container corrosion), some of these problems can also be attributed to the beta"- alumina ceramic itself. In particular, selection of raw materials, powder processing methods and sintering techniques can influence both the in-cell and out-of-cell behavior of ceramic electrolytes. Proper attention to these factors will result in excellent quality and high performance electrolytes. In this paper, recent developments on the interrelationships between raw material and processing variables and the properties and performance of lithia-stabilized beta"-alumina ceramics will be described.

Experimental Procedure

Raw Materials

Beta"-alumina powders were formulated and prepared at a composition of 8.85% Na_2O, 0.75% Li_2O and 90.4% Al_2O_3 by weight. Alpha-alumina and reagent grade sodium and lithium salts comprised the major raw materials. All materials were assayed by either gravimetric or titrimetric methods prior to processing of the ceramic.

Previous investigations have reported on a number of candidate alpha-alumina raw materials (3,11). In the earlier work, it was demonstrated that moderately high purity alpha-alumina (99.5%+) was necessary in order to fabricate high quality, conductive electrolytes. In the present investigation, a series of alumina raw materials has been carefully examined which conform to this previously set specification with respect to purity but with a range of costs. A listing of these aluminas, along with their nominally reported properties, is given in Table I. The physical and chemical properties of each raw material have been evaluated. Chemical analyses were performed using emission spectography on three replicate samples of each powder. Particle and agglomerate size measurements were performed using a Sedigraph 5000 particle size analyzer.* The phase composition and powder morphology of these raw materials were examined using x-ray diffraction and SEM techniques, respectively.

The alkali raw materials consisted of either sodium carbonate** (Na_2CO_3), sodium hydroxide**($NaOH$), sodium acetate** (CH_3COONa), or sodium aluminate*** ($NaAlO_2$), and lithium oxalate+ ($Li_2C_2O_4$), lithium hydroxide monohydrate+($Li_2OH\ H_2O$), lithium aluminate+ ($LiAlO_2$), or zeta lithium aluminate++ (zeta-$LiAl_5O_8$). The selection and use of any one of these raw materials was dependent upon the processing method employed.

Powder Preparation and Forming

Two basic processing methods were utilized in preparing beta"-alumina powders: zeta processing and spray drying.

Zeta Processing. The zeta process, first described by Youngblood et. al. (12), consists of separate mixing and calcination of two components. One component, zeta-lithium aluminate, is prepared by mixing alpha-alumina and

---

* Micromeretics, Inc., Norcross, VA.

** J. T. Baker Co., Phillipsburg, PA.

*** The $NaAlO_2$ was prepared by pre-reacting Na_2CO_3 with alpha-alumina at 1000°C for two hours. Subsequently, dry-ball milling was employed to obtain a reactive powder. This powder was milled for 19 hours at a media to powder ratio of 10:1.

\+ LITHCOA, Inc., Bessemer City, N.C.

\+\+ The zeta-$LiAl_5O_8$ was prepared by pre-reacting $Li_2C_2O_4$ with alpha-alumina at 1260°C for two hours. Dry-ball milling of this powder was also performed.

Table I

NOMINALLY REPORTED PROPERTIES OF ALUMINA RAW MATERIALS

PROPERTY	ALUMINA RAW MATERIAL					
	A	B	C	D	E	F
Chemical Properties (%)						
Al_2O_3	99.95+	99.8+	99.5+	74.2*	99.7+	99.99
Na_2O	<.008	.12	.08	.004	-	.003
SiO_2	.016	.009	.04	.008	.008	.002
CaO	.008	<.008	.01	-	.01	.001
TiO_2	.001	.001	.001	-	.0007	.0008
Cr_2O_3	.0003	.00014	.0002	-	.001	.001
Fe_2O_3	.0038	.003	.02	.005	.005	.003
K_2O	<.001	.0033	-	-	-	-
Mean Particle Size (μm)	0.55	0.8	0.5	45	5-10	3.8
BET Surface Area m^2/gm	6-8	8.0	6-8	280	13	30
Alpha Content	100%	100%	100%	0%	80%	100%

A - RCHP-DBM, Reynolds Metals Co., Richmond, VA.

B - ERCHPS-DBM, Reynolds Metals Co., Richmond, VA.

C - A-16SG, ALCOA, Pittsburgh, PA.

D - CATAPAL SB, Conoco Chemicals Division, Houston, TX.

E - ALUSUISSE BP-80, Alusuisse Metals, Inc., Fort Lee, NJ.

F - BAIKOWSKI CR-30, Baikowski International, Charlotte, NC.

* Approximately 25% H_2O

$Li_2C_2O_4$ in a 5.5 to 1 molar ratio. The material is prepared at a lithia deficient composition in order to avoid the formation of $LiAlO_2$. The second component is formed from an intricate blend of alpha-alumina and Na_2CO_3 in a molar ratio of approximately 5.24:1. Blended components are calcined separately at 1260°C for two hours and then combined in appropriate quantities to yield the correct composition. Mixing and comminution of these two components occurs concurrently during ball-milling* which is performed at media to powder ratios between 5:1 and 10:1 for 19 to 36 hours. Polyvinylbutyral** (2% by weight) is added to the milled powder in an acetone slurry as a binder. The powder is subsequently dried, deagglomerated and screened prior to isostatic compaction.

Spray Drying. In the spray drying process, batches are prepared by slurrying the active alumina raw material and semi-soluble or soluble alkali salts in a solution containing selected additives. Mixing of these constituents was accomplished for 15 to 30 minutes in either a urethane-lined SWECO mill***, using 0.5 inch cylindrical alpha-alumina media+, or by propellor action inside a stainless steel tank. Spray drying was performed in a small production dryer++, incorporating mixed air flow with two-fluid fountain-type atomization. The inlet and outlet temperatures of the dryer were varied between 200°C - 350° and 93°C - 137°C, respectively. The air pressure of the two-fluid atomizer was varied between 15 and 60 psig.

Forming. Ceramic fabrication of all powders into cylindrical tubes was accomplished by isostatic compaction at pressures between 208 MPa and 381 MPa. The dimensions of the green tubes varied between 20 and 60 mm diameter by 1 to 3 mm wall by up to 700 mm in length. These ceramic parts were subsequently bisqued in air at temperatures up to 1200°C. During bisquing, samples and tubes were heated at 1°C/min during critical temperatures where the binder or the alkali salts decompose. Heating rates at all other temperatures were 10°C/min.

Sintering

All samples and tubes were sintered by enclosing the bisqued ware inside platinum tubes or platinum-lined setter crucibles and firing at temperatures between 1550°C and 1650°C for time periods ranging between 5 and 30 minutes. Sintered ware was subjected to a post-sinter anneal at temperatures between

---

* Milling is accomplished in an Arlcite-lined Patterson Mill or in one gallon AD-99 Coors Alumina Mills (Golden, CO). In either case the milling media utilized were Sapphrox 1-1/2 inch spherical alumina balls from Sapphrox, Benton, AR.

** B-76, Monsanto, Inc., St. Louis, MO.

*** Model ML-20, SWECO, Inc., L.A., CA.

\+ Diamonite Products, Inc., Shreve, OH.

++ Bowen Laboratory Ceramic Spray Dryer #1, Bowen Eng., Somerville, N.J.

1400°C and 1475°C for times ranging between 1 and 24 hours. Several experiments were performed wherein the pre-sintered ware was subjected to an anneal prior to raising the furnace temperature to that required for rapid densification. In these experiments, samples were "pre-annealed" for time periods ranging from 1 to 4 hours at temperatures between 1400°C and 1500°C.

Characterization and Cell Testing

Post-firing ceramic property characterization involved numerous measurements. Density, ionic resistivity, diametral strength, and microstructural properties were measured using methods described elsewhere (11). Phase composition was determined using x-ray diffraction according to a previously described procedure (12). Dimensional and shrinkage measurements were made with standard calipers and gauges.

Corrosion studies and endurance testing of selected ceramics were performed in specially constructed sodium-sodium cells. Electrolyte tubes, 16 mm OD by up to 200 mm long, representing a selection from various raw materials and processes were cycled with dc current passing in and out of the tubes at constant current densities up to 1.25 A/cm^2 at temperatures around 300°C. Inspection of the electrolytes after cell testing was qualitatively performed visually.

Results and Discussion

Raw Materials

Listed in Table II are the chemical and physical properties of the various alpha-alumina raw materials. With the exception of raw materials D and E, these compounds are 100% alpha-alumina. Raw material D is a boehmite powder and material E is a partially converted gamma-alumina. Raw materials A through C are high purity Bayer-process aluminas. Raw material D is the by-product from a refinery operation and material E is prepared using a process developed and described by Ruthner et. al. (13). Alumina F is an alum derived powder. These materials range in cost from $1 - 30/kg. As might be expected, the higher cost materials possess smaller concentrations of undesirable impurities (especially Ca and Si) than do the lower cost materials. A considerable range in purity of these aluminas exists although each of them is considered to be a reagent grade chemical.

Due to the fact that beta"-alumina is an ion-conducting ceramic which operates in highly corrosive environments, excessive concentrations of specific impurities are believed to adversely affect performance and lifetime. Over the past few years, many researchers have recognized the need for evaluating the effects of impurities on the performance characteristics of beta"-alumina electrolytes. As a result, considerable literature has been generated on this topic (14-25). In particular, it has been shown that calcium and silicon are perhaps the most damaging impurities (19,20). However, systematic studies of all potential impurities have not been conducted.

Table II

PHYSICAL AND CHEMICAL PROPERTIES OF ALUMINA RAW MATERIALS

PROPERTY	ALUMINA RAW MATERIAL					
	A	B	C	D	E	F
Chemical Analysis* (ppm)						
Ca	35±12	13±4	153±15	43±2	143±6	8±3
Cr	-	-	6±1	9±1	16±1	3±0
Fe	29±1	23±3	113±28	40±1	80±2	5±1
Ni	3±1	3±1	8±3	9±1	14±2	3±1
Si	125±16	137±8	260±10	30±2	137±6	29±1
K	-	-	12±2	-	-	34±2
Mean Particle Size (μm)	≤0.5	≤0.5	≤0.5	43	5-10	≤0.3
Phase Composition	100% alpha	100% alpha	100% alpha	-	80% alpha	100% alpha

Raw Material D also contains ∿1,000 ppm Ti

Of the Bayer processed aluminas (A, B & C), considerable variation in impurity levels exist. These impurities are historical artifacts of the Bayer process itself. During one stage of the process, soda and other impurities (predominantly the alkali and alkaline earth soluble salts) can become entrapped in the gibbsite crystals during many hours of cycling in precipitators prior to their calcination to the alpha-alumina phase. Washing is not effective in removing all of these impurities and hence they eventually remain in the final product (26). Also, dry grinding of these aluminas with low purity mill linings and media can result in significant increases in the CaO and SiO_2 content as is the case for raw material C (27).

The non-Bayer processed aluminas are superior in some respects but contain other impurities (e.g., Ti, Fe) whose ultimate effects have not yet been characterized in sodium-sulfur cell applications. The alum derived raw material (alumina F) presents perhaps the most desirable properties but unfortunately it is also the highest priced. While such a material may be viable for short or intermediate term developments, it is presently priced far too high to become a practical production raw material.

Clearly, extensive work is required to determine which impurities (other than Ca and Si) are harmful to the performance characteristics and lifetime of beta"-alumina electrolytes in Na-S cells. Also, critical concentrations of these impurities need to be determined. However, it appears probable that low-cost and relatively high purity raw materials such as powders B, C and E may be eventually used in full-scale production. Nevertheless, much more information is needed concerning the long-term effects of impurities on electrolyte performance.

The alkali constituents (Na_2O and Li_2O) can be obtained or prepared in fairly high purity. The cost of various sodium oxide sources is low (i.e., $2/kg). The lithium salts, although more expensive, are not a major cost item because of their relatively small concentrations. The use of different alkali raw materials and their effect on the physical properties of the ceramic will be described.

Powder Preparation, Sintering and Electrolyte Characterization

Processing of most beta or beta"-alumina ceramics involves the reaction of the alkali raw materials with aluminum oxide. This reaction process can be effected by calcining powder mixtures prior to sintering or by firing preformed ceramic tubes containing unreacted constituents. In any case, ceramics prepared by either process pass through a series of intermediate phases prior to reaching fully stabilized beta"-alumina. (In reality, 100% conversion to the beta"-alumina phase is normally achieved following densification and post-sintering annealing.)

When powder mixtures are used, milling is required after calcination in order to maintain powder activity. Calcining and milling are basic to the so-called "zeta-process". On the other hand, spray drying utilizes insitu reaction of the alkalis with the alpha-alumina after isostatic pressing. Both of these processes have distinct advantages and potential drawbacks.

Zeta Processing. For zeta-processed powders, incorporation of the Li_2O in the form of $LiAl_5O_8$ aids in distributing the lithium ions during mixing and milling. Additionally, because zeta-$LiAl_5O_8$ crystallizes in a spinel structure, this phase may act to enhance the formation of stabilized beta"-alumina by a coherent nucleation and growth mechanism. Either case will aid in rapid conversion of the material to the beta"-alumina structure.

The zeta-process has been used to prepare powders from each of the aforementioned alpha-alumina raw materials. Listed in Table III are physical properties of zeta-process ceramics, prepared under various sintering schedules.

All of these alpha-alumina raw materials led to the production of good to excellent quality beta"-alumina ceramics. Excellent density and resistivity values were obtained. The microstructures were generally duplex in nature. While the average grain size was generally less than 50 microns, the propensity for exaggerated grain growth was prevalent with some powders. Control of runaway growth for these powders was accomplished only by engineering precise firing schedules.

It is noted that ceramics prepared from raw materials C and E, possessed the highest resistivity values. As discussed earlier, these two materials also contained the highest calcium contents. Conversion of these powders to 100% beta"-alumina was difficult. In fact, ceramics prepared from raw materials B, C and E required extensive heat-treatments ($\sim 1450^{\circ}C$ for 4 hours) in order to obtain optimum and low resistivities.

The cause of high resistivities may be attributable to one or more of several factors: (1) incomplete conversion of the beta"-alumina phase, (2) intragranular impurity effects, or (3) intergranular impurity effects.

In an effort to understand the cause of the high resistivity values, ceramics prepared from raw material C were systematically studied. X-ray diffraction analysis of ceramics prepared from raw material C indicated that they were predominantly a two phase mixture consisting of beta and beta"-aluminas. No other phases were detected. $f(\beta)$ values (12) were calculated for several samples using the integrated intensities of the (107) beta-alumina peak occurring at 33.4° of 2θ and the (01,11) beta"-alumina peak occurring at 34.6°. The range of values obtained varied between 0.01 to 0.1. From a comparison of the relative intensity data obtained from the x-ray diffraction analysis, the volume fractions of the beta- and beta"-alumina phases were estimated. For an $f(\beta)$ value of about 0.1, the volume fraction of beta"-alumina was 0.94. For an $f(\beta)$ of about 0.01, the corresponding volume fraction of beta"-alumina was 0.99*. For an $f(\beta)$ value of 0.1, utilization of a simple binary mixture rule indicated that a significant fraction of the high macroscopic resistivity can be attributed to the presence of the beta- alumina phase in the ceramic**. Conversely, for $f(\beta)$

---

*This estimation is based on a normalizing calculation of the (01,11) beta"-alumina diffraction line to the (017) beta-alumina line. The relative intensities of each of these two lines are 40% and 75%, respectively. The mass absorption coefficients for beta- and beta"-aluminas are essentially equivalent and the 100% relative peaks of each phase occur at the same 2θ value.

**Such a calculation assumes nominal resistivity values for beta- and beta"-aluminas to be 16.6 ohm-cm and 4.5 ohm-cm ($330^{\circ}C$) respectively (31).

Table III

PHYSICAL PROPERTIES OF β''-ALUMINA CERAMICS
PRODUCED USING ZETA PROCESSED POWDERS

RAW MAT'L	SINTERING SCHEDULE	PHYSICAL PROPERTIES DENSITY (g/cm^3)	% LINEAR SHRINKAGE	RESISTIVITY 300°C(ohm-cm)	MAXIMUM GRAIN SIZE (μm)
A	1550°C;7mins.	3.23±.02	15.2±.4	4.6±.7	150-200
B	1550°C;7mins.	3.21±.02	15.4±.4	4.6±.3	150-300
C	1590°C;0 min.	3.21±.01	16.0±1.0	6.3±.2	150
D	1600°C;5mins.	3.21±.02	15.0±1.0	3.8±4.3	100
E	1580°C;10mins.	3.22±.01	16.6±.2	5.1-5.7	150-300
F	1590°C;7mins.	3.21±.01	15.6±.5	4.7±.5	100-150

All samples were subjected to a post-sinter anneal at ∿1475°C for $\leq$2 hrs. Samples prepared from raw materials B and C required a pre-anneal heat treatment at ∿1475°C for $\leq$2 hours in order to obtain optimum conductivities.

values around 0.01, the apparent contribution of the beta-alumina phase to the overall resistivity seems insignificant. This assumes, however, that the nominal resistivities of the two phases are reasonably low (i.e., 4 ohm-cm and 16 ohm-cm for beta"-and beta-alumina, respectively). Such may not be the case. The presence of impurities located within the "spinel" structure or within the conduction plane itself may significantly increase these values. For instance, Boilot, et. al., have shown that the presence of M^{2+} ions with radii ≥ 0.97 Å may substitute for Al^{3+} and favor the formation of beta-alumina preferentially over beta"-alumina (18). Included in this group are Ca^{2+} (r=0.99Å), Sr^{2+} (r=1.12Å), Pb^{2+} (r=1.20Å), and Ba^{2+} (r=1.34Å). Monovalent or divalent cations with an ionic radius ≤ 0.97 tended to favor the formation of beta"-alumina. Additionally, some of these ions can be incorporated into both the beta- and beta"-alumina structures by substitution of Na^+ ions in the conduction planes if additional $O^=$ ions are also incorporated (28,29).

Several researchers have shown that the presence of certain divalent cations tends to increase the resistivity of beta-alumina type materials. For calcium containing beta-alumina (∿1% CaO), 10 to 100 fold increases in resistivity have been reported (14,30). In one instance, these higher values are attributed to the presence of secondary non-conductive phases at grain boundaries (23).

Using a method described by Virkar, et. al. (32), a comparison of the resistivities of coarse and fine grained electrolytes prepared from raw material C was used to obtain information concerning the inter and intragranular components of the resistivity. Activation energies for both the coarse and fine grained material were obtained from axial resistivity plots over a temperature range between 573°K and 648°K. The activation energy for the fine grained sample was 5.35 ± 0.14 Kcal/ mole (95% confidence limits). The coarse grained sample yielded a value of 4.76 ± 0.07 Kcal/mole. The apparent intergranular activation energy obtained from this analysis was 6.16 ± 0.20 Kcal/mole. These values compared favorably with values obtained from electrolytes prepared from higher and lower purity raw materials. The results indicate tentatively that significant quantities of impurities are not being uniformly segregated at grain boundaries.

Despite the apparent negligent effect on the grain boundary contribution to the overall resistivity, impurities such as CaO or SiO_2 may significantly alter the overall resistivity even as second phase constituents. The intergranular formation of such phases may reduce the local Na_2O or dopant content of the adjacent grains and thus aid in the formation of localized sodium depleted beta-alumina preferably over beta"-alumina. The syntactical relationships of beta- and beta"-alumina type structures suggest that the formation of either phase is quite sensitive to local changes in the dopant or Na_2O levels. The formation of a beta- alumina phase would therefore be more likely to occur near the grain boundary where these second phase compounds can segregate.

It therefore appears that the high resistivity values obtained on electrolytes prepared from raw material C are due to incomplete conversion of the material to the beta"-alumina phase. In turn, this was probably caused by the effects of impurities in destabilizing the beta"-alumina structure. Incorporation of Ca^{2+} ions either into the spinel block or within the conduction planes may have caused highly resistive beta- or beta"-alumina to form.

Spray Drying. The preparation of beta"-alumina powders by spray drying was first extensively investigated in this country by Miller, et. al. (3) Others, most notably Johnson and co-workers (33,34), have published similar work. Two approaches have been taken by these investigators in the preparation of pressable spray dried beta"-alumina powders. The first method, extensively developed by Miller, is designated the conventional method. It involves the slurrying of zeta processed powders in an appropriate quantity of water followed by spray drying. This approach mimicks more traditional synthesis methods for ceramic powders. The second approach to the preparation of beta"-alumina press powders involves the preparation of a slurry solution of alpha-alumina and soluble or semi-soluble alkali salts. The novel feature of the later preparation method is that it eliminates several energy-intensive unit operations (e.g., calcination and ball milling) that are required in conventional processing. To differentiate between the two methods, the first method will be referred to as "slurry" spray drying and the second method will be designated "slurry-solution" spray drying.

Only raw material A has been investigated by the technique of slurry spray drying. Extensive testing of this technique was not warranted because the second method of slurry-solution spray drying has a greater potential for low cost processing in mass production. Preparation of a flowable spray dried powder using this method was fairly straight forward. However, densification of such parts to acceptable mass density was more difficult than that required for standard zeta process powder (non-spray dried). This difficulty was attributed to poor compaction characteristics of the spray dried powder itself. However, with the selection of proper additives to the slurry and control of the operational parameters of the dryer, adequate densities ($3.20 \pm 0.01 g/cm^3$) and ionic resistivities at 300°C (4.5 ± 0.3 ohm-cm) could be achieved with the use of these powders. These properties were comparable to those achieved in zeta processing.

For the technique of slurry solution spray drying a comparative analysis of raw materials A, B and C has been conducted. This work involved an examination of the densification characteristics and phase conversion kinetics in ceramics prepared from these three raw materials. Water soluble NaOH and LiOH were used as the source of alkalis. Ceramics prepared from these three powder types were formed by isostatic pressing, sintering at temperatures between 1550°C and 1600°C for 10 minutes and annealing at 1475°C for time periods ranging up to four hours. The physical properties of ceramics prepared from these three powders are given in Table IV. The resistivity and phase conversion characteristics are shown in Figures 1 and 2.

During post sintering heat treatments, dramatic reductions in resistivity accompanied by substantial conversion to the beta"-alumina phase was encountered as expected. This effect was most pronounced during the first hour of annealing. Annealing for three additional hours resulted in further reductions in ionic resistivity although the change was much less dramatic than that observed in the first hour. Of some interest is the observation that after four hours of annealing, these ceramics were still not fully converted to the beta"-alumina phase. An x-ray analysis of sintered electrolytes prepared from all three raw materials revealed the presence of sodium-aluminate ($NaAlO_2$) as a tertiary phase in small quantities. The microstructures of all ceramics were significantly coarsened after post sintering heat treatments.

Table IV

A COMPARISON OF THE PHYSICAL PROPERTIES OF SLURRY SOLUTION SPRAY DRIED POWDERS

Raw Material/Property	Post-Sinter Annealing Time at 1475°C: 0 hr	1 hr	4 hr
I. A-16 SG Alumina*			
1. Density (g/cm^3)	3.18±0.00	3.19±0.01	3.18±0.01
2. % Linear Shrinkage	12.1±0.2	12.1±0.1	12.2±0.1
3. Resistivity (ohm-cm 300°C)	17.9±0.6	7.3 ±0.4	6.1 ±0.5
4. f(β) value	0.63	0.35	0.24
5. Maximum Grain Size	≤50μm	≤300μm	≤300μm
II. RCHP-DBM Alumina**			
1. Density (g/cm^3)	3.18±0.01	3.19±0.02	3.19±0.02
2. % Linear Shrinkage	11.3±0.2	11.3±0.1	11.3±0.0
3. Resistivity (ohm-cm 300°C)	15.9±2.8	6.4 ±0.3	5.3 ±0.1
4. f(β) value	0.62	0.33	0.21
5. Maximum Grain Size	≤50μm	≤300μm	≤300μm
III. ECHPS-DBM Alumina**			
1. Density (g/cm^3)	3.18±.01	3.23±0.01	3.19±0.01
2. % Linear Shrinkage	11.7±0.2	11.5±0.3	11.5±0.3
3. Resistivity (ohm-cm 300°C)	15.0±0.6	6.5 ±0.3	4.9 ±0.1
4. f(β) value	0.40	0.30	0.19
5. Maximum Grain Size	≤50μm	≤300μm	≤300μm

* Sintering Conditions for A-16SG--1585°C-10 mins.
**Sintering Conditions for the Reynolds Based Powders--1570°C-10 mins.

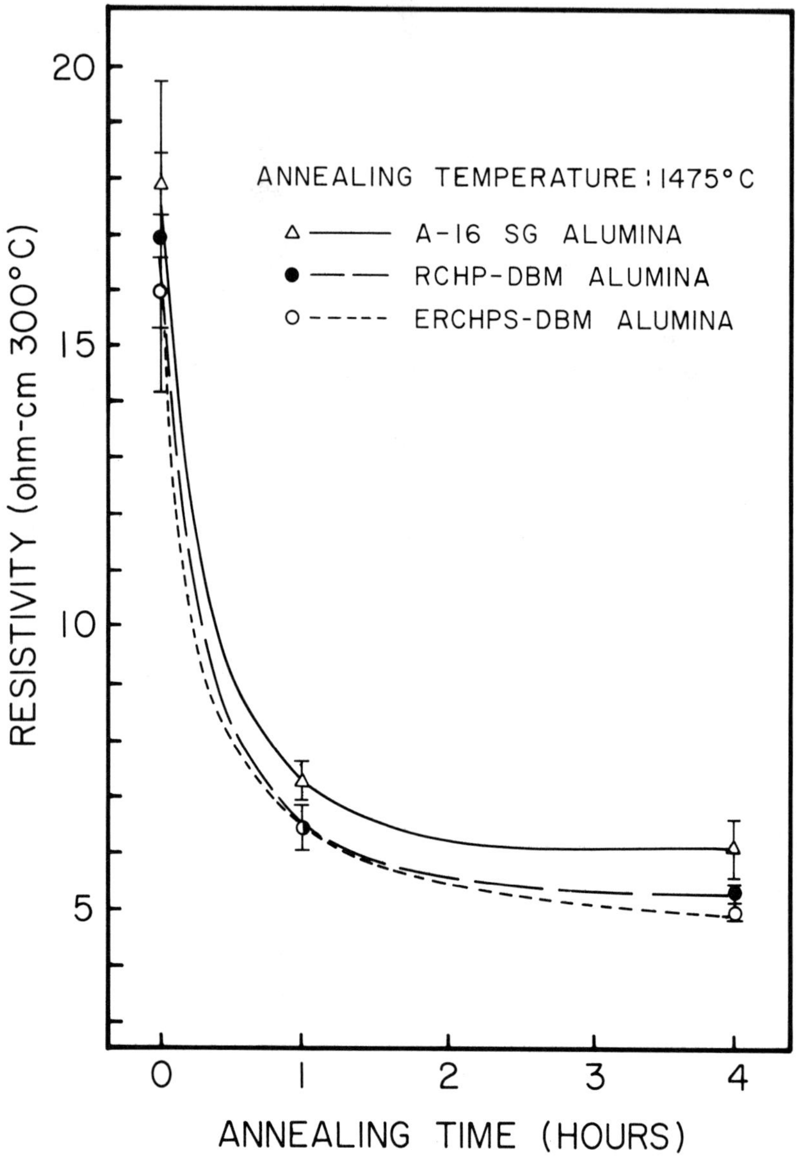

Figure 1 - Effect of Annealing on the Resistivity of Sintered Beta''-Alumina Electrolytes Fabricated from Spray Dried Powders

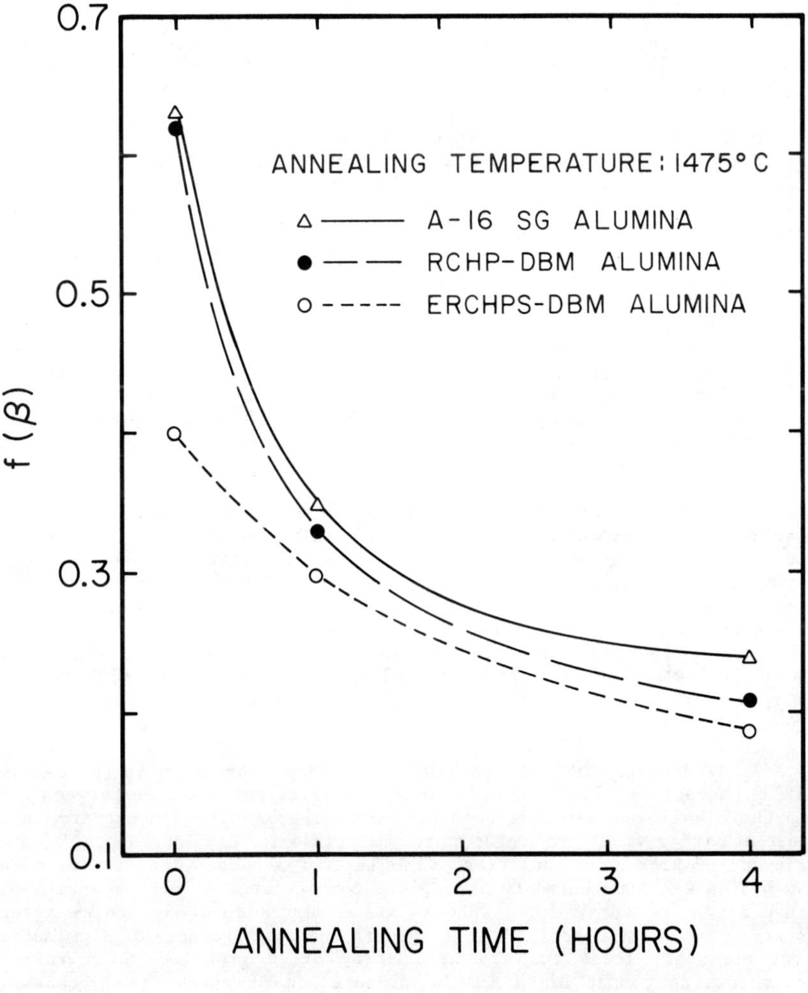

Figure 2 - Effects of Annealing on the Phase Composition of Sintered Beta"-Alumina Electrolytes Fabricated from Spray Dried Powders

Substantial property differences could be associated with the use of a particular raw material. The data clearly indicate that under all conditions raw material C led to ceramics with higher resistivities. Raw materials A and B led to electrolytes with almost equivalent properties. The higher resistivities observed with raw material C were similar to those encountered in non-spray dried (zeta process) powders. The high impurity levels in this powder are probably the cause of the high resistivities.

In addition, ceramics prepared from slurry-solution spray dried powders are more susceptible to exaggerated grain growth than electrolytes fabricated with non-spray dried zeta process powders. To overcome the problem of exaggerated grain growth, alternative processing modifications have been developed: seeding (35,36), "dual-peak" firing (37), pre-sinter annealing (38), and the use of alternative alkali raw materials.

Previous work has shown that by seeding precursor powders with fine crystals of beta"-alumina, a fully converted electrolyte with an intermediate, uniform microstructure can be produced. Seeding experiments conducted using the processing methods and raw materials presented in this report have validated this earlier work. This catalysis method of phase conversion allows for optimum and low ionic resistivities to be obtained while minimizing exaggerated grain growth.

The pre-sinter annealing process is another firing method that minimizes the formation of large idiomorphic grains. This procedure allows for homogenization of the ceramic to occur as well as the formation of many stable beta"-alumina nuclei. Experimental work using this firing technique demonstrated that this method is effective in controlling exaggerated grain growth while at the same time promoting conversion of the material to the beta"- alumina phase.

In Table V and Figures 3 and 4, the effect of alkali raw materials are summarized for various combinations of alkali raw materials in slurry-solution spray drying. The raw materials selected for use in these experiments were alpha-alumina raw material B, sodium hydroxide, lithium hydroxide monohydrate, zeta-lithium aluminate, gamma-lithium aluminate and beta-sodium aluminate. Spray drying experiments were performed for each of the following raw material combinations: (1) alpha-alumina, NaOH, and $LiOH{\cdot}H_2O$; (2) alpha-alumina, NaOH, and $LiAl_5O_8$; (3) alpha-alumina NaOH, and $LiAlO_2$; (4) alpha-alumina, $NaAlO_2$, and $LiAl_5O_8$; and (5) alpha-alumina, $NaAlO_2$, and $LiAlO_2$.

As indicated by the data in Table V, a large variation in the densification characteristics of beta"-alumina ceramics was observed between the different alkali raw material combinations. The results demonstrate that the use of either $NaAlO_2$ or $LiAlO_2$ as raw materials is very detrimental to the attainment of high mass densities. The lack of adequate densification in these bodies was attributed to incomplete compaction. A large quantity of uncrushed remnant spray dried spheres was observed in these bodies after isostatic compaction at 381 MPa. It appears that the presence of aluminates in the pressbody leads to very hard spray dried granules which resist crushing during compaction. In the preparation of spray dried powders, aluminates or similar compounds, leading to aluminum ions in solution, should be avoided. It should be noted in passing that the acetates are suitable alkali raw materials in that acceptable resistivities and mass densities can be achieved after sintering. However, the removal of large quantities of volatiles prior to sintering makes the bisquing operation difficult to control (33,34).

Table V

COMPARATIVE DENSITY VALUES FOR GREEN, BISQUED AND SINTERED SLURRY SOLUTION SPRAY DRIED ERCHPS-DBM POWDERS

Alkali Raw Material Sources	Pressed Density (g/cm^3)	Bisqued Density (g/cm^3)	Sintered Density (g/cm^3)*	% Linear Shrinkage
1. NaOH, LiOH	2.24±0.03	2.04±0.04	3.19±0.01	11.3±0.4
2. NaOH, $LiAl_5O_8$	2.23±0.02	2.07±0.02	3.19±0.01	11.2±0.4
3. NaOH, $LiAlO_2$	2.17±0.03	2.01±0.03	3.13±0.02	11.7±0.4
4. $NaAlO_2$, $LiAl_5O_8$	2.12±0.01	2.24±0.01	2.76±0.04	9.4±0.5
5. $NaAlO_2$, $LiAlO_2$	2.05±0.09	2.20±0.07	2.75±0.05	10.4±0.7

* All Samples were Sintered at 1585°C for 10 Min. Followed by a Post-Sintered Anneal of 1475°C for Four Hours.

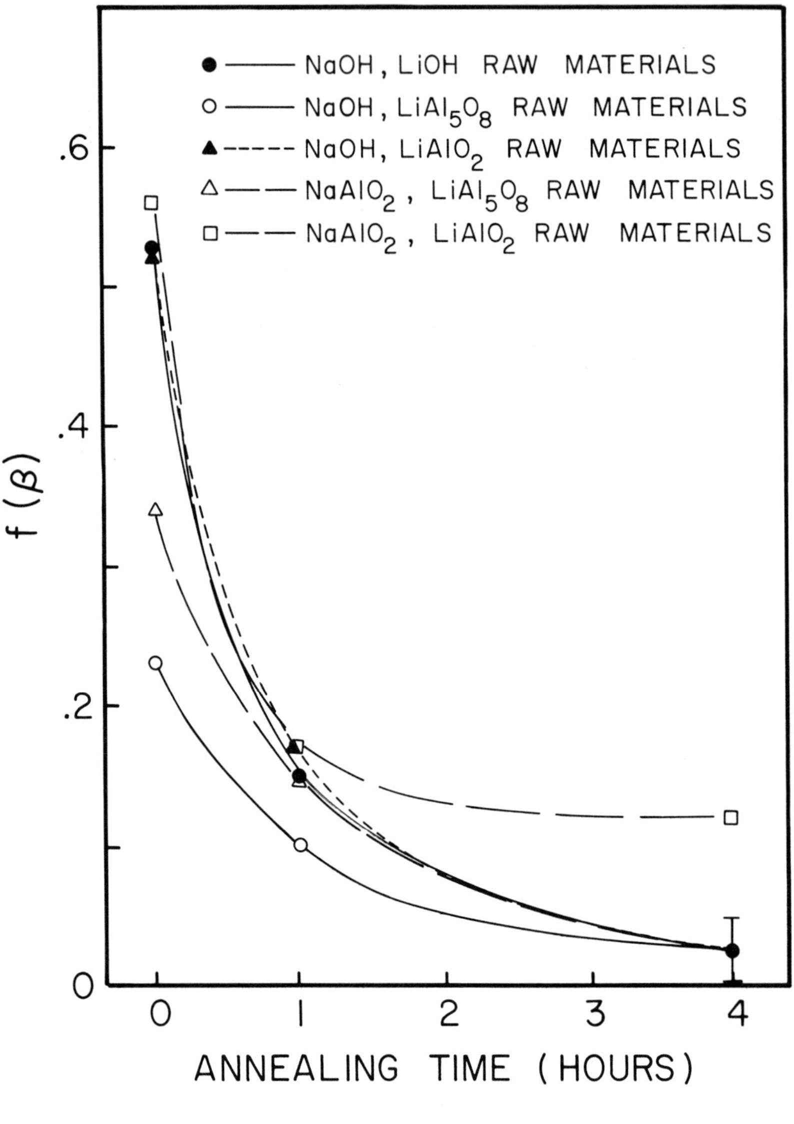

Figure 3 - Effects of Alkali Raw Material Combinations on Phase Composition of Sintered Beta"-Alumina Electrolytes Fabricated from Spray Dried Powders

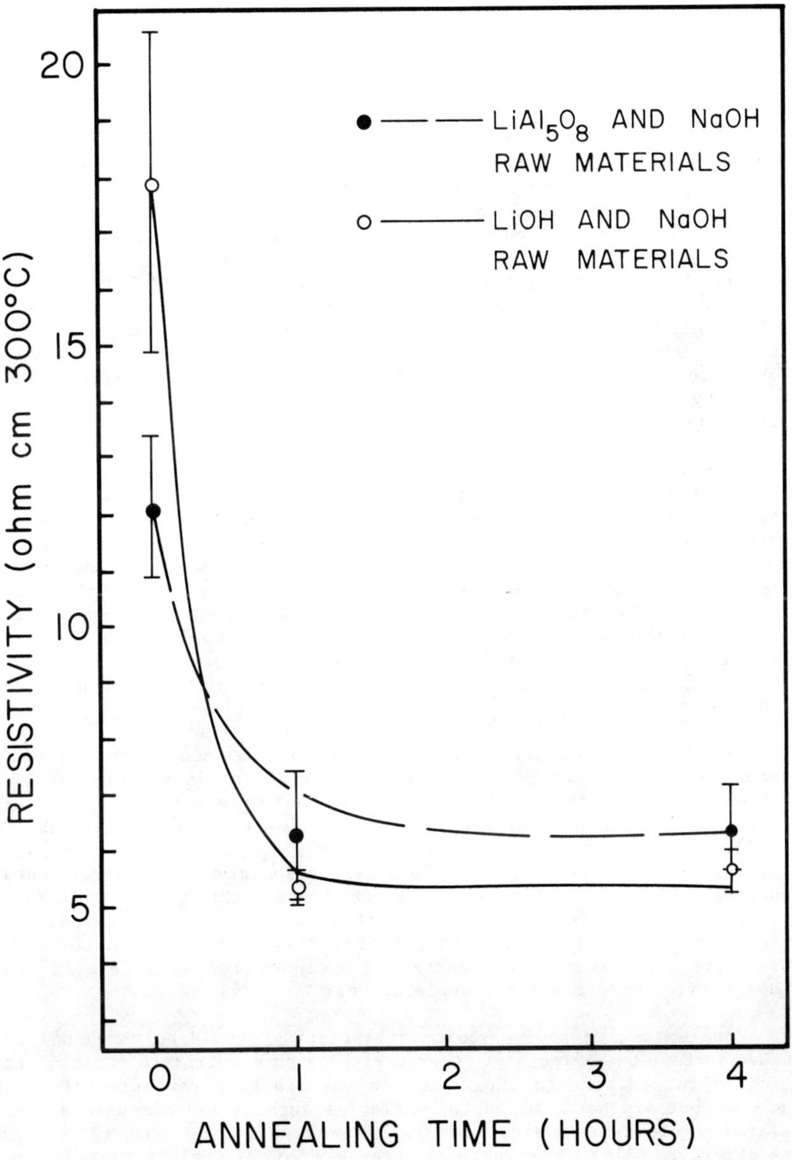

Figure 4 - Effects of Alkali Raw Material Combinations on Resistivity of Sintered Beta"-Alumina Electrolytes Fabricated from Spray Dried Powders

Sodium-Sodium Cell Testing

Beta"-alumina ceramic electrolyte tubes have been fabricated for testing in sodium-sodium test cells using the raw materials and processes described in this paper. The results of tests to date are summarized in Table VI.

A few general observations can be made. Irrespective of the method of powder processing (zeta or spray drying), electrolytes which were fabricated from alpha-alumina raw materials containing small amounts of calcium (< 50 ppm) and silicon (<150 ppm) had the lowest sodium ion resistivities and the longest lifetimes in sodium-sodium test cells operated at accelerated current densities ($\geq$0.6 A/cm^2). When the raw materials contained larger amounts of these impurities (Ca$\sim$140-160 ppm and Si 260 ppm) and even significant amounts of titanium ($\sim$1000 ppm), problems were encountered with higher ionic resistivities and/or shorter electrolyte lifetimes. Acceptable electrolyte lifetimes and attractive dc radial resistivities (< 6 ohm-cm at 300°C) can be achieved in tubular electrolytes which are fabricated by zeta and spray drying processes using reasonably economical (< \$4/kg) raw materials.

Providing the purity is reasonable with respect to calcium, silicon and possibly transition metal impurities, excellent performance which is characterized by a low and stable resistance independent of the direction of current flow (i.e., no resistance aging or asymmetric polarization) can be expected for beta"-alumina electrolytes in both sodium-sodium and sodium-sulfur test cells. Resistance rise effects (i.e., ageing) in sodium sulfur cells have been attributed to electrolytes prepared from raw material C in which the calcium impurity concentration is excessive (24). Raw material B, although a little more expensive ($\sim$\$3-4 per kilogram versus $\sim$\$1/kg), is free of these troublesome problems, probably because of its higher purity.

It has been the experience of this program that acceptable lifetime and performance characteristics of ceramic electrolytes in sodium-sodium cells at high current densities (>0.6 A/cm^2) generally translated to satisfactory performance in sodium-sulfur cell tested at lower current densities (<0.25 A/cm^2). Sodium-sulfur cell testing is more complicated and expensive than sodium-sodium cell testing due to problems associated with sulfur container (current collector) corrosion and sometimes seal failures. However, lifetimes in excess of 2000 Ah/cm^2 (both directions) in sodium-sodium cells give strong support to the proposition that electrolyte life will be sufficient in sodium-sulfur batteries designed for load leveling energy storage and electric vehicle applications.

More extensive sodium-sodium cell testing should be conducted in the future to test the effects of other variables on electrolyte performance and lifetime in addition to impurities in the starting raw materials. These include, but are not limited to, variables such as (1) dimensional factors related to electrolyte size, surface area and wall thickness, (2) sodium ion resistivity, (3) microstructural features of the electrolyte, and (4) different processing variables.

Table VI

SODIUM-SODIUM CELL TESTING
β''-ALUMINA ELECTROLYTE TUBING

RAW MATERIAL	FABRICATION PROCESS	CURRENT DENSITY (A/cm^2)	LIFE TIME TWO-WAY (Ah/cm^2)	COMMENTS
A	Zeta	1.25	>2000*	No asymmetry Good performance
B	Zeta	********************Not Tested	************************************	
C	Zeta	0.78	10	Failure caused by processing flaws
D	Zeta	0.35	12	Significant cracking of tubes
E	Zeta	0.66	>2000*	Minor degradation
F	Zeta	0.66	>2000*	No asymmetry Good performance
A	Slurry Spray Drying	0.62	>2606*	No asymmetry
B	Slurry Solution Spray Drying	0.63	>2400*	Good performance
C	" " "	1.27	≤1500	Poor performance Moderate to severe degradation

* Tubes either removed from test with no failure or cell failure caused by breakdown of the seal.

Conclusions

Low resistance, high performance, and durable beta"-alumina electrolytes can be produced by two different powder process methods: (1) zeta powder processing, and (2) spray drying (slurry and slurry-solution). Both methods can be utilized to synthesize beta"-alumina precursor powders which can be sintered to excellent properties. The technique of slurry-solution spray drying is very attractive compared to zeta powder and slurry spray drying processes in that the expensive and energy intensive operations of calcination and ball milling can be eliminated from the production process. It was shown that excessive amounts of impurities (e.g., calcium and silicon) in certain raw materials led to high electrolyte resistivity and much shorter lifetimes in sodium-sodium test cells. Selection of the correct raw material appears to be the key element in the production of high performance and durable ceramic electrolytes.

NaOH and LiOH appear to be the optimum alkali raw materials in the slurry-solution spray drying process. The use of acetates, while leading to acceptable properties, creates excessive problems in the controlled removal of volatiles during bisquing. The use of water soluble aluminates ($NaAlO_2$, $LiAlO_2$) leads to hard spray dried granules and subsequent low sintered densities.

Acknowledgements

The Department of Energy, under contract DE-AM-02-79CH10012 is acknowledged for the support of this work. Dr. M. L. Miller is acknowledged for his contribution to slurry spray drying and exploratory studies in slurry-solution spray drying. J. R. Rasmussen and L. Shaffer are acknowledged for their contributions in sodium-sodium cell testing. E. Beck, D. Prouse, S. Roberts, F. Schaaf, and R. Snow are acknowledged for their experimental assistance.

REFERENCES

1. J. T. Kummer, "β''-Alumina Electrolytes," Progress in Solid State Chemistry, 7, 141 (1972).

2. J. T. Kummer and N. Weber, "Sodium-Sulfur Secondary Battery," SAE J., 76, 1003 - 1007, (1968).

3. M. L. Miller, B. J. McEntire, G. R. Miller, and R. S. Gordon, "A Pre-Pilot Process for the Fabrication of Polycrystalline β''-Alumina Electrolyte Tubing," Am. Ceram. Soc. Bull., 58, (5) 522-526, (1979).

4. R. W. Powers, "The Electrophoretic Forming of Beta-Alumina Ceramic," J. Electrochem. Soc., 122, (4) 490-500, (1975).

5. "Fabrication and Characterization of β''-Alumina Electrolytes and Electronic-Ceramic Containers and/or Current Collectors for the Sodium-Sulfur Battery," Final Technical Report for the Period June 30, 1973 to January 31, 1978. Prepared for the National Science Foundation under Contract No. NSF-C805, R. S. Gordon - Principal Investigator, (1978).

6. G. J. Tennenhouse and R. A. Pett, "Fabrication of Thin Layer Beta-Alumina," Prepared for the National Aeronautics and Space Administration Under Contract No. NAS 3-19782, (1978).

7. R. T. Dirstine, "Fabrication of β''-Al_2O_3 Tubes from Cast Ceramic Tape," Fast Ion Transport in Solids, P. Vashishta, J. N. Mundy and G. D. Shenoy, Eds., Elsevier North Holland, Inc., 79-82, (1979).

8. M. Rivier and A. Pelton, "A New Slip Casting Technique for the Laboratory Fabrication of β-Alumina and Other Ceramics," Am. Ceram. Soc. Bull., 57, (2), 183-185, (1978).

9. S. R. Tan and G. J. May, "Production of Beta-Alumina by a Zone Sintering Process," Sci. Ceram., 9, 103, (1977).

10. B. J. McEntire, G. R. Miller and R. S. Gordon, "Sintering of β''-Alumina Electrolyte Tubing," The Sodium-Sulphur Battery, J. Sudworth and R. Tilley, Eds., Chapman and Hall, (in-press).

11. R. S. Gordon, B. J. McEntire, M. L. Miller, and A. V. Virkar, "Processing and Characterization of Polycrystalline β''-Alumina Ceramic Electrolytes," Processing of Crystalline Ceramics, H. Palmour III, R. F. Davis, and T. M. Hare, Eds., Plenum Press, 405-420, (1978).

12. G. E. Youngblood, A. V. Virkar, W. R. Cannon and R. S. Gordon, "Sintering Processes and Heat Treatment Schedules for Conductive, Lithia Stabilized β''-Al_2O_3," Am. Ceram. Soc. Bull., 56, (2), 206, (1977).

13. M. J. Ruthner and H. Krischner, "Production of Alkali-Free Alumina," ICSOBA 3rd. International Congress, Nice, 547-555, (1973).

14. I. Wynn Jones and L. J. Miles, "Production of β-Al_2O_3 Electrolytes," Proc. Brit. Ceram. Soc., 19, 161, (1971).

15. A. Imai and M. Harata, "Ionic Conduction of Impurity-Doped β-Alumina Ceramics," Japanese Journal of Applied Physics, 11, (2), 180-185, (1972).

16. J. H. Kennedy and J. R. Akridge, "Inhibition of Sintering of Beta-Alumina by V_2O_5," J. Am. Ceram. Soc., 59, (5-6), 279-280, (1976).

17. J. H. Kennedy, "Additives And Resistivity in Beta-Alumina," Superionic Conductors, G. D. Mahan and W. L. Roth, Eds., Plenum Press, 335-350, (1976).

18. J. P. Boilot, A. Kahn, J. Thery, R. Colongues, J. Antoine, D. Vivien, C. Cheverette and D. Gourier, "Influence of Foreign Ions Addition On Relative Stability and Electrical Conductivity of β- and β"-Alumina Type Phases Localization of Impurities," Electrochemica Acta., 22, 741-745, (1977).

19. M. Y. Hsieh and L. C. DeJonghe, "Silicate-Containing Sodium Beta-Alumina Solid Electrolytes," J. Am. Ceram. Soc., 61, (5-6), 185-191, (1978).

20. I. Yasui and R. H. Doremus, "Effects of Calcium, Potassium, and Iron Ions on Degradation of β"-Alumina," J. Electrochem. Soc., 125, (7), 1007-1010, (1978).

21. G. J. May, "Ionic Conductivity and Crystal Structure of Magnesium- and Cobalt-Doped Sodium-Beta-Alumina," J. Mat. Sci., 13, 261-267, (1978).

22. G. J. May, "The Influence Of Barium And Titanium Dopants On The Ionic Conductivity And Phase Composition Of Sodium-Beta-Alumina," J. Mat. Sci. Letters, 14, 1502-1505, (1979).

23. L. C. DeJonghe, "Grain Boundaries and Ionic Conduction In Sodium-Beta-Alumina," J. Mat. Sci., 14, 22-48, (1979).

24. W. G. Bugden, P. Barrow and J. H. Duncan, "The Control Of The Resistance Rise Of Sodium-Sulphur Cells," Fast Ionic Transport In Solids, J. B. Bates and G. C. Farrington, Eds., North Holland Press, 275-278, (1981).

25. W. G. Bugden and J. H. Duncan, "Effect Of Dopants On Beta Alumina Resistivity And Reliability," Sci. Ceram., 9, 348-355, (1977).

26. G. Maczura, T. L. Francis, R. E. Roesel, "Special Aluminas for Ceramics and Other Industrial Application," INTERCERAM, 3, (1976).

27. Private Communication, J. Lundenberg, Technical Service Representative-ALCOA, Bauxite, Ark., March, 1980.

28. W. L. Roth, Trans Amer. Cryst. Assoc., 11, 51, (1975).

29. Maggie Alden, John O. Thomas, and Gregory C. Farrington, "The Structure of Ba^{2+} and Ca^{2+} β''-Alumina," Fast Ionic Transport in Solids, J. B. Bates and G. C. Farrington, Eds., North Holland Press, 205-206, (1981).

30. L. C. DeJonghe, EPRI Report No. 252 (July 1975) as reported in: J. H. Kennedy, "Additives and Resistivity in Beta-Alumina, "Super-Ionic Conductors, G. D. Mahan and W. L. Roth, Eds., Plenum Press, 335-350, (1976).

31. G. R. Miller and D. G. Paquette, "Physical Properties Data Compilation Relevant to Energy Storage, III. Engineering Properties of Single and Polycrystalline Sodium Beta and Beta''-Alumina," U. S. Department of Commerce/National Bureau of Standards, NSRDS-NBS 61 Part III., (1979).

32. A. V. Virkar, G. R. Miller, and R. S. Gordon, "Resistivity-Microstructure Relation in Lithia-Stabilized Polycrystalline β-Alumina," J. Am. Ceram. Soc., 61, (5-6), 250, (1978).

33. D. W. Johnson, Jr., S. M. Granstaff, Jr., and W. W. Rhodes, "Preparation of β''-Al_2O_3 Pressing Powders by Spray Drying," Am. Ceram. Soc. Bull., 58, (9), 849-855, (1979).

34. E. M. Vogel, D. W. Johnson, Jr., and M. F. Yan, "Homogenization, Phase Formation, and Sintering of Spray Dried β''-Al_2O_3," Am. Ceram. Soc. Bull., 60, (4), 494-500, (1981).

35. G. J. May and A. Hooper, "The Effect of Microstructure and Phase Composition on the Ionic Conductivity of Magnesium-Doped Sodium-Beta-Alumina," J. Mat. Sci., 13, 1480-1486, (1978).

36. A. D. Jatkar, I. B. Cutler and R. S. Gordon, "Conversion Catalysis And Microstructure Control in the Sintering Of Lithia-Stabilized β''-Alumina," Ceramic Microstructures '76, R. M. Fulrath, and J. A. Pask, Eds., Westview Press, 414-422, (1977).

37. J. H. Duncan, "Methods of Manufacture of Beta-Alumina," U.S. Patent 4,167,550, (1979).

38. A. D. Jatkar, I. B. Cutler, A. V. Virkar and R. S. Gordon, "Microstructural Control During Sintering of β''-Alumina Compositions Through Ceramic Processing Modification," Processing of Crystalline Ceramics, H. Palmour III, R. F. Davis, and T. M. Hare, Eds., Plenum Press, 421-431, (1978).

PRODUCTION ECONOMICS-THE DRIVING FORCE BEHIND THE DEVELOPMENT OF β"-ALUMINA CERAMIC ELECTROLYTES

G. R. Miller, B. J. McEntire, and R. S. Gordon

Ceramatec, Inc.
Salt Lake City, Utah 84115
USA

The profitable manufacture of sealed β"-alumina tubular assemblies for sodium-sulfur battery applications is discussed with respect to known processing methods. The essentials of these processing steps are outlined with emphasis given to those areas which require research and development. Finally, two rather detailed cost studies are summarized which are based on different processing paths.

Introduction

The manufacturing of sodium ion conducting ceramic electrolytes and insulating seals for large scale applications such as the sodium-sulfur battery will provide some of the most challenging problems the ceramic industry has yet faced. Whether the sodium-sulfur battery eventually makes a large impact in the load leveling of electric power stations or on electric vehicles, the overall problems are similar. In either application, very large numbers of high quality, high technology ceramic electrolytes will have to be manufactured. These electrolytes will be relatively large in size, likely in closed-end tubular forms, and sealed to an insulating seal header (Fig. 1). In Table I the estimated number of required tubular electrolyte assemblies are listed for various sizes of sodium-sulfur cells if twenty-five one hundred megawatt hour (100 MWh) load leveling batteries were to be produced in a year. The numbers are based on cell and battery designs developed by the Ford Aerospace and Communications Corporation (FACC). Other developers' cell designs cause only moderate changes in these estimates.

For such applications, the allowed selling costs of assemblies are expected to be inordinately low given the fact that processing of ceramics can be expensive, especially when so many properties of the ceramic material must be optimized simultaneously, not simply because these properties are desired but are demanded by their intended use in sodium-sulfur cells. In Table II a listing is given for the properties of the β"-alumina electrolyte tubes which are required by the builders of sodium-sulfur cells. In order that an economically viable battery be constructed, it is estimated that the cost of an electrolyte assembly containing the attached seal header as sketched in Figure 1 will have to be less than or about $0.015 per square centimeter of outer electrolyte surface. For electric vehicle applications, the cost of what may be smaller electrolyte tube assemblies will have to be less than or about $0.009 per square centimeter of outer electrolyte surface area.

Given the cost limitations and a market of several million electrolytes per year, extreme pressure is placed on all aspects of the ceramic production process from the acquisition of raw materials and capital equipment, the preparation and processing of powders, forming, sintering, fabrication of seal headers, quality control programs and energy usage. Highly automated processing is demanded in order that labor costs be minimized in an industry that traditionally can be very labor intensive.

In this paper, an attempt is made to summarize the results for two of four recent cost estimates of large scale ceramic electrolyte production. Each cost estimate has assumed significantly different processing methods. In addition, some suggestions are made for research and development projects required to bring a cost effective process on line.

Process Analysis and Alternatives

The typical sodium-sulfur battery cell disclosed by Kummer and Weber (1) or a recently proposed alternative sodium-aluminum chloride cell (2) is usually cylindrical in shape enclosing a closed-end, tubular ceramic solid electrolyte separator which allows transfer of only sodium ions under the influence of an electric field. For the electrolyte, the most commonly used ceramics are in the sodium aluminate class of compounds, typically beta (β) or beta double prime (β") aluminas or some combination of these two phases.

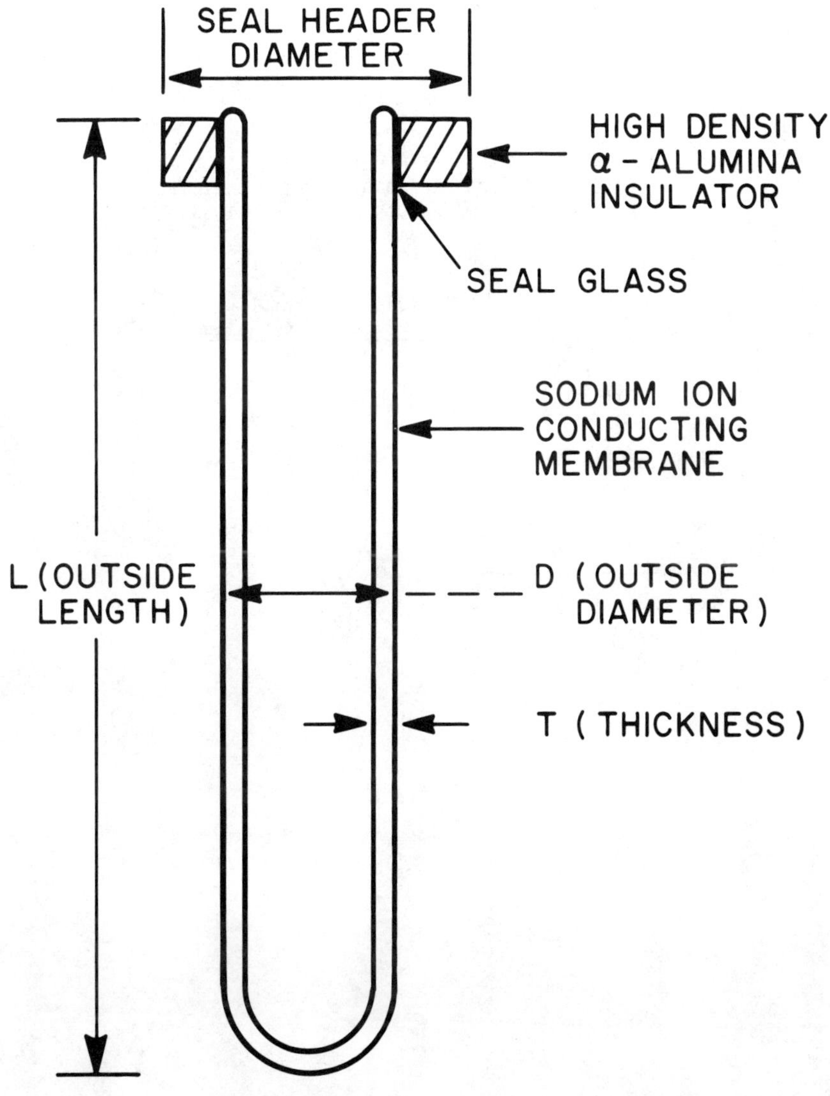

Figure 1 - Schematic of Beta''-Alumina Electrolyte Glass Sealed to an Alpha-Alumina Insulating Seal Header

Table I

NUMBERS OF β"-ALUMINA ELECTROLYTE ASSEMBLIES REQUIRED IN 25-100 MWh LOAD LEVELING BATTERIES AS A FUNCTION OF ELECTROLYTE DIMENSIONS

Dimensions (mm)*			Millions of Assemblies Required by Battery Designer
D	L	t	
34	260	2.5	13.0 (FACC)
34	460	2.5	7.1 (FACC)
50	600	2.5	3.6 (FACC)
29	448	1.45	5.7 (CGE)

*Refer to Figure 1 for definition of dimension symbols

Table II

TYPICAL PROPERTIES OF β"-ALUMINA ELECTROLYTES REQUIRED BY BATTERY MANUFACTURERS

Property	Value
Mass Density	At least 98% of theoretical at all points along tube.
Electrical Resistivity	As low as possible($\lesssim$5 ohm-cm @ 300°C)
Modulus of Rupture	>25,000 p.s.i.
Leakage to helium gas	Hermetic (<10^{-10} std cm^3 per second per cm^2 of electrolyte surface)
Surface Flaw (Crack) Density	None observable by standard dye penetrant tests
Microstructure	Can be duplex providing strength is maintained
Dimensions;	
a. Roundness	Ovality <1.01
b. Straightness	Less than 0.0015mm warp per mm of tube length
c. Diameter	Within 0.2mm of mean

The material with the higher conductivity is the β" phase. Both compounds are two dimensional conductors. The nominal compositions for each are β:$Na_2O\cdot$ 9-11 Al_2O_3 and β":$Na_2O\cdot 5Al_2O_3$. The β" phase however is unstable at high temperatures, decomposing to $NaAlO_2$ and the β phase. Thus a stabilizing element is added to maintain the β" phase at all temperatures. Common stabilizing additives are Li_2O and MgO, although additives such as ZnO among others are effective in stabilizing the β phase.

By varying the kind and amounts of stabilizing additives and the soda concentration, ceramics with widely varying properties can be produced. Some, for instance, are mixed β-β" conductors with conductivities midway between that found for the pure β or pure β" phase but with more easily controlled microstructures leading to the possibility of very high fracture strengths. Other compositions lead to high conductivity, single phase β" aluminas which are more susceptible to water degradation as for instance materials with 2-5 mole percent MgO stabilizing additives (3). Various developers have chosen many different compositions depending on their perceived needs as dictated by cell designs. As time has passed, more and more developers are concentrating on maximizing tube conductance at the expected battery operating temperatures (300-375°C). This can be accomplished by the use of thinner tube walls (causing greater problems in strength and sealing and also in dimensional control during processing) or by lowering the material resistivity for sodium ion flow--hence greater interest in the β" phase.

Several classical approaches to the processing of β or β"-alumina electrolyte tubes have been researched and costed by the four groups whose studies are available to the public. The Ford Motor Company (1978) (4), The General Electric Company (1977 and 1979) (5), Compagnie General D'-Electricite (CGE) (1976 and 1980) (6), and Ceramatec, Inc. (1980) (7) have made detailed studies of manufacturing plans and costs for producing β or β"-alumina electrolyte tubes for load leveling versions of sodium-sulfur batteries. Direct comparison of costs and processes can really only be made between the CGE and Ceramatec studies, both of which used the same costing guidelines (8). Also the CGE and Ceramatec studies had the additional benefit of an extra one or two year's process development. This paper will focus on the CGE and Ceramatec costing studies, comparing approaches to electrolyte processing and critically assessing the validity of the basic assumptions made in each study. In Figure 2 the classical steps in fabricating a ceramic product are given except that issues pertinent to the manufacture of β"-alumina tubes are emphasized in the subheadings. Some obviously uneconomical approaches such as forming by slip casting were omitted.

Raw Materials

Taking each processing step in order, consider the problem of raw material choice. Whether for any given electrolyte tube a composition is chosen which ultimately will give a β-alumina body (or mixed) has little effect on the total cost of raw materials. (Actually the β-alumina compositions require more good quality β-alumina but the β"-aluminas require a small amount of an often more expensive stabilizing element and the two effects nearly offset one another.) The determining effect on the choice of β or β"-alumina is electrolyte resistance. Since the β"-alumina electrolyte tubes possess lower resistivities than β-alumina tubes of similar dimensions, sodium-sulfur cell efficiences will be greater when constructed

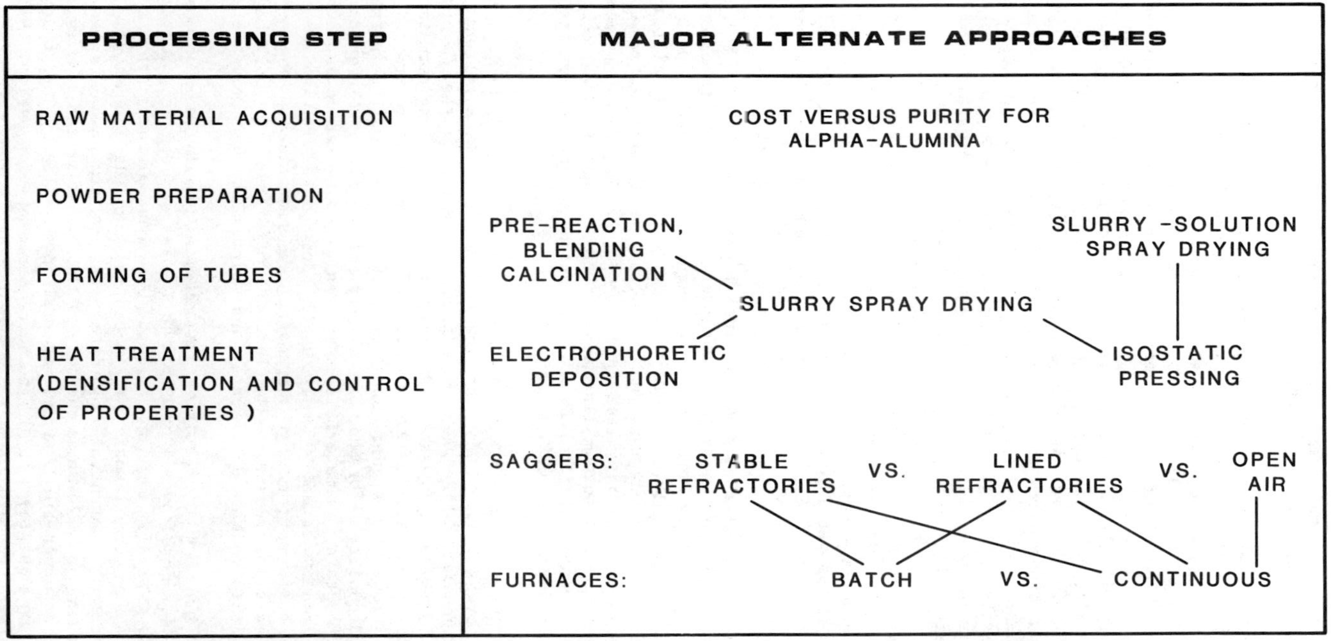

Figure 2 - Alternative Approaches to the Fabrication of Beta"-Alumina Ceramic Electrolytes with β''-alumina electrolytes and thus fewer sodium-sulfur cells will have to be built. As a result this paper will focus on β'' compositions or at least compositions which are predominately β''-alumina.

More and more battery developers appear to be recognizing the need for high purity forms of α-alumina as the chief raw material precursor to β''-alumina. Electrolytes fabricated from very high purity (>99.98%) α-aluminas tend to perform best in cells, exhibiting no resistivity rise or asymmetric polarization (9,10). Lower cost raw materials containing excessive amounts of Ca and Si have been shown to be highly detrimental to electrolyte lifetimes and the levels which are allowed may be rather minute especially in the case of calcium. Besides being associated with early electrolyte failures, excessive calcium is also associated with difficulties in completing the reactions beteween various components to form the β''-alumina phase. The phases which stabilize in a composition containing excessive amounts of calcium are not well known; it appears that levels above about 60 ppm really cannot be tolerated. Alpha-alumina raw materials containing calcium in large amounts appear to produce electrolytes which exhibit a resistivity ageing effect in sodium-sodium or sodium-sulfur cells in that the electrolyte contribution to cell resistance increases with time (passage of charge) (11,12). The rate of resistance rise can be increased by factors of at least ten when excess calcium is present in the raw material. This resistance rise, apparently due to calcium migrating to tube outer surfaces, can be removed by a chemical treatment, the cost of which has not been determined.

The effects of silica as a detrimental impurity are more difficult to trace. Its effect is apparently quite dependent on the way it is distributed in the electrolyte body and how it is combined chemically with other elements. Some electrolytes containing large amounts of silica have exhibited poor behavior in many sodium-sulfur cells and yet some cells containing large amounts of silica (probably in large globules) have performed very well. The bottom line appears to be that silica, well distributed in the ceramic electrolyte body, cannot be tolerated; yet its level of tolerance is not at all well known.

Other impurity effects are not well known either. It might be expected that certain transition metal ions could lead to some electronic conductivity in the electrolyte when present as an impurity in the β''- alumina. This contribution would decrease the battery shelf life as long term internal discharge takes place even if the electronic transference number is only, say, 10^{-3}. Obviously, a significant amount of work is required in the area of raw material quality before one can specify exactly the true requirements. At this time, certain raw materials have been given tentative approval and others have been rejected. In Table III some data from sodium-sodium qualification tests performed at Ceramatec are summarized. However, the rejected raw materials, usually the low cost materials, may only have one (unknown) impurity that is detrimental and which might be removable.

That the question of requisite raw material purity is of utmost importance is underscored by all of the cost studies undertaken to date. Raw material costs typically make up at least 30 percent of the factory cost of electrolytes (when the lowest cost potential candidate from the Bayer process is chosen). Potential candidates for α-alumina powders, however, have costs ranging up to 20 times the cost of the lowest priced candidate. In 1981 dollars, an upper limit on raw material cost is likely less than \$1.75 per pound in large quantities. Given the requirements for purity, this will provide an interesting challenge to manufacturers of alumina powders.

Table III

α-ALUMINA RAW MATERIALS
PRECURSOR - PURITY - COST - PERFORMANCE

Raw Material	Precursor	Impurities+ (ppm) Si	Ca	Approx. Cost \$/Kg Al_2O_3	Passed Na-Na Cell Test
Linde C	Alum	142-179	12-16	40-50	Yes
Baikowski CR-30	Alum	28-30	<11	13.50*	Yes
XA-139SG	Bayer	164	121	8.00	No (?)
RCHP-DBM	Bayer	109-141	23-47	7.59	Yes
RCHPS-DBM**	Bayer	129-145	9-17	4.00	Yes***
Catapal****	AlOOH	28-32	41-45	1.19	No
A-15	Bayer	81-112	130-230	0.55	?
A-15SG	Bayer	239	103	0.86	No
A-16SG	Bayer	250-270	138-168	1.06	No (?)
Alusuisse	$AlCl_3 \cdot XH_2O$	131-143	137-149	1.65	Probably

* \$ 7/Kg in Less Pure Form
** Contains 0.12% Na_2O
*** Good Endurance in Na/S Cells
**** Contains 1000 ppm Ti
+ Emission Spectrographic Analysis

Powder Preparation and Forming

The next steps in processing involve the preparation of powders of appropriate composition and homogeneity and the subsequent forming into closed-end electrolyte tubes. One cannot separately estimate costs for each unit operation in the process; instead a consistent set of unit processes must be treated group-wise. For example, one may treat as a separate powder process the pre-reacting, blending, calcining, and slurry spray drying of raw materials in combination with isostatic pressing as a method of forming the tubes. Another comparable approach is to treat the pre-reacting, blending, calcining, and spray drying steps together with electrophoretic deposition as a method of forming tubes. Although seemingly reasonable, one can come to the conclusion that the process of slurry spray drying combined with isostatic pressing is a more expensive process than slurry spray drying combined with electrophoretic deposition. In fact, since isostatic pressing does not require that the flowable powders be preconverted to the β or β'' phases, the entire process of pre-reacting, blending and calcining can be neglected in favor of only slurry-solution spray drying prior to isostatic pressing.

At the moment, costing of the powder preparation and forming steps has only been carried out for the electrophoretic and isostatic pressing methods of forming. Probably because the electrolyte tube sizes are predicted to be quite large, injection molding methods have been avoided. However, a serious question must be asked. How much is known about high volume forming of large tubes by the electrophoretic and isostatic pressing (automatic-dry bag) methods? It is the opinion of many that much work is required in both of these areas before fundamental and reliable cost data can be generated. It is quite possible that the difficulties for both of these methods may be encountered which are at least as great as injection molding of large parts.

Bisquing, Sintering, and Annealing

The last major step in ceramic tube processing may well be the source of greatest problems: heating for (a) water and organic additive release, (b) sintering to high density, and (c) phase conversion to the proper electrical conductivity while retaining proper compositional and microstructure control, mechanical strength and precise dimensional tolerances. These processes manifest themselves in (1) the choice of method for powder preparation and tube forming in that the type and quantity of organic additives is determined by these choices, (2) the choice of encapsulating saggers required for sintering at temperatures where soda loss is rapid and if lost, say at a surface, results in a useless electrolyte tube with a poor conductivity, (3) the choice of a powder preparation method which will allow conversion to the proper phases in the most economical fashion, (4) the choice of batch sintering versus continuous sintering furnaces, and (5) the method of heating the chosen furnaces.

Release of water and organic additives occurs in the bisque firing or preheat stage prior to sintering. This process may be separate from or in-line with subsequent firing processes. Often cutting or machining processes occur prior to bisquing but sometimes an advantage is found in cutting or machining after some extra green strength has been added by a higher temperature bisque firing process. The length of time and temperatures required for bisque firing depend on the nature and amounts of water and/or organic additives left in the green body from the powder preparation or forming process. In addition the bisque fire time and temperature profile required for adequate prefired strength can depend on the green density attained during forming.

Each of these considerations will favor the preparation of powders and forming of tubes in a manner in which low amounts of organics are added and high green densities are attained. The slurry-solution spray drying process will, in all likelihood, leave larger amounts of organics in the green body than a process based on electrophoretic deposition. Both forming processes can give very high green densities in the part; however, automatic dry bag isostatic pressing shows perhaps greater promise for rapid forming and tube removal from the forming mandrels. In addition, one must consider that if the bisque firing operation can be operated in-line with the sintering operation and if bisquing requires a time no greater than would normally be allowed for heating the ware to the sintering temperatures, then very little advantage of one powder preparation and forming process over another will be realized insofar as removal of the organic (water) additives is concerned.

For the unit operation of sintering to density, the question of sintering encapsulation devices or saggers has received much attention. Four methods are being considered to avoid soda loss in high temperature sintering: (1) surrounding the tube in a loose powder containing a quantity of soda sufficient to set up a local soda equilibrium between the electrolyte and the setting powder thereby inhibiting soda evaporation, (2) encapsulating the tube in a nearly impervious refractory which does not react with soda at the sintering temperatures (examples of such materials are MgO and ZrO_2, or a tube of nearly the composition of the material to be sintered), (3) placing liners or reaction barriers on the inner surface of a refractory tube which otherwise would react with soda. Examples of this method are the use of α-alumina encapsulating tubes whose inner surfaces have been pretreated with soda to form a layer of sodium aluminate or β-alumina on the inner surface. This barrier's growth is limited by diffusion of soda to the α-alumina/β''-alumina interface, the process being slow. Another example is the use of platinum foil or platinum coatings on the inner surface of, say, α-alumina to protect it from soda attack--the soda vapor is thus contained within the platinum and, if the platinum can be reused a very large number of times, the technique may be economical (4). The fourth method and probably the most desireable is to remove all encapsulating devices and sinter the tube directly in air. This has been carried out successfully and is the standard process for the Chloride Silent Power Group (13). In this process, the tubes are rapidly passed through a high temperature zone only a few centimeters long so that soda loss is minimized by virtue of the tube being at high temperatures for only a short time during which a reasonable partial pressure of soda is maintained in the hot zone (by evaporation). Tubes of very good dimensional control (in the sense of being round and straight) can be produced using this process. A high reject rate combined with a need for a long term anneal subsequent to sintering appear to be the primary disadvantages of rapid zone sintering in an air atmosphere.

One should note that methods (1), (2), and (3) for encapsulation are amenable to either horizontal or vertical sintering; dimensional control for horizontal sintering can be maintained with the use of two line support in the form of vee-setter support refractories (14). The real questions to be answered involve cost; through expenses involved with setting powders, refractories, liners, and reject rates expected from each method.

The question of converting the ceramic powders to the proper phases with the requisite properties has been closely associated with the sintering and post sintering temperature-time program. A seemingly simple method to attain requisite properties is to prepare homogeneous powders of the proper compositions and phases prior to forming and sintering--unfortunately such powders do not readily sinter. As a result, developers have resorted to preparing nearly homogeneous powders which, through the production of transient liquid phases during heating to the sintering temperature, aid in rapid (few minutes) densification at reasonable sintering temperatures. However, this approach leads to a ceramic whose phase composition is far from equilibrium. For instance, after reaching density, phases such as the β phase and other aluminates high in soda may be left even though the equilibrium (and desired) composition corresponds to 100% β"-alumina. In order to complete the reaction, several pre- or post-sintering treatments are available.

The most commonly used method is a resort to a relatively long post sintering anneal at temperatures wherein soda loss rates are not excessively high, where phase conversion can still take place in a reasonable time and yet not allow grain growth to "run away" and thus lower the electrolyte strength to an unacceptable level. If one wishes to maintain a very fine grain size while effecting phase conversion, the anneal must be very long (many hours) at a low temperature (say 1300°C). If one can tolerate larger grains, the anneal can be much shorter at a higher temperature. Obviously, this choice affects the cost of tubes both through energy use and investment in capital equipment.

A second method of promoting phase conversion is to either add seeds of size equal to or greater than the size required to accelerate heterogeneous nucleation of the desired β"-alumina phase (15) or produce such critical size nuclei in a presintering anneal (16,17).

As to choice of sintering furnace, three schools of thought have established themselves including (a) batch encapsulated sintering, (b) continuous encapsulated sintering, and (c) unencapsulated zone sintering. The proponent of each will claim that their method is chosen on the basis of lowest cost, usually because of a lack of research into the competing methods or because some other preferred portion of the fabrication process demands the particular method of conducting the heating operation. For example, a preferred process of long term, slow heating with long hold times is very compatible with batch sintering. Groups using short term sintering processes and long term annealing may prefer a combination of unencapsulated zone sintering and encapsulated batch annealing. A third group may wish to use rapid encapsulated sintering and annealing. Given that cost cutting is very important to commercial viability of the β"-alumina electrolyte, only one process giving the best cost-performance product will likely dominate in the future.

Little attention has been paid to the method of heating furnaces. Electrical resistance heating appears to be preferred by all manufacturers thus far because of the control it affords. The Chloride Silent Power Research Group has, however, used RF induction heating in their unencapsulated zone sintering method (13). As will be readily apparent in the cost analysis, energy costs play a large role in determining the economics of the fabrication process. Certainly high efficiency heating methods must be researched to include, of course, recirculated air preheat and minimizing the charge to be heated. Other methods (e.g. microwave heating) should be given importance in research planning.

Seal Header Fabrication

In most sodium-sulfur cell designs, a dense α-alumina header of cylindrical symmetry is glass sealed to the β"-alumina closed end tube. The header usually has to fit tightly to the electrolyte tube and have precise dimensions elsewhere including, often, very smooth outer surfaces. With larger and larger cells being built, the problem of producing such seal headers in an "as sintered" state with very good dimensional control such that only minor alterations to surfaces are required becomes acute; yet in order that seal headers be of low cost, say a maximum of one 1981 U.S. dollar, innovation to meet this goal is of paramount importance.

Automation

It goes without saying that any plant making low cost high quality electrolyte assemblies will have to be fully automated. The automation will have to be so extensive that from raw material unloading to finished product, no human hands touch the electrolyte or seal header. This type of automation exists in the spark plug industry but will have to be extended to the very large electrolyte assemblies envisioned for batteries. A quantum jump in automation may be necessary to devise methods for handling and moving large items during their fragile stages.

Comparison of Cost Studies

In Table IV processes are compared for major operations in the production of ceramic electrolytes which were chosen by CGE and Ceramatec in their 1980 cost studies. In Table V the estimated costs by Ceramatec for making β"-alumina tubes of various sizes from optimized processes are compared. In Table VI costs for unsealed tubes of nearly the same size are compared from the CGE and Ceramatec cost studies. Obviously some important differences exist. The study by CGE is the more wide ranging study of the two, having determined the cost of several methods of approaching the unit operations in ceramic production. Ceramatec's study was far more focused, costing only operations that had been researched to the point of pilot plant or "near pilot plant" operation in their facilities. Thus while some operations considered viable by CGE were not given much attention by Ceramatec, those operations that were chosen by Ceramatec were already reasonably well developed by Ceramatec researchers. One important item assumed by Ceramatec as an available capital asset to plant designers was extensive automatic control and delivery systems, something to be engineered by those more expert in the field. Another important assumption in the Ceramatec and CGE studies was that electrolyte tubes of large size (up to 50mm OD by 600mm length) would have the same expected lifetime in batteries and the same reject rate in processing as do smaller tubes. These assumptions are yet to be proved. Detailed comments are now in order concerning various aspects of electrolyte production.

Raw Materials

Both studies focused on the use of a relatively low cost α-alumina raw material, Alcoa A-16SG (from the Bayer Process). The choice of the A-16SG is obvious from the cost point of view in that higher purity forms of α-alumina are usually significantly more expensive. In 1980 dollars, the cost of the Alcoa A-16SG was about one half U.S. dollar per pound. Since raw material

Table IV

PROCESS COMPARISON-CGE AND CERAMATEC
COST STUDIES OF 1980*

	Item	CGE	Ceramatec
1.	Choice of α-Al_2O_3 Raw Material	Alcoa A-16SG	Alcoa A-16SG
2.	Powder Preparation for Forming	Calcining and Spray Drying	Slurry-Solution Spray Drying
3.	Forming Method	Electrophoresis**	Automatic Isostatic Pressing
4.	Choice of Saggers	Pack in β-alumina concrete with α-Al_2O_3	Platinum Lined α-Al_2O_3
5.	Sintering Method	Batch	Continuous
6.	Quality Program	Parallel Sample Batch	Multiple Point Complete Inspection
7.	Level of Automation Assumed	Moderate	High
8.	Cost per cm^2 of electrolyte surface	\$0.026	\$0.010 to 0.015

* Using guidelines prepared by A. D. Little Co. for the Electric Power Research Institute (1979) [8]

** The comparison of isostatic pressing versus electrophoresis was determined for β-alumina only.

Table V

CERAMATEC COSTS FOR ELECTROLYTE ASSEMBLIES*

Electrolyte Size (mm)	460x34x2.5	259x34x2.5	600x50x2.5
No. of sealed assemblies in 25 batteries (millions)	7.1	13.3	3.6
Total raw material costs (thousands of $)	6,500	7,900	6,900
Total variable cost (thousands of $)	2,165	3,821	1,994
Direct labor (man-years)	199	315	132
Major energy requirements (MWh)	109,300	135,400	115,600
Capital investment (thousands of $)	35,766	48,056	33,430
Plant area (square feet)	124,100	155,500	119,600
Factory Cost (thousands of $)	26,672	37,217	23,261
Selling price (thousands of $)	39,476	54,473	34,990
Selling price (per sealed electrolyte)	$5.56	$4.10	$9.72
Price per square cm of electrolyte surface	$0.0115	$0.0151	$0.0103

* 1980 $; A.D. Little guidelines, September, 1979

Table VI

FACTORY COST SUMMARY FOR
UNSEALED ELECTROLYTES

	CGE	Ceramatec
Electrolyte Size (mm)		
O.D.	29	34
Length	434	460
Wall	1.5	2.5
Production Rate in Tubes per Year	$5.7.10^6$	$7.1.10^6$
Materials Cost ($M)	16.6	10.6
Labor Force	373	199
Equipment Investment ($M)	44.6	35.8
Plant Working Area (ft^2)	161,000	124,000
Factory Cost Per Tube	$7.44	$3.81

costs play a large role in determining electrolyte costs (Figure 3), allowing the raw material cost to rise above about $1.50 per pound (as determined by the Ceramatec study) puts the economics of electrolyte production in jeopardy. The assumption that the Alcoa A-16SG can meet the sodium-sulfur battery requirements is even now being questioned by those users most enthusiastic about its use a few years ago. As previously mentioned, significant problems arise when trying to effect phase conversion with the resulting low resistivity, the phase conversion being somewhat incomplete when the A-16SG material is used as the base raw material. Furthermore, the question of ultimate lifetime in sodium-sulfur cells and performance problems related to resistance rise (calcium impurity contamination) are still issues which remain to be solved in the use of A-16SG as an electrolyte raw material. Shifting to the higher purity raw materials can require a significant cost penalty. A candidate for a compromise base material (cost vs. performance) is the Reynolds ERCHPS-DBM, an α-alumina powder containing residual soda (which would be added as an alloying compound anyway). In sodium/sodium tests at Ceramatec and sodium-sulfur tests at Ford Aerospace and Communications Corporation, β''-alumina electrolyte tubes made with this powder have performed well both in terms of low electrolyte resistance and long lifetime. Further, no problems have been encountered with resistance rise and asymmetric polarization. Thus it appears that although both the CGE and Ceramatec studies focused on the use of Alcoa A-16SG, a new raw material may have to be found which combines reasonable purity and (unfortunately) a higher, yet acceptable cost. The Reynolds ERCHPS-DBM is a possible candidate whose cost is probably borderline.

Powder Preparation and Forming

Taking the powder preparation and tube forming steps together, one finds a significant difference between the CGE and Ceramatec studies, CGE concluding that electrophoresis is a lower cost unit process than isostatic pressing (at least for β-alumina). Reasons for this include a lower equipment cost for electrophoretic deposition. However, one must recognize that the automatic isostatic pressing method of forming closed end tubes does not require converted powders--only reasonably homogeneous powders such that reactions to form the β''-alumina phase can rapidly be completed in normal sintering/annealing operations. Thus slurry-solution spray drying (a low energy, low cost preparation method) is a meaningful process for preparing powders for isostatic pressing but not for electrophoretic deposition where phase conversion and also spray drying (?) are prerequisites to forming. As a result, the combined steps of slurry- solution spray drying combined with automatic isostatic pressing can be of lower total cost than calcining and electrophoresis. Neither the Ceramatec nor CGE study directly addressed this comparison between calcining/slurry spray drying/electrophoresis and slurry-solution spray drying/ isostatic pressing.

Sintering

For the sintering processes, significantly different conclusions are again found, both for the choice of saggers and the method of moving electrolytes through the heating cycle. CGE has chosen to use setting concretes of limited life ($\sim$10 cycles) to provide support and the requisite soda atmosphere during sintering. This is apparently reflected in their rather high costs for raw materials. Ceramatec's costing study which was based entirely on the use of dense encapsulation devices suggests that reusable α-alumina setters with thin platinum metal liners are cost effective (See Table VII). As to which method is most cost effective will require long term testing.

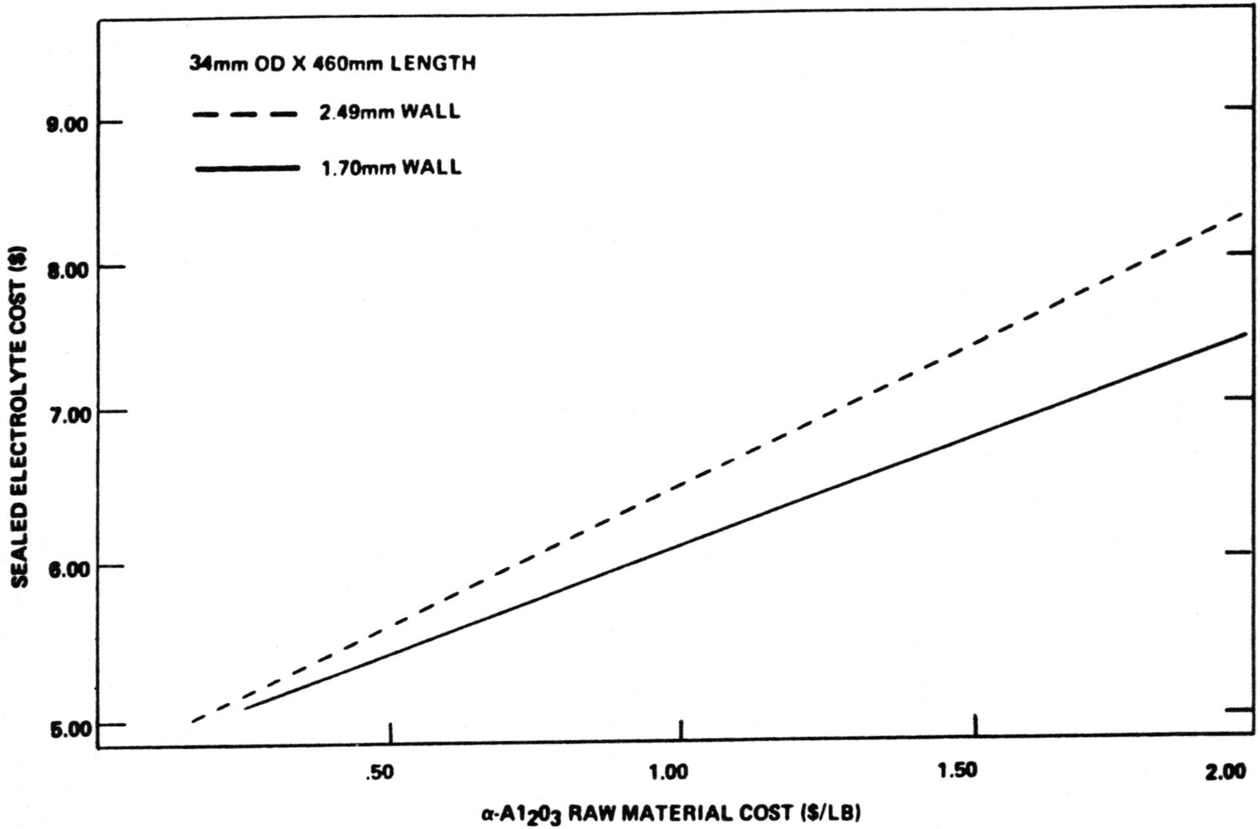

Figure 3 - Effect of Alpha-Alumina Raw Material Cost on Selling Price

Table VII

β''-ALUMINA SINTERING ENCAPSULATION COSTS*

Annual Production >10^5 Units
(33mm OD x 260mm Length)

Method	Ultimate Costs (1980 Dollars)			
	$/kWh	$/MK-II Size Electrolyte	¢/cm^2	¢/cm^2 Current
Platinum Encapsulation				
Pt Life, 100 Cycles α-Carrier Life 1200 Cycles	5.00	1.25	0.46	0.46
Platinum Lined α-Al_2O_3 Carriers				
Pt Life, 2400 Cycles α-Carrier Life 1200 Cycles	0.35	0.09	0.03	∿0.12
Pretreated α-Al_2O_3Carriers				
Carrier Life, 2400 Dycles	0.72	0.18	0.07	0.16
Inert Ceramic Refractories				
US (MgO, ZrO_2)				
Carrier Life, 240 Cycles	4.56	1.14	0.419	>1.00
British (MgO)				
Carrier Life, 360 Cycles	0.78	0.20	0.07	?

* The cost estimates in this table did not follow the A.D. Little Co. guidelines. The figures shown here should be used for comparing the relative costs of various encapsulation devices only.

One of the most significant disagreements focuses on sintering in a batch versus continuous mode. The Ceramatec study advocates the use of continuous furnace heating over batch processing based upon the costs of capital equipment, energy, labor as well as electrolyte property uniformity. CGE's study advocates the opposite with their conclusions based primarily on the cost of equipment and labor. The two groups have determined the costs of batch and continuous sintering for entirely different modes of sintering encapsulation and different requirements on the geometry of furnace loading (horizontal in the case of Ceramatec and vertical for CGE-- in both cases proper dimensional control is claimed).

Since extensive discussions have taken place which purport to justify that batch sintering of β''-alumina tubes is a lower cost operation than continuous sintering, a short comparison of the costs is in order. In order to fairly compare batch sintering with continuous sintering, most of the batch process assumptions listed in Table VIII were obtained from the proponents of batch sintering. The furnace capacity of 100 tubes per furnace results from an extrapolation of Ceramatec's experience. The assumptions regarding continuous sintering are also extrapolations of actual experience by Ceramatec.

In Table IX estimates of items directly related to cost for the two methods of sintering are compared. Taking each item in succession, the following comments can be made:

(1) The very large energy consumption per tube when a batch sintering mode is used results from the fact that considerable energy is required to heat the furnace walls: Optimal design of a batch sintering furnace requires low heat loss to the surroundings during heating and sintering operations and very rapid extractions of heat during cooling operations so that a 24 hour cycle time can be maintained.

(2) The difference between the actual costs of energy per tube for batch and continuous sintering is actually greater than the energy usage because the utilities require a premium payment for large rates of energy usage. For batch sintering, an assumption of constant rate of energy input was used over the entire 24 hours of heating and sintering. This assumption then requires that furnace operations be carefully staggered over a day's operations.

(3) The factor of 6.7 greater labor for batch sintering reflects concern over methods to automate the loading and unloading operations. Vertical sintering of tubes (to maximize the number of tubes per furnace) poses problems in automatic loading when carriers must be carefully placed around each unsintered tube. Horizontal sintering is far more amenable to automatic loading and can be accomplished in a very gentle manner, probably more gentle than can be done with manual labor.

(4) With the continuous method of sintering, the furnace throughput is far greater than that possible with a batch furnace; thus a significantly greater number of furnaces are required in batch processing. These considerations are prompting many ceramic manufacturers to move to continuous sintering.

(5) Because of the large numbers of furnaces required, the capital requirements are enormous for batch sintering, being greater than for the entire ceramics plant run with continuous sintering furnaces. For the capital requirements in continuous sintering, an automated load system is included in the cost estimate.

Table VIII

PROCESS ASSUMPTIONS
LARGE SCALE PRODUCTION ($\sim 10^7$/YEAR)
β''-ALUMINA ELECTROLYTES

Batch Sintering

1. Slurry-solution spray drying (no calcination and milling)
2. Bisquing is part of the sintering process
3. Vertical batch sintering
   a. 100 tubes per furnace
   b. 24 hour turn around of furnace
   c. Encapsulation is required

Continuous Sintering

1. Slurry-solution spray drying
2. Bisquing is continuous and in line with sintering
3. Sintering and annealing are continuous and conducted in the same furnace
   a. Six tubes, in two lines and six carriers high are passed at a velocity of about 8 cm/min.
   b. Sintering is horizontal with the use of encapsulation carriers (V setters)

Table IX

COMPARISON OF CONTINUOUS VERSUS BATCH SINTERING
LARGE SCALE (~8.5×10^6) PRODUCTION
ELECTROLYTE SIZE (34mm OD x 460mm LENGTH)

	Continuous	Batch
Energy Consumption		
1. KWh/Tube	4.4	13
2. Energy Cost ($/tube)	0.143*	0.452*
Labor (Man-Hour-Tube)	0.003	0.02
Sintering Furnaces		
1. Number	29	425
2. Capital ($10^6$$)	9	50
3. Floor Space (Sq.Ft.)	40,500	170,000
4. Annual Cost of Space ($10^3$$)	203	860
5. Daily Capacity/Furnace	~1400	~100
6. Number Encapsulation Devices	~4100	~64,000
Uniformity of Properties	Expected to be better than batch processing; some kiln to kiln variation.	Some variation tube to tube, batch to batch kiln to kiln.
Process Yield	90% (?)	89% maximum (?)

* Based on electric utility rates from Utah Power & Light Company effective 3 January 1980.

(6) The plant area required for batch sintering is very large and is reflected in very high floor space costs.

(7) The problem of sintering carriers or "saggers" has long been controversial in a β-alumina fabrication. Because so many setters must be committed to each batch furnace for such long times, the number of carriers required is very large. In this estimate it has been assumed that at anytime, approximately 33% of the setters must be out of the furnaces and being loaded. In continuous sintering with automated loading and unloading, combined with rapid return mechanisms, the number of carriers not in the furnace can be held to a minimum. Given these estimates, it is clear that much greater flexibility in the choice of carriers is available to those processes utilizing a continuous sintering approach.

(8) Finally it is Ceramatec's experience that property uniformity and process yield percentages are greater with continuous sintering. Using a "head to toe" arrangement of tubes and setters, each tube should see the same time temperature profile. Thus the tubes are not subjected to varying heat up or cool down rates which can vary with time of day or operator vagaries. Given that fewer furnaces are required, the furnace to furnace variations are also more limited.

In summary, the authors can find no reason to give batch sintering any consideration whatever in large scale production of beta alumina electrolyte tubing.

Quality Control

There is an interesting, different approach to the development of a quality program for both groups. Ceramatec suggests a part by part quality inspection program which inflates the labor costs tremendously, whereas the CGE process would use the complete processing of a parallel sample batch for quality control evaluation other than leak testing. Given that the quality of the ceramic must be high and expected reject rates are expected to be high (25% in this study), it is not clear as to which program should be instituted. It is likely that significant pilot plant production experience will have to be generated to determine the most technologically effective and cost effective approach. In any case the part by part inspection program, if found to be required, will require new sophistications in automated inspection if costs are to be significantly reduced. Most methods available to the metals industry (X-ray, Eddy current methods, etc.) are not useful in the ceramics industry.

Automation

A final comment concerns the level of automation assumed. Minimal labor in the fabrication area is required in the process envisioned by Ceramatec. However, significant additional labor is required in their quality inspection program. Whether such labor minimization can be truly effected depends entirely on the development of proper computer controlled equipment for loading and unloading fragile parts in continuous firing furnaces. Certain aspects of computer controlled processing of ceramics are already available but, to our knowledge, have not been applied to the large ceramic items projected for the sodium-sulfur battery. Ceramatec's study envisioned a completely automated process for loading and unloading tubes from encapsulation saggers in continuous, horizontal sintering furnaces whereas the CGE (and all other studies) have projected manual procedures.

Projected Costs

With the foregoing comments on the widely variant process conclusions in mind, the selling costs of an electrolyte tube as summarized in Tables V and VI show the expected wide diversity. The very fact that both studies predict costs per electrolyte of less than about \$0.03 per square centimeter of outer electrolyte surface is in itself amazing. However, the discrepancy between the Ceramatec cost of about \$0.014 per cm^2 and the CGE cost of \$0.026 per cm^2 for comparable tube sizes is quite significant and is probably assignable to the infancy of the technology and the inherent errors in projecting into the unknown.

Summary

The rather large discrepancy between the electrolyte costs projected by the CGE and Ceramatec cost studies strongly suggests that development proceed rapidly in several areas. Should there have been general agreement on overall costs, one could simply assign small discrepancies between certain unit operations to individual group preferences. Real questions arise concerning the many assumptions which each group was required to make in these cost studies. Major questions and required development areas can be summarized as follows:

(1) What purity is required for the raw material? Specifically what impurities at which tolerance levels can be used for electrolytes in long life, low resistance sodium sulfur cells?

(2) Given answers to (1), can one develop methods of manufacture using readily available raw materials to achieve the requisite impurities at low cost? With the anticipated use of millions of pounds of these raw materials per year and with the demonstrated success of the sodium-sulfur battery, sufficient economic challenge should be available.

(3) Large scale demonstrations of the various alternate methods of preparing powders and forming tubes should be made, not just to end up with tubular shapes but tubes which are sintered easily with requisite out of cell properties and demonstrable performance in large sodium-sulfur cells.

(4) True cost data should be generated for the various methods of encapsulating electrolyte tubes for sintering and annealing using methods actually faced in industrial size furnaces. Such data should arise from carrier lifetime studies in furnaces where real thermal gradients exist which might decrease carrier lifetimes.

(5) If expected overall acceptance rates for large electrolyte tubes are only about 75 percent as required in the 1979 A.D. Little guidelines, the development of an adequate quality program is imperative. If for instance the program envisioned by Ceramatec is required and actual factory costs as seen by CGE are combined, the cost of an electrolyte tube may easily become prohibitive. It would appear that efforts to find ways of making acceptable those tubes now considered rejects will have to be made--not an easy problem. If reject rates can be cut to say less than five percent, the CGE program of producing a control sample is probably viable and costs are within striking

range. Until large numbers of large tubes are made, economic projections will be on loose grounds.

(6) Energy costs are expected to be quite high, so high that real efforts should be made toward developing energy efficient processes. The attempt to make tubes via open air (unencapsulated) rapid zone sintering is a start in the right direction. Other methods of heating such as plasma arc and microwave sintering are possibly attractive.

(7) Automation is a must; the problem of handling relatively weak greenware, the automation of presses and tube delivery systems must be brought on line to keep labor costs at a minimum.

Acknowledgements

The Department of Energy under Contract DE-AM-02-79CH10012 is acknowledged for support of this work.

References

1. Kummer, J.T. and Weber, N., "A Sodium Sulfur Secondary Battery", Trans. S.A.E. 76, 1003 (1968).

2. Auborn, J.J. and Granstaff, S.M. Jr., "Sodium-Sulfur-Aluminum Chloride Cells", Proc. 15th IECEC (Seattle), 575 (1980).

3. Haar, W., Fischer, W., Kleinschmager, H. and Weddigen, G., Report from the Central Research Laboratories, Prepared at DOE Battery Contractor's Workshop, Argonne National Laboratory, Brown, Boveri and Cie, Heidelberg (1976).

4. "Sodium Sulfur Cell Cost Study-Mark I-AL Design", Prepared by Manufacturing Feasibility Department, Engineering and Research Staff, Ford Motor Co., (1978).

5. "Manufacturing Planning and Cost Study for 100 MWH Sodium Sulfur Battery", prepared for the Electric Power Research Institute by the General Electric Co., (1977, 1979).

6. "Economic Assessment and Comparison of Alternative Beta-Alumina Electrolytes", prepared by Compagnie General D'Electricite Marcoussis, for the Electric Power Research Institute (1976, 1980).

7. "Manufacture of β''-Alumina Electrolytes, α-Alumina Seal Headers and Sealing Operations", prepared for the Aeronutronic Division of Ford Aerospace and Communications Corporation by Ceramatec, Inc. (1980).

8. "Revised Guidelines for Estimating the Capital Costs of Advanced Battery Systems for Utility Energy Storage", prepared for the Electric Power Research Institute by Arthur D. Little, Inc. (1979).

9. Communicated by Yuasa Battery Company at the Fourth DOE/EPRI Sodium Sulfur Battery Workshop at Heidelberg, Germany, September (1980).

10. Breiter, M.W. and Dunn, B., "Time Dependence of the Asymmetric Resistance of Polycrystalline Beta-Alumina", to be published in Electrochimica Acta (1981).

11. Demott, D.S., "Resistance Rise in Sodium-Sulfur Cells", J. Electrochem. Soc., 122, (10) 2312 (1980).

12. Bugden, W.G., Barrow, P., and Duncan, J.H., "The Control of the Resistance Rise of Sodium-Sulphur Cells", Fast Ionic Transport in Solids, J. B. Bates and G. C. Farrington, eds., (Amsterdam: North Holland Publishing Co., 1981) pp. 275-278.

13. Tan, S.R. and May, G.J., "The Production of Beta-Alumina by a Zone Sintering Process", Sci. Ceram. 9, 103 (1977).

14. Miller, Mark L., McEntire, B.J., Miller, G.R., and Gordon, R.S., "A Pre-Pilot Process for the Fabrication of Polycrystalline β"-Alumina Electrolyte Tubing", Am. Ceram. Soc. Bull., 58 (5), 522 (1979).

15. Jatkar, A.D., Cutler, I.B., and Gordon, R.S., "Conversion Catalysis and Microstructure Control in the Sintering of Lithia-Stabilized β"-Alumina", Ceramic Microstructures '76 R.M. Fulrath and J.A. Pask, Eds., Westview Press Boulder, Colorado, p. 414 (1977).

16. Jatkar, A.D., Cutler, I.B., Virkar, A.V., and Gordon, R.S., "Microstructure Control During Sintering of β"-Alumina Compositions Through Ceramic Processing Modification", From Processing of Crystalline Ceramics Mat. Science Res. Vol. II, H. Palmour III, R.F. Davis and T.M. Hare Eds., Plenum Press (1978) p. 421.

17. Duncan, James H., "Methods of Manufacture of Beta-Alumina", U.S. Patent No. 4,167,550.

PROCESSING EFFECTS ON TOUGHNESS FOR

LIQUID PHASE SINTERED W-Ni-Fe

R. M. German and J. E. Hanafee*

Materials Engineering Department
Rensselaer Polytechnic Institute
Troy, New York 12181
USA

*Chemistry and Materials Science Department
Lawrence Livermore National Laboratory
Livermore, California 94550
USA

A two-phase heavy alloy composite based on 95W-3.5Ni-1.5Fe (wt%) was fabricated from elemental powders by liquid phase sintering. The composition was tested for toughness variations with test temperature and cooling rate from post-sintering anneals. Past reports disagree concerning cooling rate effects on toughness. This composition undergoes a ductile to brittle transition with decreasing test temperature; the transition temperature is close to room temperature. The cooling rate from post-sintering anneals over 1000°C has a large influence on toughness; rapid quenching gives superior toughness. These findings support an impurity segregation explanation for embrittlement in the heavy alloys.

Introduction

The heavy alloys are tungsten based two-phase composites used in applications requiring high density.[1] The alloys are liquid phase sintered from blended elemental powders. After sintering, the microstructure consists of a rounded tungsten phase (typically 50μm in diameter) surrounded by a matrix phase containing dissolved tungsten. The typical chemical composition ranges from 80 to 98 percent W with either Ni-Cu, Ni-Fe, or Ni-Fe-Co additions. Understandably, the mechanical properties are variable with microstructure, chemistry, and processing. Yield strengths in excess of 500 MPa are fairly common; however, ductility and toughness tend to be unpredictable. Generally, the Ni-Fe alloys exhibit superior mechanical properties and a 7:3 ratio of nickel to iron is observed to be optimal.[2-4]

In spite of numerous studies on the heavy alloys dating back to the 1930's, there is still uncertainty as to the sources of toughness variation. Considering the large number of parameters associated with this material the observed variability in toughness is not surprising. Generally the factors influencing toughness can be divided into three categories. First are those factors which produce differing results between studies such as composition, sintering temperature, test geometry, sintering atmosphere, and heat treatment. Second are those factors which give differing properties between similarly processed heats. This group includes density, pore size, impurities and part size. Third are the factors which contribute to property variations within a single heat of heavy alloy such as thermal and gravitational gradients. This multitude of variables is summarized in Figure 1 and it is emphasized that these factors are interrelated. Hence, studies aimed at optimizing specific properties like toughness must be performed carefully to avoid confusing results from the other factors.

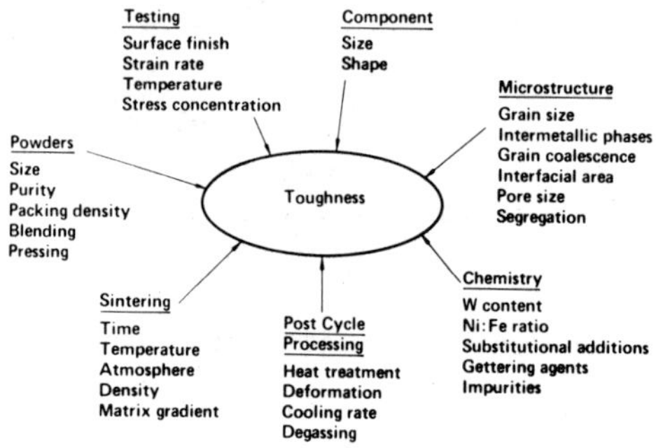

Figure 1 - Factors known to influence toughness of liquid phase sintered tungsten-nickel-iron.

Many past studies have optimized mechanical properties of the heavy alloys through either rapid quenching or slow cooling from temperatures above 1000°C. The conflicting results from such investigations are summarized in Table I and clearly illustrate the complexity of the problem. For example, the studies by Jones and Munnery,[3] Edmonds and Jones,[4] Kuzmick,[5] Dzykovich et al.,[6] and Muddle and Edmonds[7] found high toughness or high ductility in rapidly cooled heavy alloys. Alternatively, Krock,[2] Krock and Shepard,[8] Churn and Yoon,[9] Minakova et al.,[10,11] and Bazhenova et al.,[12] obtained excellent properties with slow cooling. The latter three papers give specific attention to the cooling rate effects and show increased tensile elongations with slower cooling rates. The proposed explanations for the cooling rate sensitivity include intermetallic phase formation, matrix phase saturation, hydrogen embrittlement, altered ductile-brittle transition temperature, and impurity segregation. Most likely each of these proposed processes contributes to the embrittlement. However, how dominance shifts with alloy composition, material purity and material processing is unclear. A recent study[13] has shown that rapid cooling from the liquid phase (sintering) temperatures gives embrittlement due to matrix supersaturation in tungsten and the formation of matrix solidification porosity. In wrought tungsten, brittle intergranular failure is commonly associated with impurity segregation.[14] Similarly, segregation of impurities is a possible cause of embrittlement in the heavy alloys as well.[4,7,12]

It is probable that toughness variations associated with heavy alloys represent several effects. The obvious contradictions between investigations can not be resolved without greater experimental detail. The purpose of this study was to determine the cooling rate effect on toughness for the 95W-3.5Ni-1.5Fe alloy. Past experience on this alloy has demonstrated considerable heat-to-heat variation in toughness. Hence post sintering anneals up to 20 hours at temperatures over 1000°C with an air cool are used to minimize the variations. In this condition, the ductility and toughness are improved. Typical properties for the annealed material are given in Table II.

Table I. Cooling Rate Necessary for High Ductility or High Toughness in Tungsten Base Alloys

Authors	Composition	Cooling Rate	Proposed Explanation
Krock[2]	80W, 85W, 90W, 95W Ni:Fe ratio 13:7	Slow	--------------------
Krock & Shepard[8]	80W, 85W, 90W, 92W Ni:Fe ratio 7:3	Slow	--------------------
Jones & Munnery[3]	90W-7.5Ni-2.5Cu	Rapid	prevent formation of WNi_4 intermetallic
	90W-7Ni-3Fe	Rapid	--------------------
	95W-3.5Ni-1.5Fe	Rapid	--------------------
Kuzmick[5]	90W-6Ni-4Cu	Rapid	prevent formation of brittle phase
Dzykovich et al.[6]	90W-10Ni 90W-7Ni-3Fe 90W-5Ni-5Fe 90W-3Ni-7Fe	Rapid	prevent formation of intermetallic WNi_4
Minakova, et al.[10,11]	88W, 90W, 93W, 95W, 97W Ni:Fe ratio 7:3	Slow	allow for matrix phase aging
Edmonds & Jones[4]	90W-5Ni-5Fe	Rapid	prevent formation of intermetallic W(Ni,Fe)
	90W-7Ni-3Cu	Rapid	prevent impurity segregation
Bazhenova et al.[12]	90W-7Ni-3Fe	Slow	allow for aging of matrix
Muddle & Edmonds[7]	92.8W-7.2Cu	Rapid	prevent impurity segregation

Table II. Nominal Mechanical Properties for 95W-3.5Ni-1.5Fe Heavy Alloy

Yield strength, MPa	620
Ultimate tensile strength, MPa	900
Elastic modulus, GPa	375
Reduction in area, %	26
Elongation, %	23
Charpy impact energy (unnotched), J	30

Experimental

Material for this investigation, 95W-3.5Ni-1.5Fe, was fabricated from blended elemental powders. The tungsten was minimum 99.9 percent pure (General Electric type UB 5.0) with a Fisher subsieve size between 3 to 4μm, and a mean sedimentation size of approximately 7μm. Both the nickel and iron powders were carbonyl types (INCO and GAF, respectively) with minimum purities of 99.5 percent, -325 mesh (less than 44μm), and an average size less than 10μm. The powders were blended for 30 minutes without lubricant or binders and loaded into polyvinylchloride bags. The bags were evacuated, sealed and cold isostatically pressed at 200 MPa. The compacts were induction sintered at 1470 ± 5°C for two hours in a dynamic hydrogen atmosphere with a subsequent 1350°C, 0.5 hour vacuum anneal with an air cool from 1000°C. More complete processing details for the 95W-3.5Ni-1.5Fe alloy are given by Northcutt,[15] Dickinson et al.,[16] Zukas and Levinson,[17] and Zukas and Sheinberg.[18]

The resulting material had a density of 18.15 g/cm^3 (99.9% of theoretical) and a total impurity content of less than 500 ppm by weight. Figure 2 shows the typical microstructure of the two-phase material.

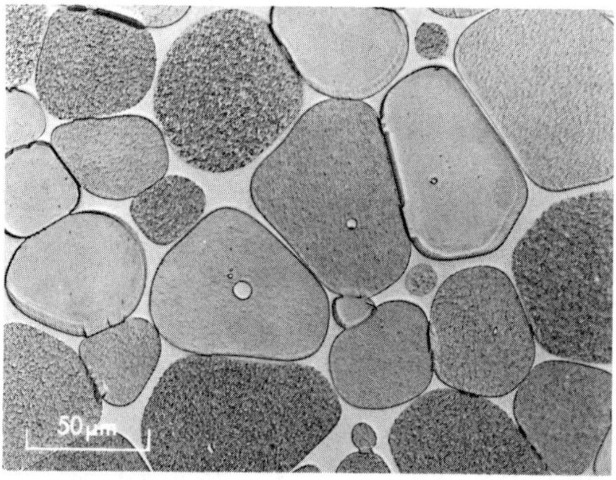

Figure 2 - Optical micrograph of 95W-3.5Ni-1.5Fe showing the coarse tungsten grains dispersed in the matrix phase (modified Murakami's etch).

In this study attention was given to impact behavior. Previous testing determined the material to be highly sensitive to notches. Thus unnotched and keyhole notched Charpy specimens were used for these experiments. The keyhole specimens were 10x5x55mm with a 2mm diameter hole and 1.2mm wide slot giving a fracture area of 5x5mm, while the unnotched specimens were 10x10x55mm. The keyhole specimens were tested at -100°, 22°, and 300°C. The unnotched specimens were fabricated from two additional heats of alloy and were tested at 20°C. The samples were polished to a 0.8μm (RMS) surface finish. The unnotched specimens were subjected to vacuum heat treatments to determine cooling rate effects on toughness. The treatments were 1000°C for two hours, water quench; 1300°C, for two hours, vacuum cool; and 1000°C, for two hours, slow furnace cool. The cooling curves are given in Figure 3.

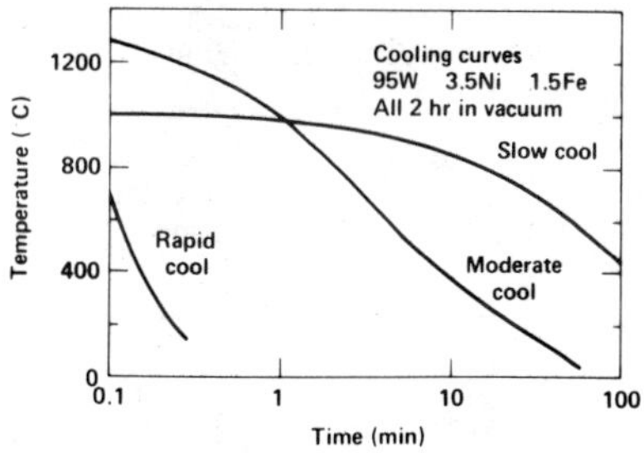

Figure 3 - Cooling curves for the three heat treatments.

In the critical temperature range of 400°C to 800°C,[4,10] the total exposure times are estimated as 0.1, 10, and 100 min. respectively. Hereafter, these cycles will be referred to as rapid, moderate and slow cools. After impact testing, examinations were conducted to help explain the observed toughness changes. The examinations were by standard techniques and included x-ray diffraction, chemical analysis, hardness, microhardness, optical metallography, scanning electron microscopy, transmission electron microscopy, Auger spectroscopy, electron microprobe, and quantitative metallography.

Results

The keyhole Charpy toughness data is shown in Figure 4 as a function of test temperature. Previous studies have noted that the heavy alloys go through a ductile-brittle transition at approximately room temperature and the test temperatures were selected to span this range.[12,15] The results shown in Figure 4 indicate a transition temperature below 22°C is characteristic of the 95W-3.5Ni-1.5Fe alloy.

The impact energies for the unnotched Charpy bars are plotted in Figure 5 for both heats of material as a function of the characteristic cooling times. The rapidly cooled specimen from heat 2 did not break, thus its impact energy was greater than the 81J machine setting. For comparison, the impact energies for the starting materials are indicated by arrows in the lower right corner of this figure. Obviously rapid quenching provides for the highest toughness for both heats of material.

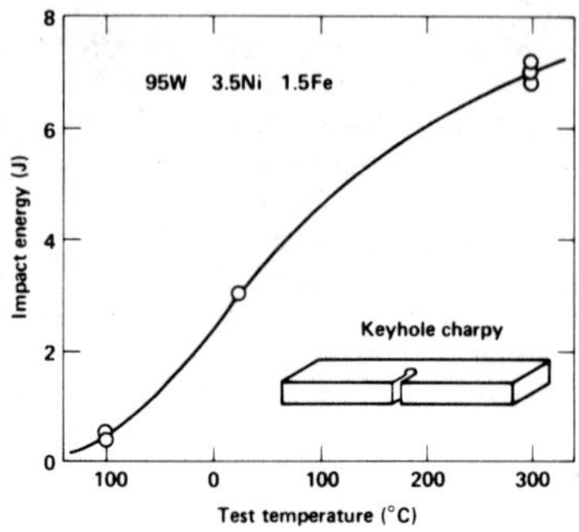

Figure 4 - Keyhole Charpy impact energy data versus test temperature for 95W-3.5Ni-1.5Fe heavy alloy.

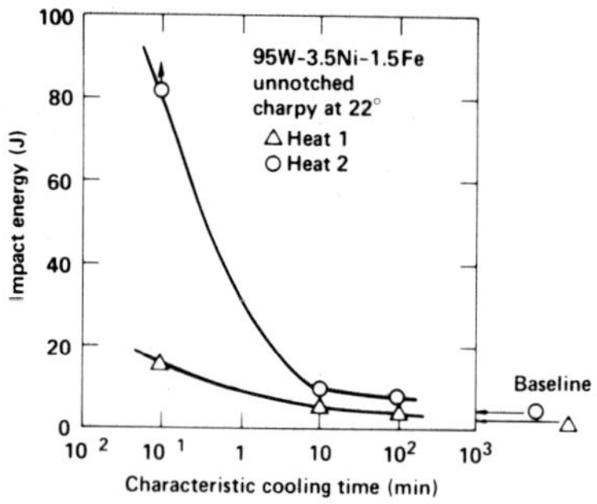

Figure 5 - Charpy impact energies for unnotched specimens from two heats of 95W-3.5Ni-1.5Fe heavy alloy. The characteristic cooling times were taken from the cooling curves in Figure 3.

Bulk hardnesses were determined for both heats in all four conditions and are shown in Table III. There were no significant differences in bulk hardness; all specimens exhibited essentially the same hardness and thus

Table III. Rockwell D Hardnesses
(Typical standard deviation = 1.2)

Condition	Heat 1	Heat 2
Starting condition	47.9	46.4
Rapid cool	46.7	47.3
Moderate cool	47.5	47.6
Slow cool	47.0	44.9

hardness properties do not correlate with the toughness data in Figure 5. The samples from heat 1 were subjected to microhardness measurements to determine individual phase responses to cooling rate. The data listed in Table IV shows a small but systematic decrease in matrix hardness as the cooling rate decreases. Quantitative metallography failed to measure any

Table IV. DPH Microhardnesses on Matrix (25g) and Tungsten (100g) Phases

(Typical standard deviation errors ±18 on the tungsten phase ±25 on the matrix phase)

Condition	Matrix Phase	Tungsten Phase
Starting condition	224	380
Rapid cool	254	373
Moderate cool	230	370
Slow cool	215	387

changes in the amount of matrix phase (9.6 vol. %) with cooling rate. Additionally, microprobe scans determined the matrix chemistry to be unchanged by cooling rate. The matrix chemistry is 19.5 ± 2.4% W, 58.8 ± 1.8% Ni and 21.7 ± 0.6% Fe. Optical metallography failed to detect any differences between samples cooled at different rates. Thus, phase morphology, quantity of matrix phase and matrix chemistry are ruled out as possible explanations of the observed toughness differences.

X-ray diffraction of the slow cooled material failed to identify any intermetallic phase involving W and Ni. The matrix phase was indexed as FCC with a lattice constant of 0.3581 ± 0.0006nm.

Scanning electron microscopy was used to determine the fracture path. Four possible paths exist as diagrammed in Figure 6; (1) cleavage of the tungsten grains, (2) failure in the matrix phase, (3) failure at the tungsten-matrix interface, and (4) failure at the tungsten-tungsten interfaces. The backscatter images shown in Figure 7 are the fracture surfaces of the Charpy impact specimens tested at three temperatures. The degree of tungsten particle cleavage (path 1) decreases with increasing test temperature, consequently failure through the matrix (path 2) increases with test temperature. The amount of tungsten-tungsten grain boundary failure is fairly uniform over the three test temperatures (path 4). It appears that this is the weakest boundary since all fracture surfaces exhibit approximately the same amount of failure by this path.[9] With increasing test temperatures, the amount of failure along the tungsten-matrix path (path 3) increases. However, as Figure 8 demonstrates, the different cooling rates cause a major change in the amount of tungsten-matrix failure. The slow cooled material shows very little tungsten cleavage and considerable tungsten-matrix (path 3) failure. Alternatively, the rapidly cooled material exhibits tungsten cleavage and less tungsten-matrix failure. Cooling rate does not appear to have a significant influence on the amount of tungsten-tungsten grain boundary (path 4) failure. The rapidly cooled material has considerable matrix phase failure in constrast with the slow cooled and moderate cooled material.

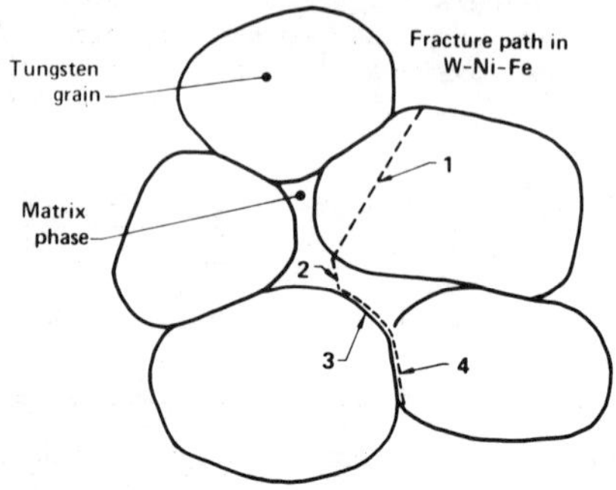

Figure 6 - Schematic diagram of the fracture path possibilities in the liquid phase sintered alloy.

Discussion

The ductile to brittle transition with decreasing test temperature has previously been noted for the heavy alloys.[2,4,8,12] Krock and Shepard[8] clearly demonstrated this transition resulted from differences in the flow stress temperature dependence of the two phases. The tungsten phase is more temperature dependent and hence there is a shift to tungsten cleavage at lower temperatures. Additionally, the heavy alloys have more tungsten-tungsten interfacial area and less matrix phase (which acts to arrest crack growth) as the tungsten content increases. Thus the 95W-3.5Ni-1.5Fe has more sensitivity to test temperature than the 90W-5Ni-5Fe (ref. 4), 90W-7Ni-3Fe (ref. 12), and 85W-10.5Ni-4.5Fe (ref. 8) alloys. Hence, the observed test temperature effect on impact energy is attributed to the lower matrix phase content and larger interfacial area found with the 95W alloy.

In the absence of other changes, it would be expected that decreases in hardness in simple systems would be associated with increases in toughness. Thus, even the microhardness changes are small, they indicate that mechanical properties of the matrix are not a factor in the toughness variations with cooling rate.

The cooling rate effect on toughness is attributed to interfacial segregation. In this investigation, rapid cooling from a post-sintering anneal resulted in improved toughness. For the heavy alloys there is substantial disagreement as to the cooling rate effect as demonstrated in Table I. Several possible explanations exist for the toughness sensitivity to cooling rate. These include impurity segregation to interfaces,[4,7] compositional and heat treatment effects on the matrix phase and tungsten grain chemistries,[6,19,20] hydrogen embrittlement of the matrix phase,[21,22] formation of intermetallic compounds,[3,5,6,10,11] changes in the defect (pore) structure,[9] and a ductile-brittle transition temperature close to room temperature.[12,16] In wrought tungsten there is a strong impurity effect on ductility.[14,23] Considering the change in fracture path shown

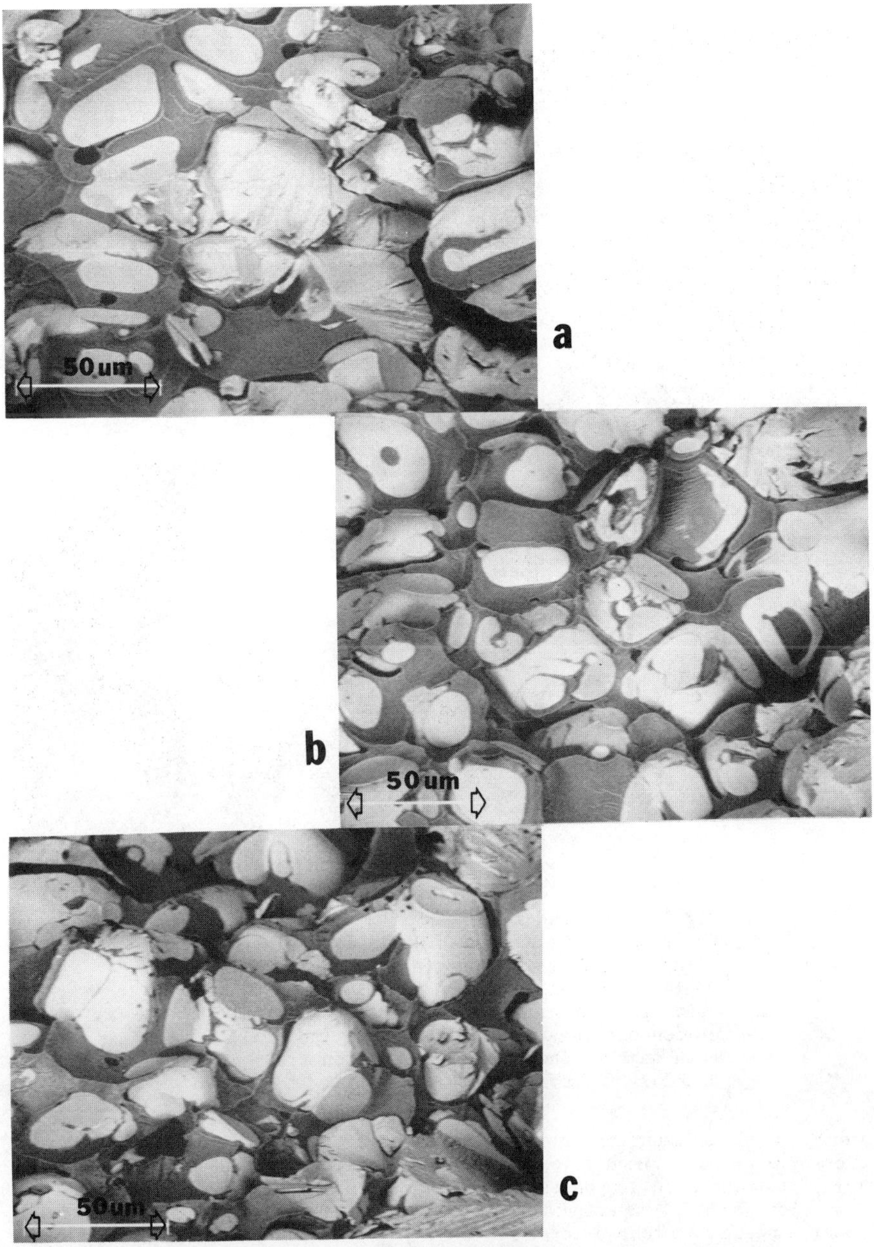

Figure 7 - Backscatter electron SEM images of the 95W-3.5Ni-1.5Fe heavy alloy fracture surfaces (the tungsten phase is light and the matrix phase is dark). The test temperatures were (a) -100°C, (b) 22°C, and (c) 300°C. All four fracture modes can be seen in these micrographs.

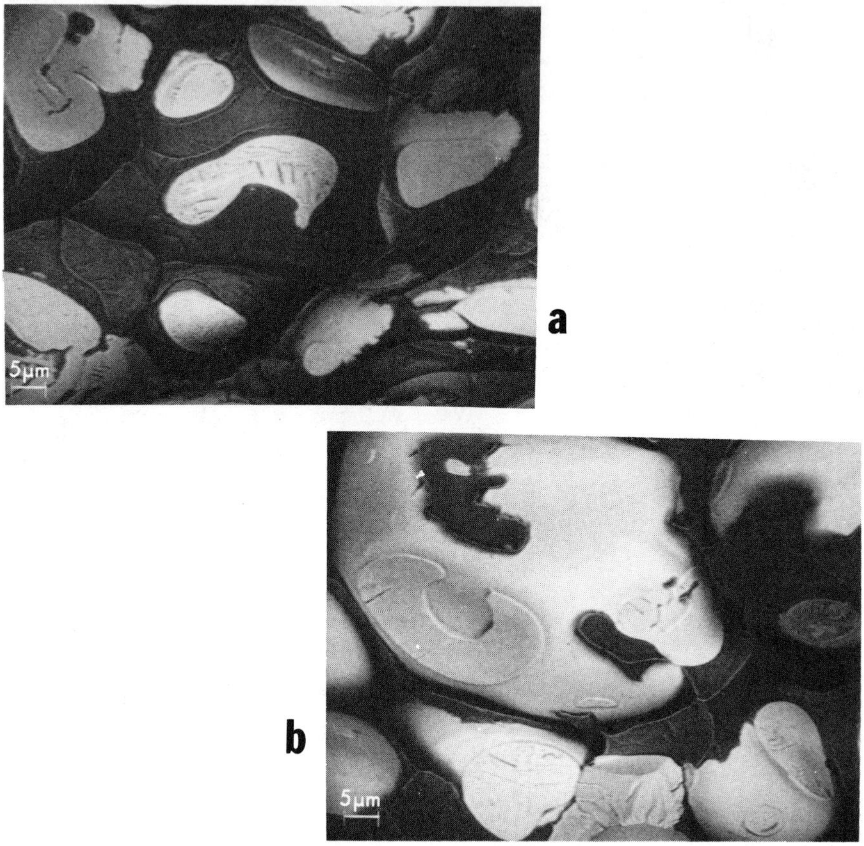

Figure 8 - Backscatter electron SEM images of (a) rapidly cooled, and (b) slowly cooled heavy alloy heat 2 fracture surfaces. The rapidly cooled material evidences matrix phase failure (path 2) whereas the slowly cooled material evidences tungsten-matrix failure (path 3). Both samples show tungsten-tungsten grain boundary failure (path 4).

in Figure 8, an impurity role appears to be operative in the heavy alloys. The segregation of impurities to interfacial areas on slow cooling would be more detrimental to toughness as the matrix content is decreased. Thus, the 95W alloy would be expected to be more sensitive to cooling rate than the lower tungsten content alloys.

The most recent experimental findings support the impurity segregation and intermetallic formation explanations for embrittlement.[4,7,12] Edmonds and Jones[4] and Minakova et al.[10,11] give evidence of intermetallic formation in 90W alloys. However, they disagree as to the effects of slow cooling as noted in Table I. With respect to impurity effects Edmonds and Jones[4] and Muddle and Edmonds[7] report Auger data showing impurity segregation in embrittled heavy alloys. In these cases, slow cooling from above 1000°C results in decreasing impurity solubility and adequate diffusive mobility so that segregation to interfaces is observed. The difference in purity, composition, processing, and microstructure between the various studies possibly explains some of the apparent conflict. In this study a comparison between impurity segregation and intermetallic formation explanations was performed. Unlike Edmonds and Jones[4], x-ray diffraction failed to detect any intermetallic phases; which confirms the earlier negative finding by Jones and Munnery.[3] Transmission electron microscopy was performed on two samples (300°C keyhole and slow cooled heat 2). Several interfaces were examined in detail and as Figures 9 and 10 demonstrate no interfacial intermetallic phases could be detected. Material given a slow cool cycle showed very low toughness. Based on Edmonds and Jones,[4] this would be the prime candidate for forming an intermetallic phase. For the 95W-3.5Ni-1.5Fe heavy alloy there is no evidence for an intermetallic; thus specific evidence for an impurity segregation effect was sought.

Impurity analysis on heats 1 and 2, as presented in Table V, demonstrate a sizeable impurity concentration. Heat 2 which exhibited the

Table V. Impurity Analysis in ppm
(nd = not detected)

Element	Heat 1	Heat 2
Al	15	15
Be	4	1
Ca	40	25
Cb	10	nd
Cr	25	25
Cu	15	3
K	15	nd
Mg	4	1
Mn	40	25
Mo	120	120
Na	15	4
P	<10	<10
S	*	*
Ti	5	3
Zn	8	8

*No reliable value available.

higher impact energies also has the lower impurity content. Fracture surfaces from the Charpy bars were subjected to Auger, ESCA, and SIMS analysis in an attempt to establish evidence of an impurity effect. Unfortunately, the complex fracture path and rough surface topography hindered a clear identification of an impurity role. Both the Auger and SIMS analyses indicate sulfur on the fracture surface. However, efforts

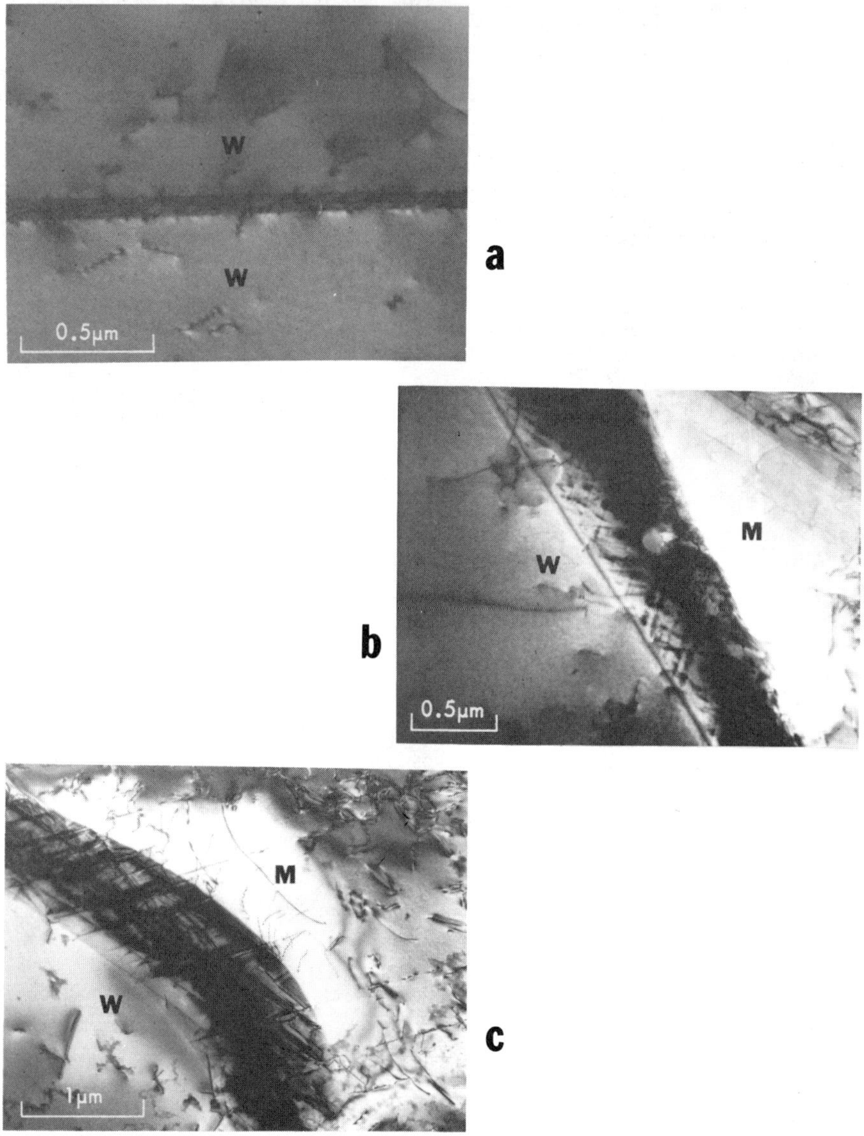

Figure 9 - Transmission electron micrographs of interfacial regions in 95W-3.5Ni-1.5Fe heavy alloy; sample taken from keyhole Charpy specimen tested at 300°C. The tungsten phase is indicated by W and the matrix phase by M: (a) tungsten-tungsten grain boundary, (b) and (c) tungsten-matrix interfaces.

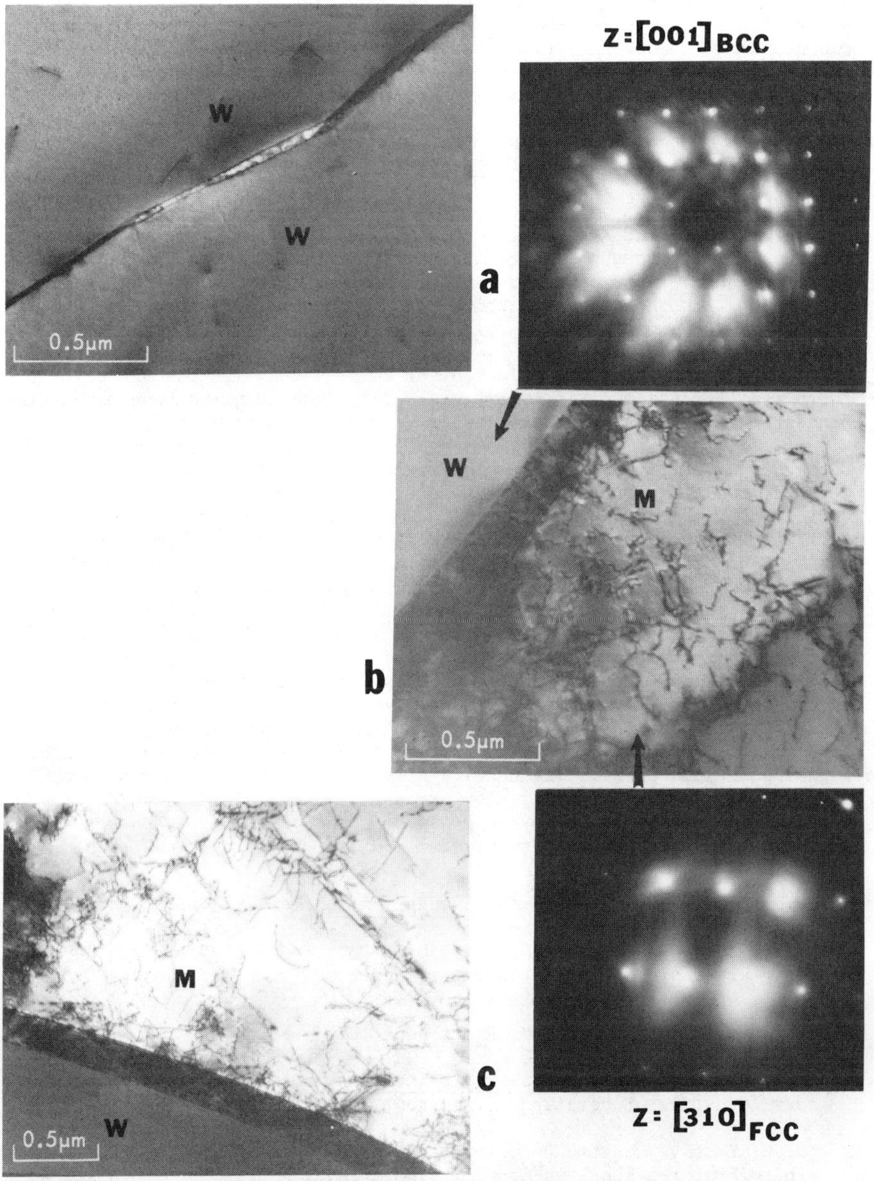

Figure 10 - Transmission electron micrographs of heat 2 material given a slow cooling cycle. Again the matrix and tungsten phases are indicated by M and W, respectively. The interfaces are (a) tungsten-tungsten, (b) and (c) tungsten-matrix. No evidence of interfacial precipitation could be found in the transmission microscopy.

to-date have failed to establish a clear correlation between the toughness and sulfur concentration on fracture faces. In part, this may be due to the complex fracture and sulfur partitioning to the tungsten phase. Thus sulfur is present on the fracture surface whether failure is through the grains or along segregated matrix-tungsten interfaces. Microprobe examinations showed that the sulfur content was highest in the tungsten phase, and that the impact energy did appear to correlate with the bulk sulfur level. Efforts are in progress to further clarify the possible role of sulfur in this alloy system.

From these findings it is concluded that impurities are responsible for the observed toughness variations with cooling rate in 95W-3.5Ni-1.5Fe. Microstructural features are essentially unchanged by the differing heat treatments. Furthermore, variables such as composition, hardness, and density do not explain the ductile-brittle toughness transitions with test temperature and cooling rate. Past suggestions of intermetallic formation and matrix phase aging are rejected for this system. In the 95W alloy there is a large amount of interfacial area. The tungsten-tungsten grain boundaries are known to be embrittled by impurities. In the present case the role of impurities is very strong. Slow cooling promotes interfacial segregation of impurities, thus the fracture path is predominately along the tungsten-tungsten and tungsten-matrix boundaries. The impurity content correlates with the impact energies, showing the detrimental role of impurities on toughness. Thus, the 95W alloy exhibits the highest toughness when rapidly cooled from a homogenization temperature of approximately 1000°C. On the other hand, slow cooling gives a decreasing impurity solubility coupled to a high diffusive mobility. Consequently, the material is embrittled by impurity segregation to interfacial boundaries. Past conflicting reports concerning the cooling rate effect are probably due in part to different impurity contents. Based on these findings, it is probable that high purity heavy alloys will exhibit high toughnesses and less sensitivity to cooling rate. However, the sensitivity to test temperature as demonstrated in this study can not be totally eliminated through use of higher purity material. The ductile-brittle transition with test temperature is due to the differing flow stress and ductility dependencies on temperature for the two alloy components. Hence, lower toughnesses are expected at lower test temperatures.

Acknowledgements

This study was supported by the U.S. Department of Energy through the Lawrence Livermore National Laboratory under Contract W-7405-Eng-48.

References

1. F. V. Lenel, Powder Metallurgy Principles and Applications, Metal Powder Industries Federation, Princeton, N.J., 1980.

2. R. H. Krock, "Elastic and Plastic Deformation of Dispersed Phase Liquid Phase Sintered Tungsten Composite Material", Metals for the Space Age, F. Benesovsky (ed.), Metallowerk Plansee, Reutte, Austria, 1965.

3. D. J. Jones and P. Munnery, "Production of Tungsten Alloy Penetration Radiation Shields", Powder Met., 10 (1967) pp. 156-173.

4. D. V. Edmonds and P. N. Jones, "Interfacial Embrittlement in Liquid-Phase Sintered Tungsten Heavy Alloys", Met. Trans., 10A (1979) pp. 289-295.

5. J. F. Kuznick, "Development of Ductile Tungsten-Base Heavy-Metal Alloy", pp. 166-172 in Modern Development in Powder Metallurgy, Vol. 3, H. H. Hausner (ed.), Plenum Press, New York, N.Y., 1966.

6. I. Y. Dzykovich, R. V. Makarova, O. K. Tedorvich, and I. N. Frantsevich, "Distribution of Elements During the Formation of Sintered Alloys of the System W-Ni-Fe", Soviet Powder Met. Metal Ceram., 4 (1965) pp. 655-660.

7. B. C. Muddle and D. V. Edmonds, "Interfacial Segregation and Embrittlement in Liquid-Phase Sintered Tungsten Alloys", Phil. Trans. Royal Soc. Lond., A295 (1980) p. 129.

8. R. H. Krock and L. A. Shepard, "Mechanical Behavior of the Two-Phase Composite, Tungsten-Nickel-Iron", Trans. TMS-AIME, 227 (1963) pp. 1127-1134.

9. K. S. Churn and D. N. Yoon, "Pore Formation and Its Effects on Mechanical Properties in W-Ni-Fe Heavy Alloy", Powder Met., 22 (1979) pp. 175-178.

10. R. V. Minakova, A. N. Pilyankevich, O. K. Teodorovich, and I. N. Frantsevich, "Nature of the Strength of Tungsten-Nickel-Iron Alloys Part I. Precipitation Hardening Characteristics of W-Ni-Fe Alloys", Soviet Powder Met. Metal Ceram., 7 (1968) pp. 396-399.

11. R. V. Minakova, A. N. Pilyankovich, O. K. Toodorovich, and I. N. Frantsevich, "Nature of the Strength of Tungsten-Nickel-Iron Alloys Part 2. Rupture Mechanism of W-Ni-Fe Alloys", Soviet Powder Met. Metal Ceram., 7 (1968) pp. 473-477.

12. L. G. Bazhenova, A. D. Vasilev, R. V. Minakova, and V. I. Trefilov, "Nature of the Low-Temperature Brittleness of a W-Ni-Fe Alloy", Soviet Powder Met. Metal Ceram., 19 (1980) pp. 34-38.

13. T. K. Kang, E.-Th. Henig, W. A. Kaysser and G. Petzow, "Effect of Cooling Rate on the Microstructure of a 90W-7Ni-3Fe Heavy Alloy", pp. 189-203 in Modern Developments in Powder Metallurgy, Vol. 14, H. H. Hausner, H. W. Antes and G. D. Smith (eds.), Metal Powder Industries Federation, Princeton, N.J., 1981.

14. A. Joshi and D. F. Stein, "Intergranular Brittleness Studies in Tungsten Using Auger Spectroscopy", Met. Trans., 1 (1970) pp. 2543-2546.

15. W. G. Northcutt, Jr., "Fabrication Development of Tungsten Alloy Penetrators", Union Carbide Corp., Y-12 Plant, Oak Ridge, Tenn., Report No. Y-1994, August 1975.

16. J. M. Dickinson, R. E. Riley and J. M. Taub, "Strengthening of Liquid Sintered Tungsten Alloys", pp. 329-352 in Modern Developments in Powder Metallurgy, Vol. 8, H. H. Hausner and W. E. Smith (eds.), Metal Powder Industries Federation, Princeton, N.J., 1973.

17. E. G. Zukas and L. S. Levinson, "Matrix Motion of Low-Volume Matrix W-Ni-Fe Composites Sintered in a Temperature Gradient", J. Mater. Sci., 10 (1975) pp. 863-869.

18. E. G. Zukas and H. Sheinberg, "Sintering Mechanisms in the 95%W-3.5%Ni-1.5%Fe Composite", Powder Tech., 13 (1976) pp. 85-96.

19. T. K. Kang and D. N. Yoon, "Coarsening of Tungsten Grains in Liquid Nickel-Tungsten Matrix", Met. Trans., 9A (1978) pp. 433-438.

20. R. V. Makarova, O. K. Teodorovich, and I. N. Frantsevich, "The Coalescence Phenomenon in Liquid-Phase Sintering in the Systems Tungsten-Nickel-Iron and Tungsten-Nickel-Copper", Soviet Powder Met. Metal Ceram., 4 (1965) pp. 554-559.

21. W. G. Northcutt, Jr.: private communication, Union Carbide Corp., Y-12 Plant, Oak Ridge, Tennessee.

22. M. Yodogawa, "Effects of Cold Rolling and Annealing on the Mechanical Properties of 90W-7Ni-3Fe", presented at the Fifth Round Table Conf. on Sintering, Portoroz, Yugoslavia, Sept. 1981.

23. A. W. Funkenbusch, F. Bacon and D. Lee, "The Influence of Microstructure on Fracture of Drawn Tungsten Wire", Met. Trans., 10A (1979) pp. 1085-1091.

DEVELOPMENT OF CRYSTALLOGRAPHIC TEXTURE IN

OXIDE DISPERSION STRENGTHENED ALLOYS

M. P. Anderson, J. Y. Koo and R. Petkovic-Luton

Exxon Research and Engineering Company
P.O. Box 45
Linden, New Jersey, 07036
USA

Crystallographic textures and microstructures developed in oxide dispersion strengthened alloys have been examined as a function of thermo-mechanical treatments. The alloys studied were Fe-20Cr-4.5Al-0.5Ti-0.5Y_2O_3 and Ni-16Cr-5Al-1Y_2O_3. The Fe-base alloy is characterized by a deformation type texture consisting of {112} <110> and {114} <110> in the warm worked state. This texture is stable with heat treatments up to 1300°C, where a change in preferred orientation is observed. The new texture is described by $(4\bar{7}1)$ [113] and $(\bar{1}10)$ [112]. In contrast, the Ni base alloy is characterized by a recrystallization type texture in the warm worked state. This preferred orientation is best described as a combination of a [001] fiber texture and a {100} <001> rolling texture. The microstructural evolution with heat treatments is also discussed with respect to the development of specific textures.

Introduction

Oxide Dispersion Strengthened alloys have potential technological importance as structural materials for high temperature applications. This is because these alloys possess good high temperature strength and creep resistance achieved by particle strengthening the matrix with a fine dispersion of thermally stable oxides (1-4). Because the powder metallurgy techniques used in fabrication are relatively new (5,6) the metallurgical variables which are important in controlling the microstructure and properties are not completely known. One parameter which is important in the deformation and creep characteristics of other alloy systems is crystallographic texture (7,8). It has been suggested by Sellars and Petkovic-Luton(9) that strong texture in ODS alloys could be responsible for the hypothesized threshold stress for diffusional creep. To explore the influence of texture upon the properties of ODS alloys it is first necessary to characterize the development of preferred orientation in these materials and to understand the mechanisms whereby changes of texture occur. Previous work has been largely confined to investigations of thoria or yttria dispersed Ni and Ni-base alloys (10-12), and the reported results are contradictory. Singer did not observe preferred orientation in Ni-base samples consolidated by hot extrusion (11). In contrast, Burton and Tien have reported both an extrusion texture and annealing textures in a similar system (12). The objective of this study is to examine the texture developed by thermo-mechanical treatment of the commercially available Fe-base alloy MA956 and an experimental Ni-base alloy to shed light upon the relationship between texture, microstructure, and the progress of recrystallization.

Experimental

The Fe-base alloy MA956 was supplied by the International Nickel Company and the Ni-base alloy Ni503 was supplied by Special Metals Corporation. Both alloys were prepared by the mechanical alloying process (5,6). Powders having the compositions given in Table I were mixed in a high energy attritor, and consolidated by canning, evacuation, and extrusion to rectangular bar. The Fe-base alloy was in addition warm rolled.

Table I. Compositions of two ODS Alloys, wt. Pct.

Alloy	Fe	Ni	Cr	Al	Y_2O_3	Ti
MA956	74.5	-	20	4.5	0.5	0.5
Ni503	-	78	16	5	1	-

MA956 was examined both in the as-worked state and after several high temperature heat treatments. Heat treatments were carried out for 14 hours in air at the the temperatures listed in Table II. The samples consisted of small coupons cut from the bar with dimensions of 3 cm x 3 cm x 1 cm.

Table II. Heat Treatments

Alloy	Symbol	Temperature, °C
	A0	as-worked
	A1	1000
	A2	1100
MA956	A3	1200
	A4	1300
	A5	1400
Ni 503	B0	as-extruded
	B4	1300

Ni503 was examined in the as-extruded state and also after a single heat treatment at 1300°C for 10 hours as shown in Table II.

Texture measurements were made on a Rigaku DMAX II-A diffractometer using an automated pole figure device. Data were collected for the (110) reflection in the case of the alloy MA956, and for the (200) reflection in the case of the alloy Ni503. Mo radiation was used with a Zr filter. The Decker method was employed in transmission and the Schultz method in reflection so that the entire pole figure could be obtained (13).

Microstructural characterization was carried out using optical metallography, TEM and STEM. Samples for transmission electron microscopy were cut from the central portions of the coupons. Thin foils were prepared initially by mechanical methods followed by jet polishing using an electropolishing solution consisting of 300 ml methanol, 175 ml butanol, and 30 ml perchloric acid. Prior to TEM observations, the jet polished foils were slightly ion milled on both sides to minimize possible surface effects. The specimens were examined using a Philips 400T electron microscope equipped with STEM attachment, field emission gun, and energy dispersive X-ray analyzer at an accelerating voltage of 100 KV. STEM images were used to determine submicron grain sizes following the planimetric method. Microanalysis on the dispersoids was conducted using a 10 nm electron probe. The data were analyzed by the thin foil ratio method (14) using the calculated k values given by Goldstein (15) for standardless thin film analysis.

Results

Fe base Alloy

As-worked State. The (110) pole figure of sample A0 is given in Figure 1. The intensities are contoured in fractions of the maximum peak intensity. The orientation of the pole figure is such that

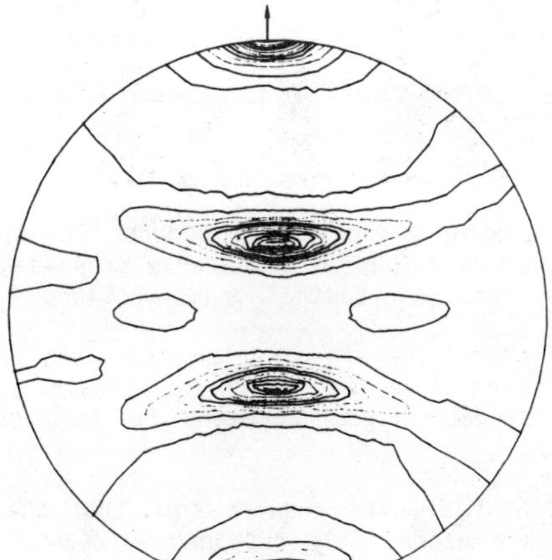

Figure 1-(110) pole figure of the as-worked Fe-base alloy. 15 contours per maximum peak intensity

the extrusion axis (EA) runs vertically and the long transverse direction (LT) horizontally. The short transverse direction (ST) is perpendicular to the plane of the figure.

The data may be characterized as showing that the preferred orientation has the symmetry characteristic of rolling (mm). One mirror contains the EA-ST axes. The second mirror is perpendicular to the first and contains the LT-ST axes. The texture is strong with (110) pole peaks parallel to the EA and inclined at 60° to the EA in the EA-ST plane. (110) poles are not only confined to these peaks but also exhibit a spread of directions in a 60° band, and a 90° band, which correspond to rotations about the EA. Two smaller maxima are also present in the 90° band.

The microstructure of the as-worked state (Figs. 2(a) and 2(b)) consists of a fine grained matrix with a significant dislocation density. The grains are elongated in the EA direction with the smallest diameter parallel to the ST direction. The mean grain diameter is

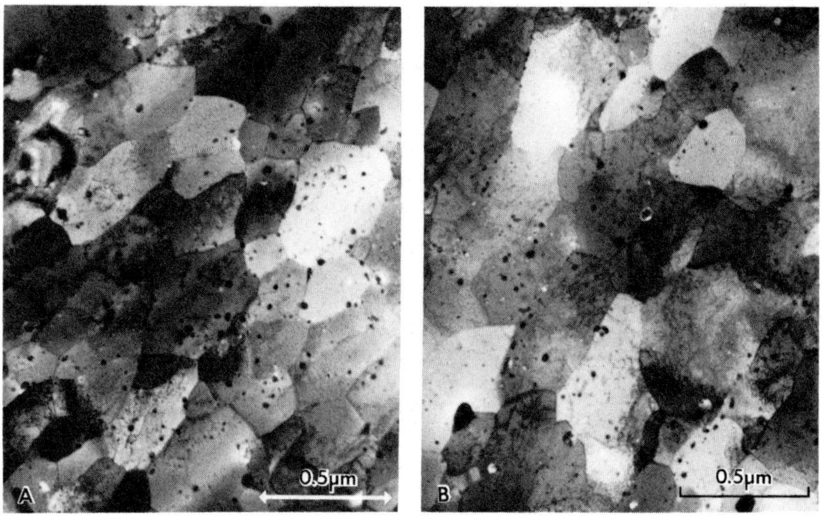

Figure 2-STEM micrographs showing grain morphology and dislocations in as-worked MA956 (a) Transverse cross-section, (b) Longitudinal cross-section

0.4 micron. The dislocations are distributed non-uniformly in that some grains are dislocation free while others contain a high dislocation content. Some dislocation rearrangement to form sub-boundaries has occurred.

Annealed States. The results show that the main features of the initial state are essentially retained with heat treatments up to 1200°C. The (110) pole figures of A2 and A3 are given in Figs. 3(a) and (b) respectively. However, the low level intensity bands which occur in A0 are not sharply defined in A2 and A3. Also, the density outside of the main peak locations (parallel to the EA direction and inclined at 60° to the EA direction) has slightly increased.

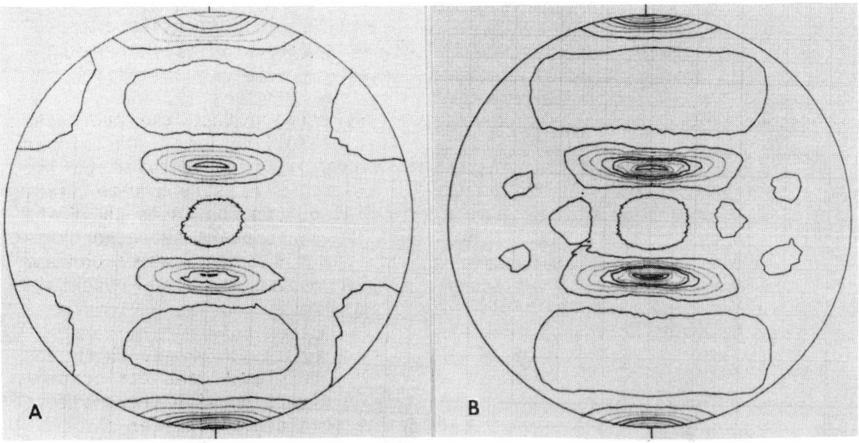

Figure 3-(110) Pole figures of Fe-base alloy heat treatment 14 hrs. at (a) 1100°C and (b) 1200°C. (a) 10 and (b) 15 contours per maximum peak intensity

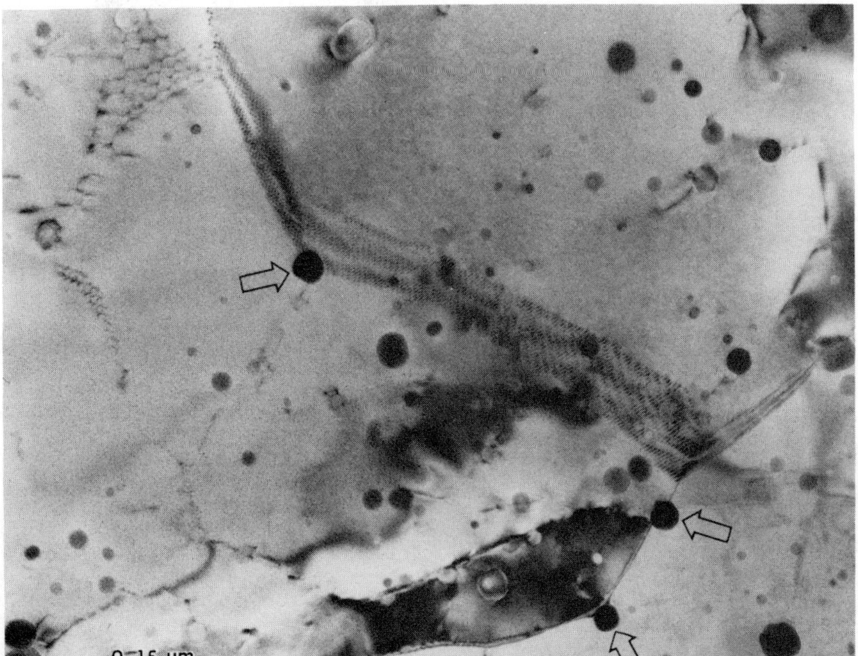

Figure 4-TEM Micrograph showing small angle boundaries developed in MA956 heat treated at 1200°C for 14 hrs (transverse cross-section). Very fine dispersoids ($\lesssim$ 200Å) are Y_2O_3 oxides and the relatively larger ones (arrow marked) are mixed oxide $AlYO_3$

The microstructures of A1, A2, and A3 are similar to that in the as-worked state. The morphology of the grains remains enlongated, with a significant average dislocation density. Careful examination of A3 reveals that polygonization has occurred in many grains, and dislocations which are present are pinned by dispersoids (Figure 4).

In A4 and A5 discontinuous grain growth has occurred which is characterized by an abnormally large grain size (Figures 5(a) and (b)). The morphology of the grains is columnar with the long diameter parallel to the extrusion axis direction. TEM micrographs show that the grains have a very low dislocation density, and that the substructures seen at lower temperature are completely eliminated. The grains are sufficiently large that the (100) pole distribution could be determined from Laue back-reflection photographs. The complete pole figure for

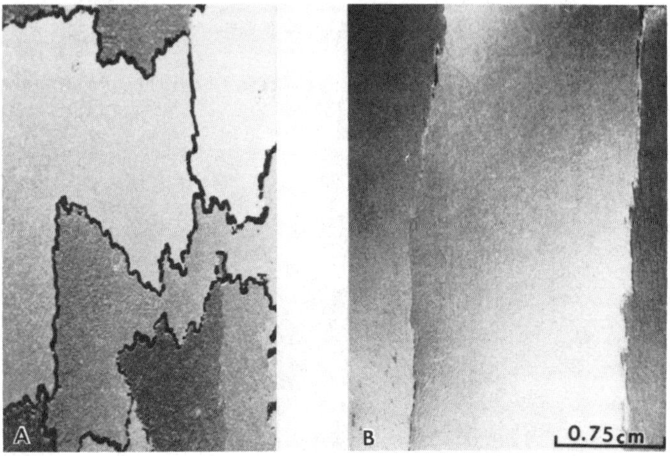

Figure 5-Optical micrograph revealing grain morphology in (a) transverse cross-section and (b) longitudinal cross-section of MA956 heat treated at 1400°C for 14 hrs.

sample A5 is given in Figure 6. The projection has been made with the ST axis vertical, the LT axis horizontal, and the EA perpendicular to the plane of the figure. The pole distribution is of lower symmetry than the deformation texture. Mirror symmetry is maintained across the EA-ST plane. However, mirror symmetry does not occur across the EA-LT plane.

Ni-base Alloy

As-extruded state. The (100) pole figure of the experimental Ni-base alloy is shown in Figure 7(a). The orientation of the pole figure is such that the EA direction is vertical, the LT direction is horizontal, and the ST direction is perpendicular to the plane of the figure. As in the case of MA956 the texture is strong and the intensity distribution has mm symmetry. (100) pole peaks occur parallel to the EA and at 90° to the EA axis in the EA-ST plane. A weaker band of poles is also present at 90° to the EA.

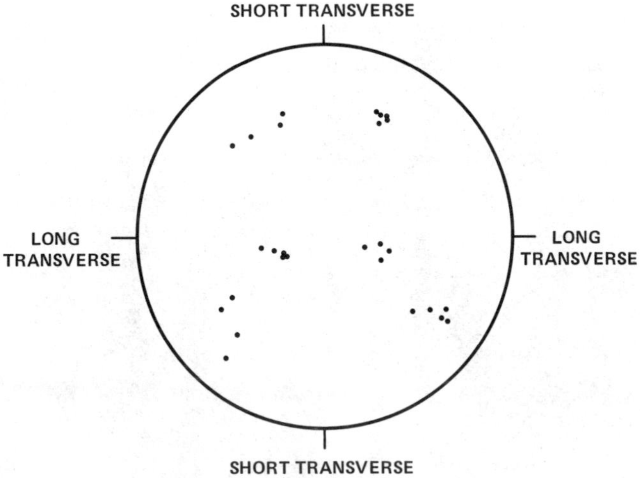

Figure 6-(100) Pole figure of MA956 heat treated at 1400°C for 14 hrs.

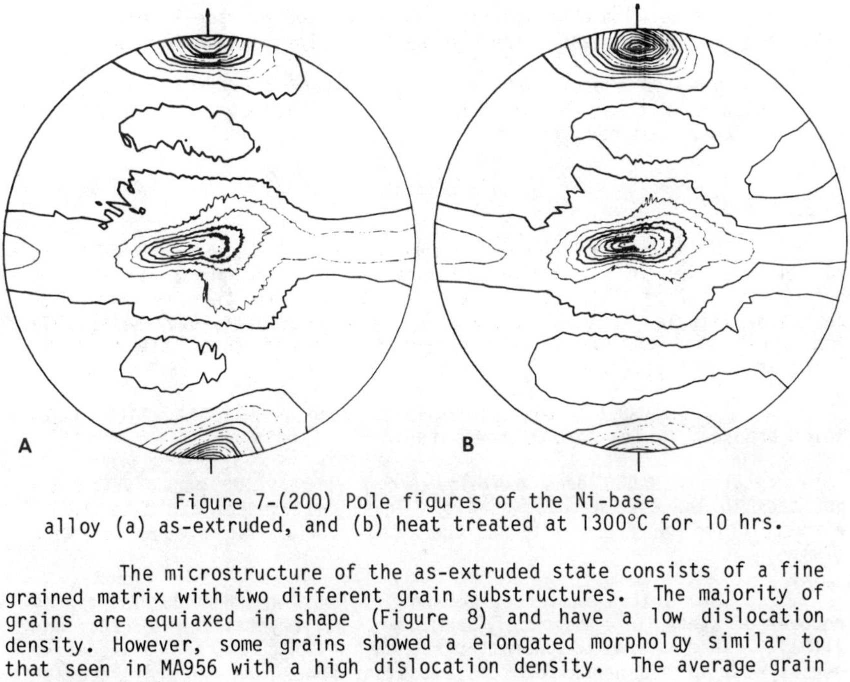

Figure 7-(200) Pole figures of the Ni-base alloy (a) as-extruded, and (b) heat treated at 1300°C for 10 hrs.

The microstructure of the as-extruded state consists of a fine grained matrix with two different grain substructures. The majority of grains are equiaxed in shape (Figure 8) and have a low dislocation density. However, some grains showed a elongated morpholgy similar to that seen in MA956 with a high dislocation density. The average grain size is 0.5 micron.

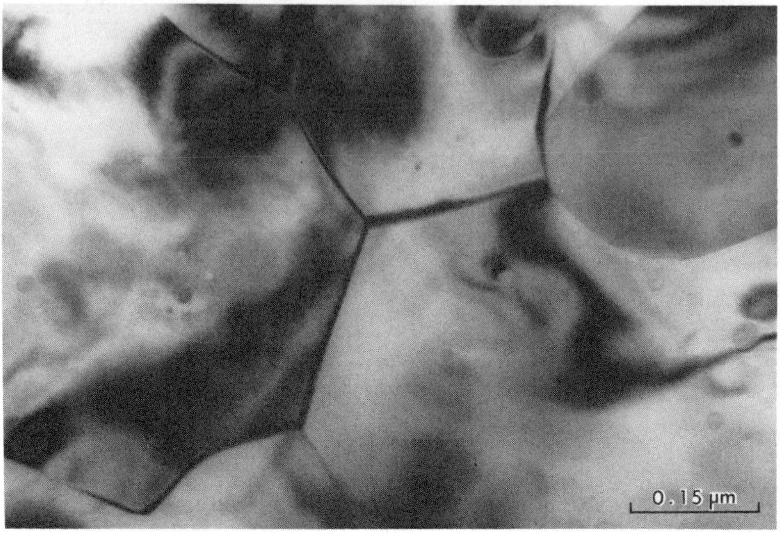

Figure 8-TEM micrograph of the as-extruded Ni-base alloy. Majority of grains present are equiaxed with low dislocation density.

Annealed state. No change in preferred orientation was observed with heat treatment at 1300°C (Figure 7 (b)). Unlike MA956 the microstructure remained fine grained.

Discussion

Fe-base Alloy

The crystallographic texture of the as-worked state of MA956 can be analyzed into a number of components. The principle contributions consist of {112} <110> and {114} <110>. This texture is similar to the deformation texture exhibited by other bcc metals such as Mo (16) and V(17), and the warm rolling texture seen in sheet steels (18). There are, however, two differences:

(i). MA956 does not exhibit pronounced {001} <110> peaks which are seen in low carbon sheet steels.

(ii). MA956 does exhibit strong {112} <110> peaks which are not seen in low carbon sheet steels. These differences could be due to a restriction on cross slip on the {110} <111> slip system in MA956 (19).

The initial microstructure is similar to the partially recovered state of moderately deformed polycrystalline Fe(20) and Al(21). This state is characterized by a large average dislocation density but with a nonuniform distribution; some grains are dislocation free, some grains contain high dislocation density, and others are fragmented into subgrains of different dislocation content. The microstructure in conjunction with the texture suggests that the conditions required for primary recrystallization, viz. high stored energy, strong lattice curvature, and steep spatial gradient of stored energy, are satisfied.

The retained texture and polygonization which are observed with heat treatments between 1000°C to 1200°C imply that recovery processes are operating but that primary recrystallization does not occur. This behavior is similar to the "recrystallization in situ" process seen in precipitation strengthened systems such as Al-Cu (22,23). In this case inhibition of primary recrystallization is attributed to particle pinning of both dislocations and high angle boundaries, which slows dislocation rearrangement and annihilation and prevents high angle boundary migration.

Heat treatment at 1300°C and above leads to an annealing texture which in the case of the particular sample examined here can be broken down into two different average grain orientations. The first average grain orientation is described by:

[113] parallel to extrusion axis

[$21\bar{1}$] parallel to long transverse direction

[$4\bar{7}1$] parallel to short transverse direction

The second grain orientation is approximately related to the first by reflection through the EA-ST mirror. The EA-ST mirror corresponds to a (112) type plane, which is a twin plane in the bcc system. The second average grain orientation corresponds to:

[112] parallel to extrusion axis

[$\bar{1}\bar{1}1$] parallel to long transverse direction

[$\bar{1}10$] parallel to short transverse direction

The relative misorientation between like symmetry grains ranges from 0° to 15°. The relative misorientation of unlike grains is ~ 160° (about the extrusion axis). The two different grain orientations are shown with respect to the bar in Figure 9. These components are within the spread of orientations predicted in the primary recrystallization of Fe (24).

The elimination of the dislocation substructure by the growth of new grains characterized by a recrystallization texture suggests that MA956 undergoes a process similar to primary recrystallization at temperatures above 1300°C. Further investigations are in progress to more clearly understand the mechanism of nucleation.

Ni-base Alloy

The preferred orientation of the as-extruded state has the mm symmetry characteristic of rolling (the alloy was extruded to bar shape) but the texture is not a typical fcc rolling texture. The pole distribution contains {100} <001> and {110} <001>, but is better described as a combination of a [001] fiber texture and a {100} <001> rolling texture. The [001] fiber texture has been observed in the extrusion of Al

(25) and has been attributed to recrystallization during deformation. Similarly the {100} <001> rolling texture is a recrystallization texture encountered in a number of fcc systems - Ni, Cu, Au, Al etc. (26,27).

The microstructure of the as-extruded state consists predominently of fine equiaxed grains with low dislocation content. This

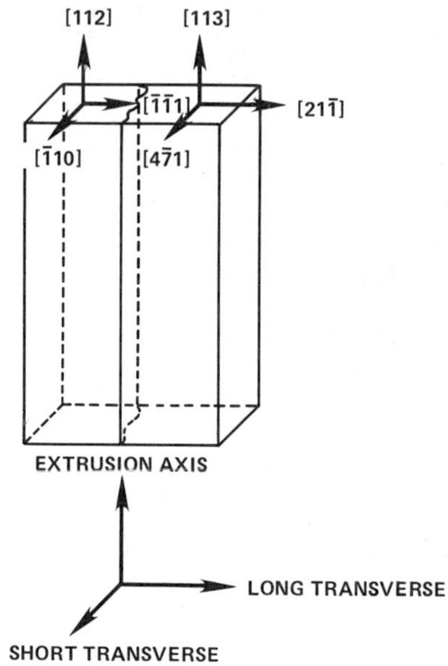

Figure 9-Grain orientations in the Fe-base alloy MA956 heat treated at 1400°C for 14 hrs.

information in conjunction with the texture suggests that the alloy has undergone partial recrystallization either during or following extrusion, but has not undergone grain growth.

This hypothesis is partially supported by heat treatment at 1300°C. Unlike MA956 which exhibits discontinuous grain growth in this temperature range, no change in texture or microstructure is observed. Little driving force for grain growth is present in the case where no dislocation density gradient exists and the grains are uniformly sized.

Summary

The crystallographic textures and microstructures developed in two oxide dispersion strengthened alloys were examined in the as-consolidated states and after several high temperature heat treatments.

The Fe-base alloy MA956 is characterized by a deformation type texture in the warm rolled state. The principle contributions can be described as {112} <110> and {114} <110>. The microstructure consists of sub-micron size grains with significant dislocation content. The main features of the preferred orientation of the initial state are retained with heat treatments up to 1200°C. The microstructures also remain fine grained, but there is evidence of some dislocation rearrangement to form cells. Above 1300°C discontinuous grain growth is observed which is characterized by an abnormally large grain size. The dislocation substructures seen in the initial state are completely eliminated. The texture is described by $(4\bar{7}1)$ [113] and $(\bar{1}10)$ [112] which are components predicted in the primary recrystallization of Fe.

The elimination of the dislocation substructure by the growth of new grains characterized by a recrystallization texture suggests that MA956 undergoes a process similar to primary recrystallization at temperatures above 1300°C. Further investigations are in progress to more clearly understand the mechanism of nucleation.

The Ni-base alloy is characterized by an unusual preferred orientation which is best described as a combination of a [001] fiber texture and a {100} <001> rolling texture. The microstructure consists of sub-micron size grains with low average dislocation content. No change in preferred orientation or grain size is observed with heat treatment at 1300°C. The microstructure in conjunction with the texture suggests that the alloy has partially recrystallized either during or following extrusion.

Acknowledgment

The authors would like to thank Lydia Massey for her help in typing this manuscript.

References

1. N. J. Grant, Strengthening Mechanisms in Metals and Ceramics, Proceedings of the 12th Sagamore Army Materials Research Conference, p.75, Syracuse University Press, 1966.

2. G. M. Ault and Burte, Oxide Dispersion Strengthening, p.14, Gordon and Breach Science Publishers, New York, 1968.

3. B. A. Wilcox and A. H. Clauer, Superalloys, C. T. Sims and W. C. Hagel eds., p.200, John Wiley and Sons, New York, 1972.

4. K. H. Kramer, "Dispersion Hardened Superalloys," Powder Metallurgy International, 9 (3) (1977) pp.105-112.

5. J. S. Benjamin, U.S. Patent 3,738,817.

6. S. Baranow, U.S. Patent 4,010,024.

7. W. F. Hosford, Textures in Research and Practice, J. Grewen and G. Wassermann eds., p.414, Springer-Verlag, Berlin, 1969.

8. G. Y. Chin, The Inhomogeneity of Plastic Deformation, R. E. Reed Hill ed., p.83, American Society for Metals, 1973.

9. C. M. Sellars and R. Petkovic-Luton, "Creep of Dispersion-Strengthened Alloys," Materials Sci. and Engineering., 46 (1980) pp.75-87.

10. W. B. Hutchinson and B. A. Wilcox, "The Influence of a Thoria

Dispersion on Preferred Orientation in Nickel Alloy," Metal Science Journal, 7 (1973) pp.6-11.

11. R. F. Singer and G. H. Gessinger, "Superplastic Deformation of an Oxide-Dispersion Strengthened Superalloy," Proceedings of the 2nd Risϕ International Symposium on Metallurgy and Materials Science, N. Hansen, A. Horsewell, T. Leffers, and H. Lilholt eds., Risϕ National Laboratory, Roskilde, Denmark, 1981.

12. C. J. Burton, S. Baranow, and J. K. Tien, "Recrystallization and Texture Development in Dispersion-Strengthened Nickel-Base Superalloys," Met. Trans. A, 10A (1979)

13. B. D. Cullity, Elements of X-ray Diffraction, pp.285-295, Addion-Wesley Publishing Co. Inc., Reading, MA, 1967.

14. J. I. Goldstein, J. L. Costley, G. W. Lorimer, and S. J. B. Reed, Scanning Electron Microscopy, O. Johari ed., 1 (1977) p.315.

15. J. I. Goldstein, Introduction to Analytical Electron Microscopy, J. J. Hren, J. I. Goldstein, and D. C. Joy eds., p.83, Plenum Press, New York (1979).

16. N. Semchyshen and G. A. Timmons, Trans. Amer. Inst. Min. Met. Eng., 194 (1952) p.279.

17. C. J. McHargue and J. P. Hammond, Trans. Amer. Inst. Min. Met. Eng., 194 (1952) p.745.

18. Horoshi Tekechi, Hiroshi Kato, and Shin'ichi Nagaghima, Trans. of the Met. Soc. of A.I.M.E., 242 (1968), p.56.

19. I. L. Dillamore and W. T. Roberts, Acta Metallurgica, 12 (1964) p.281.

20. C. J. E. Smith and I. L. Dillamore, Metal Science, 4 (1970) p.161

21. P. Faivre, D. Phil. Thesis, University of Sussex, 1973.

22. E. Hornbogen and U. Koster, 5th Intl. Leichmetalltagung Leoben 1968 Aluminium-Verlag, Dusseldorf 1968, p.5.

23. U. Koster and E. Hornbogen, Z. Metallkde., 59 (1968), p. 792.

24. I. L. Dillamore, "Recrystallization Textures in Bcc Metals," Trans. of the Met. Soc. of AIME, 233 (1965), pp.702-707.

25. C. J. McHargue, L. K. Jetter, and J. C. Ogle, Trans. AIME, 215 (1959), p.831.

26. C. S. Barrett, Structure of Metals, p.570, McGraw-Hill, New York, 1952.

27. I. L. Dillamore and W. T. Roberts, "Preferred Orientation in Wrought and Annealed Metals," Met. Reviews 10 (1965), pp.329-345.

HIGH DENSITY FERROUS COMPONENTS BY ACTIVATED SINTERING

K-S. Hwang and R. M. German

Materials Engineering Department
Rensselaer Polytechnic Institute
Troy, New York 12181

High density powder metallurgy components have been made by activated sintering, with notable success in the refractory metal systems. Systematic attempts were made in this work to promote high densities in a simple ferrous system based on the theoretical backgrounds previously developed for the refractory metals. Twelve alloying compounds were used as binary additives to a water atomized iron powder in the initial experiments. Additionally, selected ternary systems were tested to determine synergistic effects. The resulting mechanical properties, physical properties and microstructures are presented and explained in terms of the main criteria for enhanced sintering. In addition, the search for a non-embrittling sintering enhancement aid is discussed with respect to current grain boundary cohesion models.

Introduction

Work in activated sintering has been conducted widely in different fields; various levels of success have been detailed in iron-based materials, but the best success has been in refractory metals. As summarized by Rashamwala and Tendolkar (1) the techniques for achieving rapid sintering (activated sintering) can be categorized into either physical or chemical methods. Among these methods, only a few are industrially viable. This work concentrates on chemical additions since this is the most cost-effective method and an interesting one in theory.

Studies (2-10) on activated sintering of several refractory metals have been very successful, and a fairly complete theoretical model has evolved in the past five years (11). The process is basically enhanced grain boundary transport by a heterodiffusion process, relying on the presence of a segregated grain boundary phase which provides a short-circuit diffusion path.

Under the assumption that activated sintering is of substantial merit for ferrous systems, the first concern is with the nature of the grain boundary sintering enhancement layer. An additive is desired which will remain enriched at the grain boundaries. Additionally, the solubility of the matrix (iron) should be high in this additive phase. The reverse, solubility for the additive in the matrix is not necessary. This segregated layer should then provide a high heterodiffusion flux path for matrix atoms, thereby giving high shrinkage rates and densities. Since the intent is to have good mechanical properties, we impose an additional criterion that the additive not be embrittling to iron. There are three groups of elements which raise the grain boundary cohesion of iron (12-15): 1) the small interstitials like carbon, boron and nitrogen; 2) the refractory metals like tungsten, molybdenum and tantalum; and 3) the noble metals like ruthenium, rhodium and platinum. The high cost of these latter elements eliminates them from serious consideration at this point. With this embrittlement restriction, the candidates were then limited to potential activators which are either neutral or increase grain boundary cohesion, and compounds which consist of the above elements. These additives, such as B_4C, BN and WC, have the potential of a superimposing activation effects from the individual elements as the compound dissociates during sintering. Some well known densification enhancement elements such as phosphorous and silicon were included in this study in order to understand their functions in sintering compacts. However, the phosphorous and silicon addition are predicted by the grain boundary cohesion model as being unsuitable segregated second phases.

Based on these concepts, the phase diagrams and the past work, compositions and additives were selected (Table I). In the screening experiments, every additive was evaluated as a function of mechanical properties and densities. After the screening experiments, ternary systems (Table II) were designed based on the individual effect of the additive in order to reach even better properties.

TABLE I: Alloying Compositions of Specimens in the Screening Experiment

Alloying Element	Percentage
B	.2%, .5%, .8%, 1.1%
B_4C	.2%, .5%, .8%, 1.1%
BN	.2%, .5%, .8%, 1.1%
Co	1%, 2%, 3%, 4%
Mo	1%, 3%, 6%, 10%
Ni	1%, 3%, 5%, 5%
$Ni(NO_3)_2 \cdot 6H_2O$	.2%, .8%,1.4%, 2.0%
P	.2%, .6%,1.0%, 1.4%
Si	1%, 3%, 5%, 7%
Ta	.5%,1.0%,2.0%, 3.0%
W	2%, 4%, 6%, 8%
WC	1%, 3%, 5%, 7%
None (pure iron)	

NOTE: Two samples were prepared for each condition.

TABLE II: Alloying Compositions of Specimens in the Second Experiment

Base Material	Third Additive	Sintering Temperature °C
Fe-1.1% B	.5% Mo	1200
	1% Mo	1200 & 1100
	2% Mo	1200
	3% Mo	1200 & 1100
	4% Mo	1200
	6% Mo	1200 & 1100
	10% Mo	1200 & 1100
	2% Ni	1200 & 1100
	4% Ni	1200 & 1100
	6% Ni	1200 & 1100
	8% Ni	1200 & 1100
	10% Ni	1200 & 1100
	.2% P	1200 & 1100
	.5% P	1200 & 1100
	1.0% P	1200 & 1100
	1.5% P	1200 & 1100
	1% Si	1200 & 1100
	3% Si	1200 & 1100
	5% Si	1200 & 1100
	7% Si	1200 & 1100
	None (Fe-1.1% B only)	1200 & 1100

NOTE: Three tensile bar specimens were prepared for each condition.

Materials

The effect of activators can be confounded by the impurities in the base material. Hence, Hoeganaes Ancorsteel 1000B atomized iron powder was selected as a starting material because of its high purity and high compactibility. The characteristics and morphology of the as-received powder are shown in Table III and Figure 1, respectively.

The additive powders were supplied by Alfa or Cerac. They were all more than 98% pure and were all -325 mesh except nickel nitrate powder which was -250 mesh.

Experiment

The experimental procedure is outlined in Figure 2. According to the designed compositions, 1000B iron powder was mixed with alloying elements in a double shell mixer for one hour. Powders were then compacted in a 100-ton hydraulic press. Slugs of 1.25 cm in diameter and 1 cm in height were prepared for screening experiments. For the second part of the experiments, tensile bars having the shape specified by ASTM standard E8 were produced in a floating die. All samples were compacted under a pressure of 475 MPa. The die walls were lubricated by a suspension of 100 gm of zinc stearate in 1 liter of 1,1,1-trichloroethane.

TABLE III: Characteristics of Ancorsteel 1000B Powder

Apparent Density: 2.84g/cm^3
Flow Rate: 23.8 sec/50g
H_2 Loss: 0.09%

Screen Analysis:

mesh	+60	+80	+100	+140	+200	+250	+325	-325
%	0	4.6	8.9	22.8	23.9	3.9	21.0	14.9

Composition:

	C	S	N	O	P	Si	Sn	Cr	Ni
%	.01	.009	.001	.092	.009	<.01	.008	.04	.05

	Mn	Cu	Mo
%	.10	.03	.02

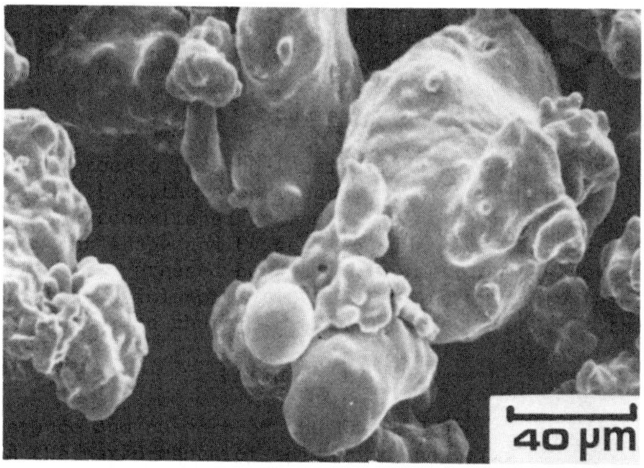

Fig. 1: A SEM photograph of the Hoeganaes Ancorsteel 1000B iron powder.

In the screening experiment, compacts were sintered at 900, 1100, and 1200°C in a molybdenum batch furnace under a dynamic atmosphere of pure, dry hydrogen. The fluctuation of the temperature in the hot zone was controlled to within ±3°C. After one hour of sintering, specimens were cooled in the cooling zone for 40 minutes.

Hardness and density measurements were conducted for every specimen. Microstructures of selected materials were examined to help understand the sintering phenomena. Moreover, compacts of interest were axially drilled through and radially crushed by a hand press. The fracture surface was then examined using a scanning electron microscope (SEM) to assess for transgranular failure.

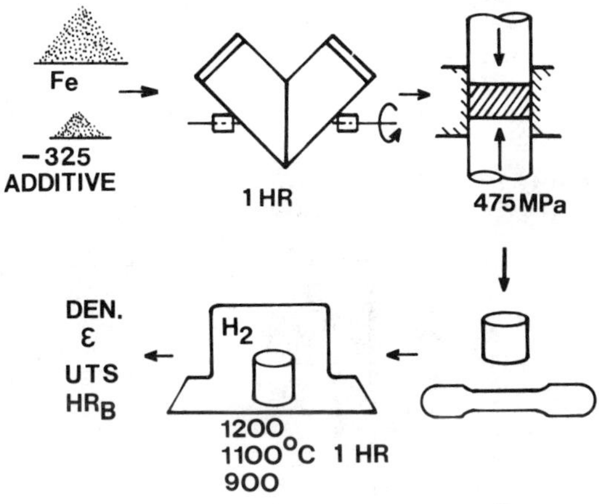

Fig. 2: A schematic diagram showing the experimental procedure.

Based on these results the second set of experiments were conducted with a 1.1% B base addition, mixed with a third additive— Ni, Mo, P, or Si. These powders were compacted to tensile bars and sintered at 1200 and 1100°C for one hour. The tensile strength, elongation, hardness and densities were measured with microstructures and fracture surfaces being examined for selected specimens.

Results and Discussions

Screening Experiments

Figure 3 shows that when B, P, Mo, and WC were added to pure iron pow-

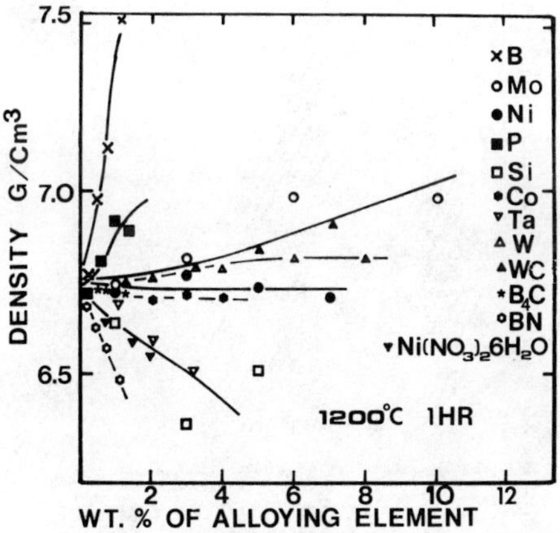

Fig. 3: Effect of additives on sintered densities during the screening experiments at 1200°C, 1 hr.

der, they enhanced densification. However, since all samples were compacted at the same pressure, different green densities resulted due to the alloying; hence, the densification effect of the additives were better judged by the densification parameter ΔD:

$$\Delta D = \frac{\text{sintered density - green density}}{\text{theoretical density - green density}}$$

From this parameter, it is shown in Figure 4 that B has the highest densification effect at 1200°C, followed by P, B_4C, Si and Mo.

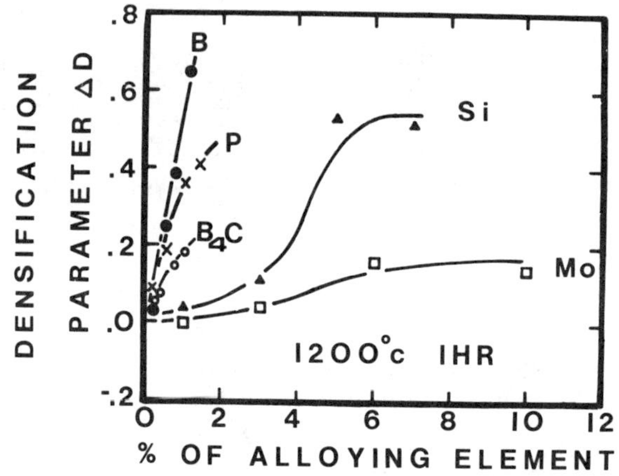

Fig. 4: Effect of additives on the densification parameter ΔD at 1200°C.

Although only a few additives can increase the density of the compacts, most alloying elements can improve the hardness. Figure 5 shows that eight

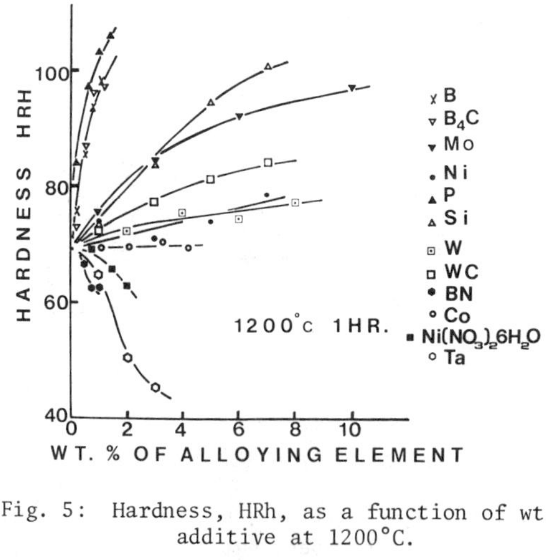

Fig. 5: Hardness, HRh, as a function of wt. % additive at 1200°C.

out of twelve additions improve the hardness. Among these, B, B_4C and P are the most effective, which correlates directly with the density increases. In this presentation no effort is made to normalize for specimen density differences with respect to the sintered properties.

The microstructure of Fe-1.1% B as shown in Figure 6(a) consists of iron particle and a matrix phase which was liquid at the sintering temperature of 1200°C. The latter was formed during sintering which amounts to 23 vol. % under image analysis. This results in 96% theoretical density and thereby increase the hardness significantly. The matrix phase has a eutectic structure of iron and Fe_2B as identified by x-ray diffraction. Figure 6(b)

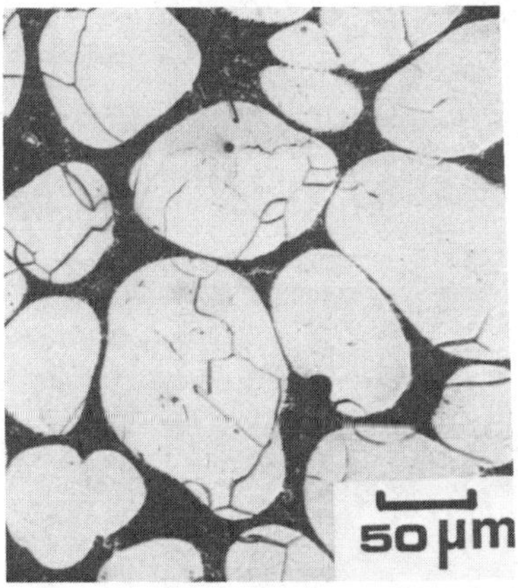

Fig. 6(a): Microstructure of Fe-1.1% B sintered at 1200°C for 1 hr. showing a liquid phase sintered microstructure.

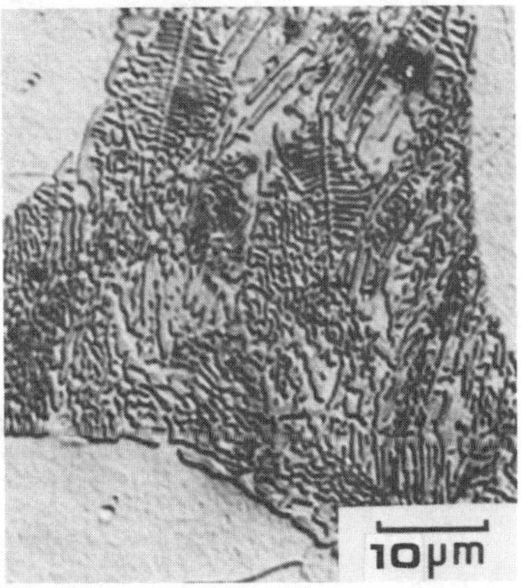

Fig. 6(b): The eutectic structure of the inter-grain liquid phase matrix.

provides a high magnification micrograph of the eutectic matrix phase. The intrinsic hardness of the lamella structure was measured as 177 DHP (100g), as compared with 80 DHP hardness for the iron grains, thus contributing to the increase in the bulk hardness. Scanning electron microscopy was used to examine the fracture surface of this material. As shown in Figure 7, ductile, transgranular type fracture was observed in accord with the preliminary analysis based on grain boundary cohesion.

In contrast to the Fe-B system, ductility was not anticipated in either the Fe-P or Fe-Si alloys even though enhanced densification was possible. Figure 8 shows the microstructure of Fe-1.4% P. Contrasted to Figure 9 which gives the pure iron structure, the Fe-P alloy shows a larger grain size and a rounder, coarser pore structure. As the amount of phosphorus increases both the density and hardness increase. However, its embrittling effect also increases as shown in Figure 10(a) with 100% intergranular type fracture surface. The small droplet-like particles on the fracture surface are complex compounds of Fe-P-Si-Mn as identified by energy dispersive x-ray analysis (Figure 10(b)). These particles might have once formed as a thin layer of liquid phase along the grain boundaries. As this liquid phase solidified the droplets formed and apparently impaired the mechanical properties, especially ductility.

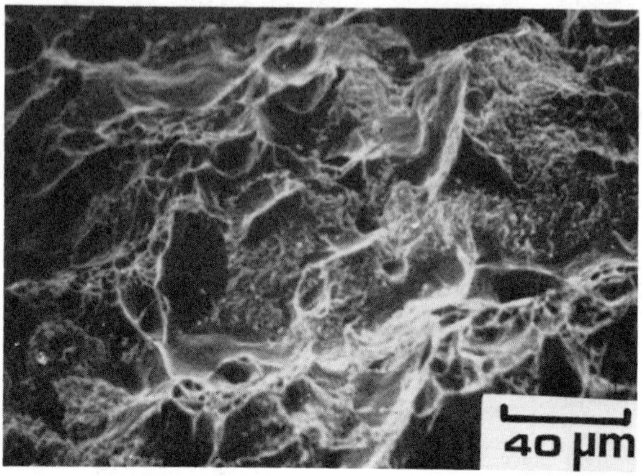

Fig. 7: A SEM photograph of Fe-1.1% B fracture surface, showing the predicted transparticle fracture path.

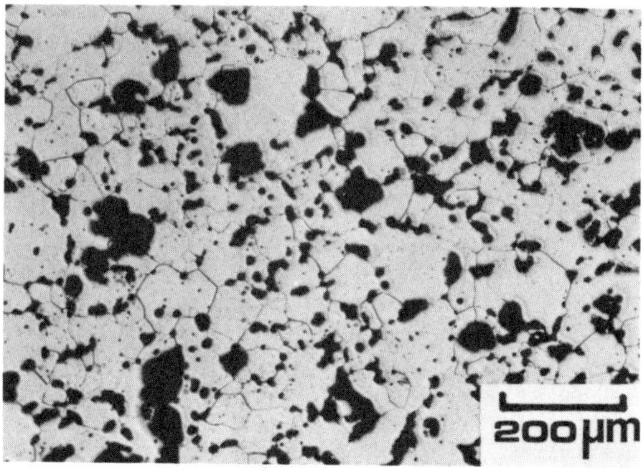

Fig. 8: Microstructure of Fe-1.4% P sintered at 1200°C for 1 hr.

Other additives such as BN, Co, nickel nitrate, and Ta, decrease the hardness of the iron compacts as shown in Figure 5. This is mainly due to the low sintered densities, possibly reflecting poor compactability.

Figures 11 and 12 show the densities and hardness of specimens sintered at 1100°C and 900°C respectively. None of these densities are greater than 6.9 gm/cm^3. Nevertheless, hardness shows significant changes for both sintering temperatures; B, B_4C and P give the greatest improvements to hardness, similar to the results observed at 1200°C.

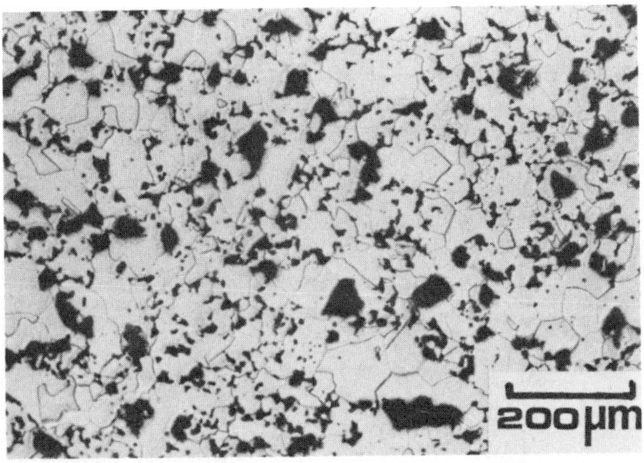

Fig. 9: Microstructure of pure 1000B compact sintered at 1200°C for 1 hr.

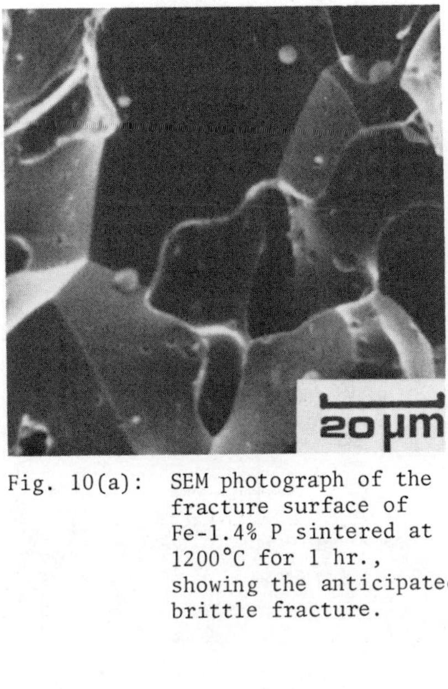

Fig. 10(a): SEM photograph of the fracture surface of Fe-1.4% P sintered at 1200°C for 1 hr., showing the anticipated brittle fracture.

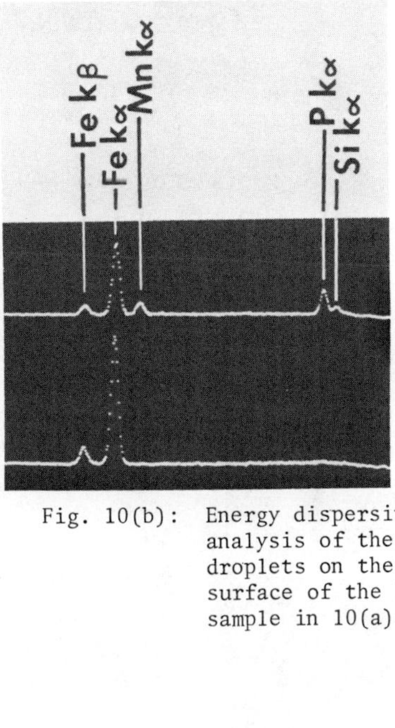

Fig. 10(b): Energy dispersive x-ray analysis of the small droplets on the fracture surface of the Fe-1.4% P sample in 10(a).

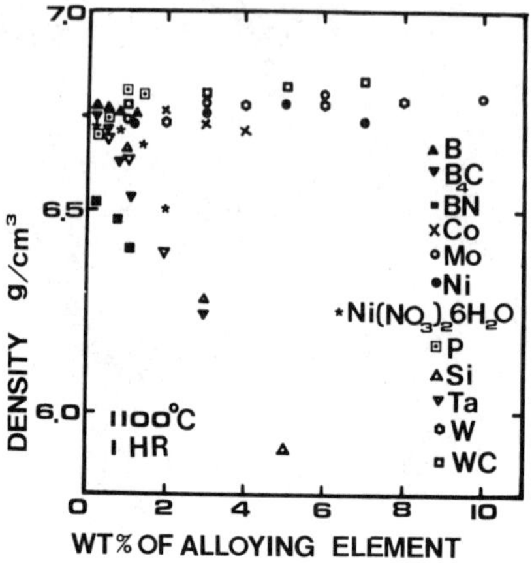

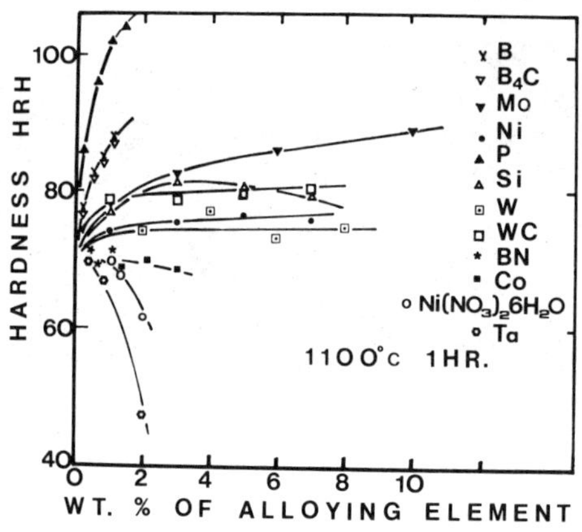

Fig. 11: Density and hardness as a function of wt. % of additives for a 1100°C sintering temperature.

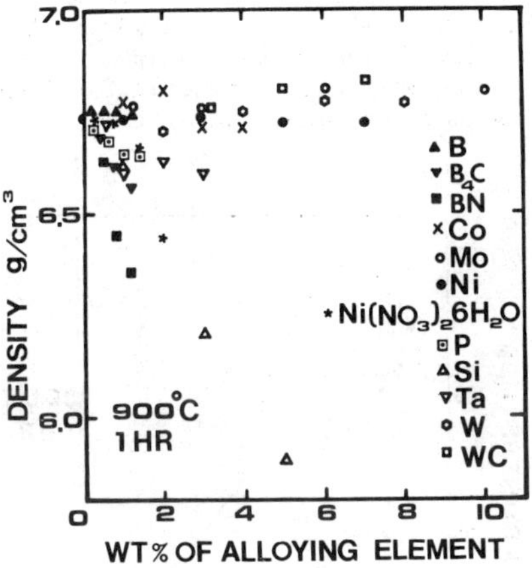

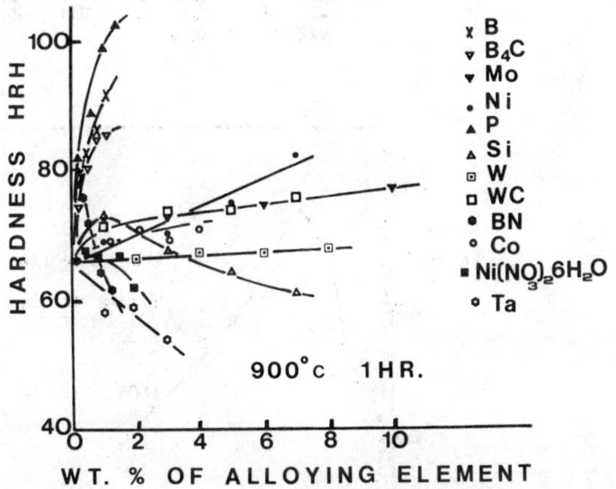

Fig. 12: Density and hardness as a function of wt. % of additives for a 900°C sintering temperature.

Ternary Systems Based on Fe-1.1% B

Four different ternary systems (as summarized in Table II) were sintered at 1100 and 1200°C. Only the Fe-B-Mo alloy survived the 1200°C temperature, other systems lost shape due to the formation of low melting point eutectics. Hence, property measurements were only conducted for Fe-B-Mo alloys sintered at 1200°C. In Figure 13, both hardness and ultimate tensile strength show a

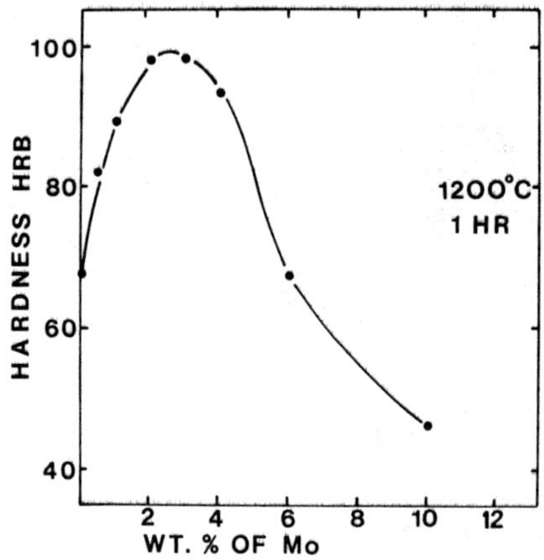

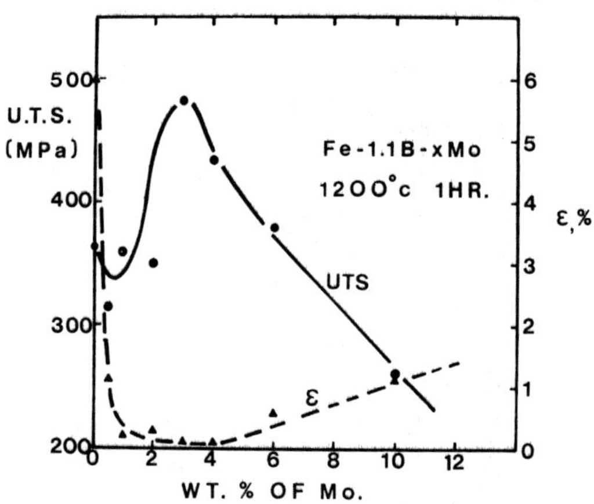

Fig. 13: Hardness, tensile strength and elongation as a function of wt. % of Mo for ternary Fe-1.1% B-Mo alloys sintered at 1200°C.

peak value at 3 to 4% Mo. The elongation, on the other hand, has the lowest value at this composition. The microstructure given in Figure 14(a) shows that a liquid phase provided the high densification. However, as the Mo content increased a matrix compound formed in increased concentration; with

10% Mo all liquid phase disappeared and was replaced by $FeMo_2B_2$ (Figure 14 (b)), as identified by x-ray diffraction. Figure 14(a) also shows that a

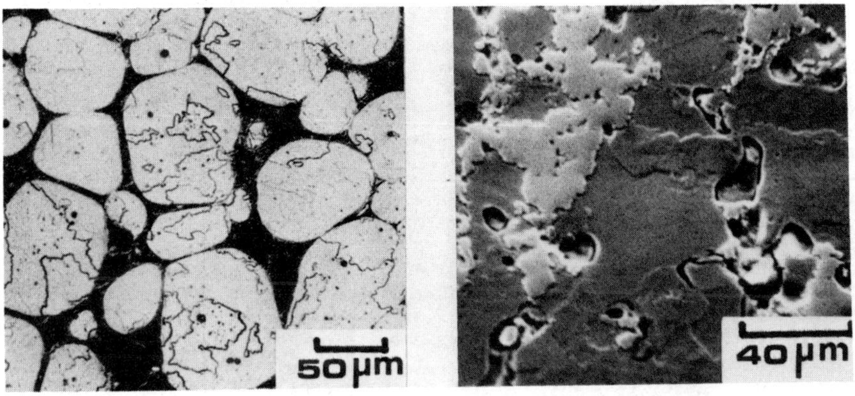

Fig. 14: Microstructure of Fe-1.1% B-Mo alloy (a) 3% Mo (b) 10% Mo. Both sintered at 1200°C for 1 hour.

massive transformation occurred inside the grain. However, this does not contribute to the strengthening as was proved by Knoop hardness measurements conducted at areas both inside and outside the transformation region. Microhardness measurements were also conducted on specimens of different Mo content. Based on these findings, given in Figure 15, it is felt that the peak

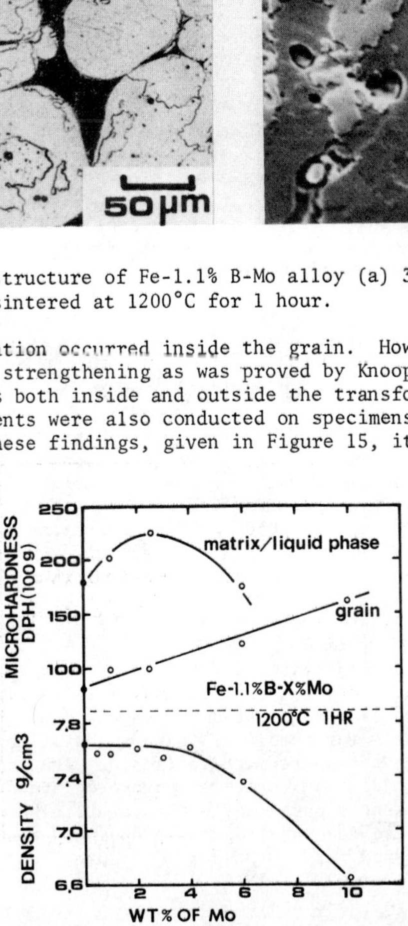

Fig. 15: Microhardness of grain and matrix phases, and density of Fe-1.1% B-Mo as a function of wt. % Mo.

value of tensile strength and hardness is a result of three factors: (1) As the Mo content increases over 4%, the amount of liquid phase decreases drastically and thereby the sintered density decreases. (2) As the Mo content increases, the hardness inside the grain increases due to the solid solution strengthening effect. (3) As the Mo content increases, the hardness of the liquid phase increases and reaches a peak near 4% Mo. This is possibly due to the formation of finely dispersed $FeMo_2B_2$. With larger amounts of Mo, the amount and size of the compound increase and hence there is a loss of the strengthening effect.

Figure 16 shows the properties of specimens from all four ternaries based on Fe-1.1% B sintered at 1100°C. In the Fe-B-Ni system, a good combination of strength and ductility was obtained at 6-8% Ni. In the Fe-B-Mo system, no significant improvement could be observed with increasing Mo content. For the Fe-B-P and Fe-B-Si systems, although the hardness and the tensile strength could be improved, the sintered material proved to be extremely brittle. The Fe-B-Ni alloys are most impressive at the 1100°C sintering temperature. In the Fe-B-Mo ternary, there was little evidence of liquid phase sintering. Consequently, densification suffered and as noted earlier, the mechanical properties were depressed by the lower sintered densities. The basic contention is that elements like B, Mo and Ni are beneficial second phase additions to Fe since they promote sintering densification while avoiding grain boundary embrittlement. Alternatively, both P and Si have been identified as elements which lower the grain boundary cohesion of Fe. Thus in spite of the positive densification and hardening effects of P and Si, they give rise to inferior ductilities. The presence of 1.1% B can only partially offset the embrittling effect even though rapid densification is obtained with these ternary compositions. Future studies will further refine these concepts to provide a basis for identifying sintering enhancement additions which raise all mechanical properties.

Conclusions

The ductility results for various iron powder systems agree fairly well with the preliminary predictions based on grain boundary cohesion models (12). Compacts up to 96% theoretical density were obtained in the Fe-B system by liquid phase sintering. The Fe-1.1% B-4% Mo ternary alloy, when sintered at 1200°C for 1 hour, reached the 500 MPa strength level but with a small ductility. The Fe-1.1% B - 8% Ni ternary alloy, when sintered at 1100°C for 1 hour reached the 390 MPa strength level with 2% elongation. However ternary Fe-B-P or Si alloys gave high densities but poor ductilities. This detrimental embrittling effect overshadows the strengthening benefit.

The criteria for enhanced sintering of ferrous metal powders are following the concepts developed for the refractory metals. The ideal sintering activator will remain as a segregated second phase at grain boundaries, will have a high solubility for iron, will have a low solubility in iron, and will provide a higher transport rate (diffusion) than normal iron lattice diffusion. Unique additional criteria have been added in this study since a non-embrittling second phase has been sought. The data contained in this paper support these concepts regarding enhanced mechanical properties in activated sintered ferrous powder metallurgy components.

Acknowledgement

The authors wish to thank Remington Arms Company for providing financial support and to Mr. Kenneth H. Moyer of Hoeganaes Corporation for supplying the iron powder.

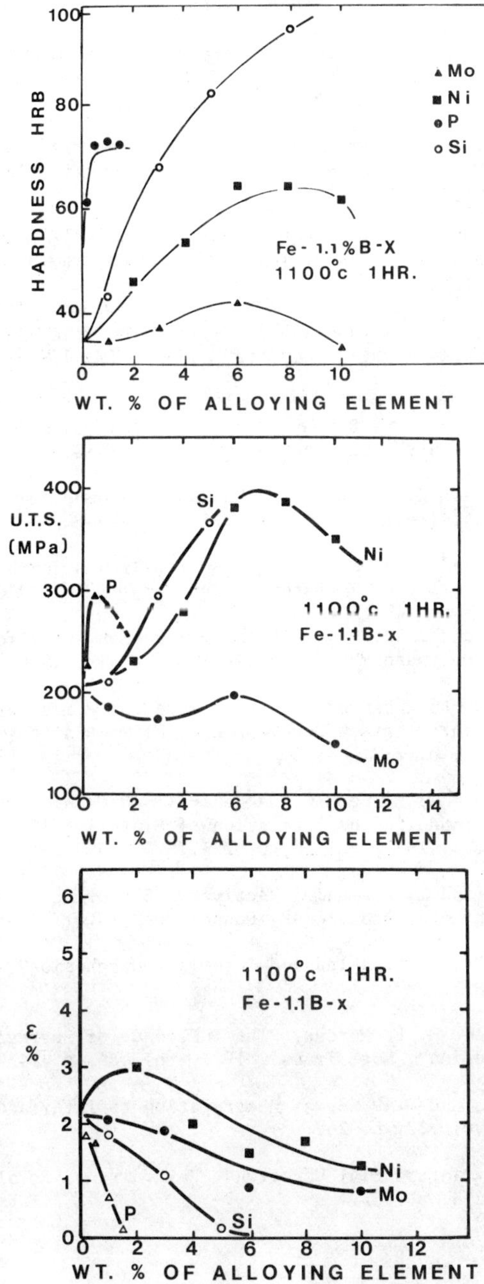

Fig. 16: Hardness, tensile strength and elongation of Fe-1.1% B base alloy as a function of wt. % of third additive at 1100°C.

References

1. A. S. Reshamwala and G. S. Tendolkar, "Activated Sintering I and II", Powder Met. Inter., 1969, vol. 1, no. 2, p. 58 and 1970, vol. 2, no. 1, p. 15.

2. J. H. Brophy, L. A. Shepard and J. Wulff, "The Nickel-Activated Sintering of Tungsten", Powder Metallurgy, W. Leszynski (ed.), Interscience, New York, 1961, p. 113.

3. J. H. Brophy, H. W. Hayden and J. Wulff, "The Sintering and Strength of Coated and Co-Reduced Nickel Tungsten Powder", Trans. TMS-AIME, 1961, vol. 221, p. 1225.

4. J. H. Brophy, H. W. Hayden and J. Wulff, "The Final Stages of Densification in Nickel-Tungsten Compacts", Trans. TMS-AIME, 1962, vol. 224, p. 797.

5. H. W. Hayden and J. H. Brophy, "The Activated Sintering of Tungsten with Group VIII Elements", J. Electrochem. Soc., 1963, vol. 110, p. 805.

6. R. M. German and Z. A. Munir, "Enhanced Low-Temperature Sintering of Tungsten", Met. Trans., 1976, vol. 7A, p. 1873.

7. R. M. German and Z. A. Munir, "Systematic Trends in the Chemically Activated Sintering of Tungsten", High-Temp. Sci., 1976, vol. 8, p. 267.

8. R. M. German and Z. A. Munir, "Heterodiffusion Model for the Activated Sintering of Molybdenum", J. Less-Common Metals, 1978, vol. 58, p. 61.

9. R. M. German, "An Enhanced Diffusion Model of Refractory Metal Activated Sintering", Sintering-New Developments, M. M. Ristic (ed.), Elsevier, New York, 1979, p. 257.

10. Z. A. Munir and R. M. German, "A Generalized Model for the Prediction of Periodic Trends in the Activation of Sintering of Refractory Metals", High-Temp. Sci., 1977, vol. 9, p. 275.

11. R. M. German and Z. A. Munir, "Activated Sintering of Refractory Metals by Transition Metal Additions", Powder Met. Inter., 1982 in press.

12. M. P. Seah, "Adsorption-Induced Interface Decohesion", Acta Met., 1980, vol. 28, p. 955.

13. J. P. Stark and H. L. Marcus, "The Influence of Segregation on Grain Boundary Cohesion", Met. Trans., 1977, vol. 8A, p. 1432.

14. E. D. Hondros and M. P. Seah, "Segregation to Interfaces", Inter. Metals Rev., 1977, vol. 22, p. 262.

15. J. P. Hirth, "Interfacial Cohesion", Phil. Trans. Royal Society London, 1980, vol. A295, p. 139.

THE EFFECT OF COMPOSITION AND CONSOLIDATION PARAMETERS ON THE THERMAL CONDUCTIVITY OF B_4C-C CERMETS*

J. E. Smugeresky
Materials Development Division I
Sandia National Laboratories
Livermore, California 94550
USA

H. J. Rack**
Sandia National Laboratories
Albuquerque, New Mexico 87185

B_4C-Cu cermets are candidates for thermally conductive neutron shields. For an optimum combination of neutron opacity and thermal conductivity, a uniform dispersion of B_4C in a continuous copper matrix is desired. Using powder processing, variations in B_4C content, particle size and consolidation parameters were studied to determine their effect on thermal conductivity of these cermets. B_4C content ranged from 20 to 80 vol. pct. and particle sizes were less than 150 microns for various hot isostatic pressing temperatures, times and pressures. The thermal conductivity can be explained by a continuous copper matrix model for low B_4C content and by an alternate B_4C and Cu layer model for high B_4C contents despite a uniform distribution of B_4C in each case. Good thermal conductivity and full density were readily achieved by this technique.

*This work supported by U.S. Department of Energy, DOE, under Contract Number DE-AC04-76DP00789.

**Present address: EXXON Enterprises, Greer, SC 29651

Introduction

In order to provide increased protection from radiation exposure after a severe accident during the transportation of radioactive spent nuclear reactor fuel, boron carbide-copper cermets have been proposed as neutron shields in shipping casks (1). The good neutron absorption of B_4C combined with the high thermal conductivity of Cu, make the B_4C-Cu cermet suitable for protection against neutron radiation while providing a heat path for dissipation of heat from the hot spent fuel. Furthermore, being a solid rather than a liquid the B_4C-Cu cermet cannot leak out in the event of an impact with a sharp or pointed object and thus the cermets provide an increased measure of safety over present day liquid neutron shields.

The target composition for the neutron shield is 75 vol. pct. B_4C-25 vol. pct Cu, which requires a uniform distribution of the B_4C in a continuous copper matrix and a fully dense solid. Cermets which meet these requirements should have optimum neutron capacity and thermal conductivity.

Powder metallurgy processing should be an ideal fabrication technique to provide a uniform mixture of such different materials; boron carbide being quite light and very hard, with copper being quite heavy and very soft. By combining suitable blending techniques for the powders with consolidation at suitable high pressures by hot isostatic pressing, fully dense uniform cerments can be expected. This report contains the details of an exploratory study to determine the feasibility of achieving such cermets and the effect of composition and consolidation parameters on the thermal conductivity and density of B_4C-Cu cermets.

Experimental Procedure

The investigation consisted of two phases: I. Evaluation of the effect that B_4C content had on the density, strength, thermal conductivity, and continuity of the copper matrix, and II. Evaluation of the influence that consolidation parameters had on the density and thermal conductivity. In both parts, technical grade boron carbide powder was used along with commercially pure copper powder. As seen in Figure 1, the boron carbide powder was irregular while the copper powder was spherical. The boron carbide particles were screened and only those particles less than 149 microns were used. Similarly the copper was screened but only those particles less than 44 microns were used. The resultant size distributions are plotted in Figure 2 and show the extent of particle size overlap. The powders were blended and loaded

a b

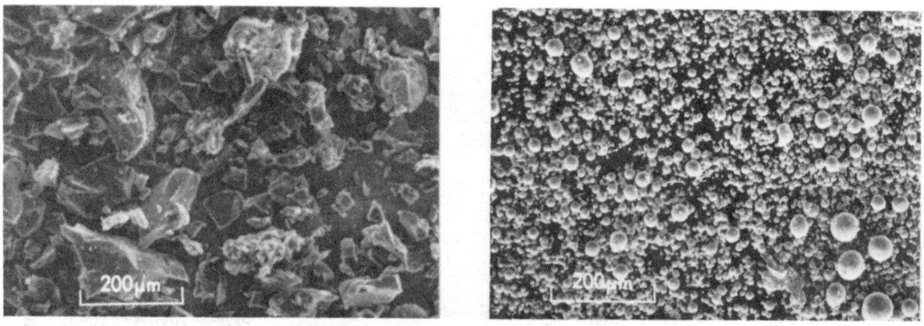

Figure 1. As-Received powders; (a) Technical grade B_4C, < 149 microns; (b) commercial purity copper, < 44 microns.

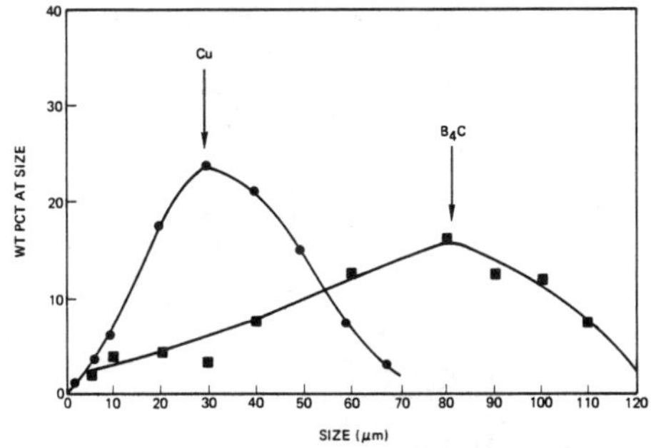

Figure 2. Particle size distribution for the as-received powders.

into stainless steel cans, which were then evacuated and sealed. Consolidation was done by hot isostatic pressing (HIP). In Phase I, the composition varied from zero to 80 vol. pct. B_4C with the compaction parameters fixed at 800°C, 100 MPa and 3 hr. at temperature and pressure. In Phase II, consolidation temperatures of 700, 800 and 1000°C, consolidation pressures of 35, 103 and 138 MPa, and the times at temperature and pressure of 1, 3, and 6 hr. were examined for both the 70 and 80 vol. pct. B_4C compositions.

Results

Phase I

Optical microscopy revealed that a continuous copper matrix existed for samples with B_4C contents of 50 vol. pct. or less (see Figure 3).

a b c

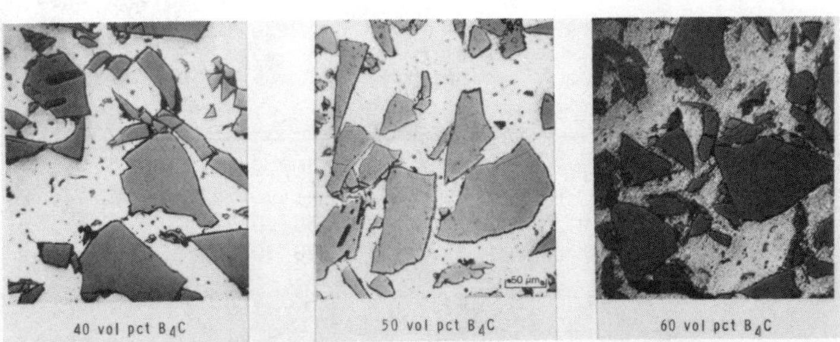

Figure 3. Optical microstructure: (a) 40 vol. pct B_4C; (b) 50 vol. pct. B_4C; and (c) 60 vol. pct B_4C.

The density of the compacted B_4C-Cu cermet was calculated as a percentage of the theoretical density and is plotted as a function of composition in Figure 4. Full theoretical density was achieved for B_4C contents of 60 vol. pct. or less. At the higher B_4C contents, variations in particle size distribution was seen to have some effect on the density, but it was difficult to single out which size distribution was optimum. The size variations are listed in Table I. Simple screening operations as well as more sophisticated operations like gap-grading (2) were used. The coarse particle size range was -74+53 microns and the small particle size range was -44 microns.

Thermal conductivity, measured by a comparator technique, ranged from 4 watts/m-K for 80 vol. pct. B_4C to ∿ 360 watts/m-K for zero vol. pct. B_4C (all copper) and is plotted in Figure 5.

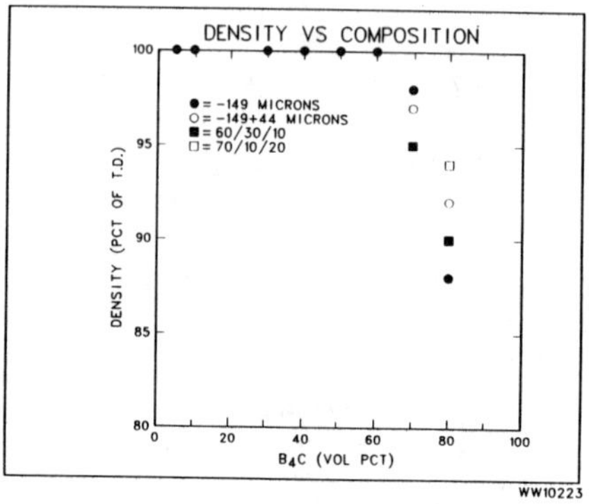

Figure 4. Density as a function of composition showing full density achieved for ≤ 60 vol. pct. B_4C. Note effect of particle size at higher B_4C contents.

Table I. Density of B_4C-Cu Cermets for Various Particle Size Distributions

B_4C Content (vol. pct.)	ρ (pct. T.D.)	B_4C Size (microns)	Cu Size (microns)
70	97.5	-149	-44
70	97.0	-149+44	-44
70	95.6	-149	-44
70	95.4	- 88	-44
70	94.9	60/30/10*	-44
70	91.0	- 44	-44
70	90.6	- 44	-44
70	89.9	-149	-44
80	93.6	70/10/20*	-44
80	92.0	-149 + 44	-44
80	90.3	50/20/30*	-44
80	89.5	60/30/10*	-44
80	88.2	-149	-44
80	87.7	-149	-44
80	84.7	- 44	-44
80	84.5	-149	-44
80	83.6	0/70/30*	-44
80	82.6	- 44	-44
80	81.0	- 44	-149 + 44
80	79.4	- 44	-149 + 44
80	74.3	- 88	-44

*Percentages of coarse, medium and fine particle sizes: see text for details.

THERMAL CONDUCTIVITY VS COMPOSITION

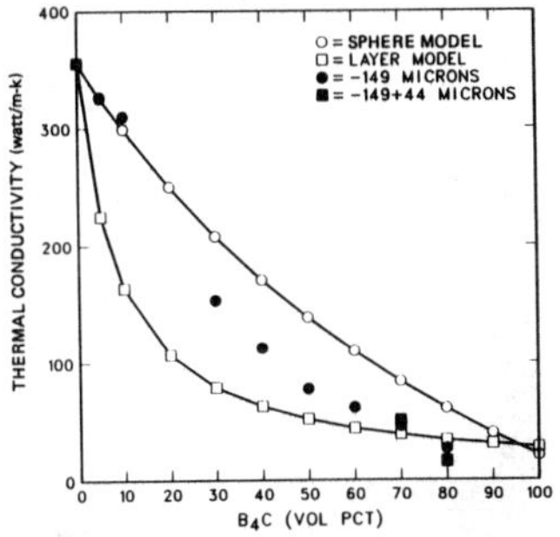

Figure 5. Thermal conductivity as a function of composition. Comparison of data with calculations for two models.

Phase II

The density and thermal conductivity after consolidation is shown in Table II. Densities ranged from 81 to 95 pct. of theoretical density for the 70 vol. pct. B_4C but only 70 to 89 pct. of theoretical density for the 80 vol. pct. B_4C. Thermal conductivity ranged from 2.4 to 8.4 watts/m-K for the 70 vol. pct. B_4C samples and from 1.1 to 4.6 watt/m-K for the 80 vol. pct. B_4C. The highest values of density and thermal conductivity are indicated with an asterisk.

Table II. Density and Thermal Conductivity of B_4C-Cu Cermets for Various Consolidation Conditions

B_4C Content (vol. pct.)	HIP T T(°C)	HIP P (MPa)	HIP t (hr)	ρ (pct. T.D.)	K (watt/m-K)
70	700	35	1	82	3.6
70	700	35	6	81	3.7
70	700	138	1	92	8.4*
70	700	138	6	90	7.2*
70	1000	35	1	81	3.3
70	1000	35	6	82	2.4
70	1000	138	1	95*	6.2
70	1000	138	6	94*	6.9
70	800	103	3	90	2.8
70	800	103	3	98	14.0
80	700	35	1	70	1.1
80	700	35	6	73	1.5
80	700	138	1	89*	2.7
80	700	138	6	85	1.9
80	1000	35	1	74	1.7
80	1000	35	6	75	1.5
80	1000	138	1	88*	3.2*
80	1000	138	6	88*	4.6*
80	800	103	3	85	2.8
80	800	103	3	88	4.0
80	800	103	3	88	-

Discussion

These results show that fully dense B_4C-Cu cermets can be fabricated with up to 60 vol. pct. B_4C. For 70 vol. pct. B_4C the best density was 98% of T.D. using the as-received particle size distributions while for the 80 vol. pct. B_4C the best density was only 94% of T.D. In contrast to the 70 vol. pct. B_4C, the 80 vol. pct. B_4C had the highest density when gap grading for optimizing packing density was used (an increase from 88 pct. T.D. to 94 pct. T.D. was observed). However, the simple screening of the fine B_4C (those particles less than 44 microns) also had a significant impact on the density. This simple operation is considered to be the least time consuming and thus should be the least costly method of increasing the density of 80 vol. pct. B_4C cermets.

The measured thermal conductivity values were relatively low for the 50 to 80 vol. pct. B_4C sample, being more like the thermal conductivity of B_4C than that of Cu (see Figure 5). Two different models of the influence of volume fraction B_4C on the thermal conductivity of the B_4C-Cu cermets have been examined in an attempt to explain these results. In Case I, a homogeneous distribution of B_4C spheres in a continuous matrix of copper is assumed. The thermal conductivity, k, is then given by (3)

$$k = k_c \left[\frac{1 + 2V_B\left(\frac{1-k'}{1+2k'}\right)}{1 - V_B\left(\frac{1-k'}{1+K'}\right)} \right] \quad (1)$$

where k_c = thermal conductivity of copper

$k' = \frac{k_c}{k_B}$

k_B = thermal conductivity of B_4C
V_B = volume fraction of B_4C.

A plot of the thermal conductivity predicted by this model is given by the open circles in Figure 5. It can be seen that this model only applies for low contents of B_4C, suggesting that a continuous conduction path has not been achieved even for the fully dense compositions at 30, 40 and 50 vol. pct. B_4C. In Case II. alternating parallel layers of B_4C and Cu are assumed. This case represents the opposite extreme of Case I, in that it assumes no connection whatsoever of the more conducting species (copper). The thermal conductivity in this latter instance is then given by (3)

$$k = \frac{1}{\left(\frac{1-V_B}{k_c}\right) + \left(\frac{V_B}{k_B}\right)}$$

which is plotted with the open squares in Figure 5. This plot suggests that for 60 to 80 vol. pct. B_4C, the material behaves more like unconnected layers of Cu than like a continuous copper matrix. At the intermediate B_4C contents, the measured values lie between the two calculated values suggesting transitional behavior. Clearly there is room for improving the thermal conductivity at all B_4C contents greater than or equal to 30 vol. pct.

In addition to the lack of copper matrix continuity, the presence of porosity can also reduce the thermal conductivity. This is most evident for the 70 and 80 vol. pct. B_4C samples which have considerable porosity. However, increasing the density of the 80 vol. pct. B_4C samples did not result in an increase in thermal conductivity although an increase was observed for increased density of 70 vol. pct. B_4C. Furthermore, the porosity cannot be used to explain the reduced thermal conductivity for samples in the 30 to 60 vol. pct. B_4C range since all samples were found to have achieved full theoretical density. Additional effort needs to be made to determine the extent of continuity and/or uniformity of the copper matrix. The important points are that the measured thermal conductivity values might be considered as typical and that the calculations based on the two models provide anticipated scatter in the data. For design purposes the measured values are sufficient while the development work to improve the values can be carried out. One additional point is that the predicted thermal conductivity drops off rather quickly with increased B_4C content even for the ideal case of a continuous matrix, and at 40 vol. pct. B_4C is less than half that of copper. In the 50 to 80 vol. pct. B_4C range there appears to be little extra benefit derived from the use of copper as a conducting matrix.

In the parametric study (phase II) of this investigation, the temperature pressure and time were varied and samples were prepared for both the 70 and 80 vol. pct. B_4C compositions. Only these compositions were studied because full density was not achieved for consolidation at the nominal (phase I) 800°C, 103 MPa, 3-hour conditions. The parametric variables were chosen using the strategy of experimentation based on statistical analysis techniques (4). Examination of Table II shows that although full theoretical density was not achieved for either composition, an indication of the recommended consolidation conditions was obtained. Generally, the thermal conductivity is directly related to the percent of theroetical density, although there is a good deal of scatter in the data. For each composition, samples compacted at the highest pressure had the highest thermal conductivity. For the 70 vol. pct. B_4C sample 700°C provided the best results whereas for the 80 vol. pct. B_4C, 1000°C provided the best results. The parametric study suggests that time is of minor importance and only one hour is sufficient to achieve the density and thermal conductivity representative of each temperature-pressure combination.

Conclusions

1. The thermal conductivity of B_4C-Cu cermets depends upon the composition more than upon the consolidation parameters.

2. For low B_4C contents, the cermets had thermal conductivities nearly the same as pure copper but for high B_4C contents the thermal conductivity was nearly the same as that of pure B_4C.

3. For cermets near the target 75 vol. pct. B_4C composition for spent nuclear fuel shipping cask neutron shields, the thermal conductivity is most strongly dependent upon the relative density of the compact, which in turn is dependent on the consolidation pressure.

4. Although the thermal conductivity of the 70 to 80 vol. pct B_4C cermets can be enhanced by increasing the density, achievement of full density will still not provide material with a thermal conductivity significantly greater than that of B_4C.

References

1. George C. Allen, Jr., R. G. Eakes, J. M. Freedman, R. B. Pope, S. A. Dupree and W. P. Schimmel, Jr., "Conceptual Designs for LMFBR Spent Fuel Shipping Casks", SAND77-1483, Sandia National Laboratories, Livermore, March 1978.

2. R. K. McGeary, Journal of the American Ceramic Society 44, p. 513 (1961).

3. W. D. Kingery, Introduction to Ceramics, (J. Wiley & Sons, 1960).

4. "Strategy of Experimentation", DuPont Technical Publication, E. I. DuPont de Nemours and Co., Wilmington, Delaware, October 1975.

FABRICATION OF COPPER INFILTRATED TUNGSTEN JET VANES

J. H. Peterson and K. E. Meiners
Battelle
Columbus Laboratories
505 King Avenue
Columbus, Ohio 43201
USA

M. Ripley-Lotee
Naval Weapons Center
China Lake, California 93555
USA

Battelle-Columbus Laboratories has developed a net-shape powder metallurgy production process to reduce the cost of producing copper-infiltrated tungsten jet vanes. By pressing the vane to near net shape, the material required is held to a minimum. In addition, by machining in the green condition using an abrasive belt, the machining time for most of the surfaces of the vane can be reduced to 10-15 minutes. Relatively high pressure is used during cold-pressing-to-shape to impart sufficient green strength to the vane to allow machining in the green condition. The as-pressed density is high enough so that the shrinkage occurring during sintering is very predictable. Only limited conventional machining is required after the sintering and infiltration operations.

Inspection of the finished parts indicated that high quality vanes could be produced by this process. Actual firing tests of these vanes indicated that their performance is equal to or better than that of those made from commercially available copper infiltrated tungsten material.

Introduction

Several naval rocket motors use jet vanes as the thrust vector control technique. To decrease production costs it was necessary to develop a lower cost manufacturing process for these copper infiltrated tungsten jet vanes. The vanes, shown in Figure 1, must survive the extremely high temperature of the propulsive gas (5000F) during the nominal firing time of four to five seconds. These vanes are fabricated from an 80 percent dense tungsten skeleton which is infiltrated with copper. During development work at the Naval Weapons Center to obtain the optimum vane design, commercially available copper infiltrated tungsten in the form of a rectangular block was used. This material then had to be machined into the vane configuration with considerable loss of material and high machining expense. The objective of the work at Battelle was to develop a process which could produce these vanes at a much lower cost.

Cost reduction of the vane hinges squarely on the ability to minimize the amount of tungsten used in the vane and the amount of machining required, particularly on the sintered part. The process developed at Battelle utilizes simple rapid machining of a near-net, green-pressed, powder shape prior to sintering and infiltration. Very little conventional machining is required to complete the jet vane to meet final dimensions after sintering and infiltration.

Experimental Work

The experimental work involved not only the selection and processing of the tungsten powder but the development of a unique machining approach for the tungsten jet vane.

Powder Selection and Processing. In the initial development phase three different particle sizes of tungsten powder and blends of these powders were evaluated for use as the starting material for the jet vane. To assess the suitability of these powders; their isostatic compaction characteristics, sintering behavior, and ease of infiltration of the sintered skeleton with copper were evaluated.

The target specifications for the copper infiltrated tungsten material were a density range of 75-83 percent of theoretical for the porous tungsten skeleton and a tensile strength of greater than 80,000 psi after infiltration. The tungsten powders evaluated are listed in Table 1. Neither the vibratory packing density nor the green density of the 0.5 - 1.5 micron powder appeared sufficiently high to be further considered. The 3.5 - 4.0 micron, the 5 - 6 micron and both of the blends showed satisfactory vibratory densities and green density after cold pressing at 85,000 psi. The green densities at 60,000 psi were too low; there would be too much shrinkage during sintering. At 150,000 psi the density was at the upper limit of the desired range for the pressed density. In addition, a high pressure such as 150,000 psi is not practical from a production standpoint. Blend B, which consisted of all three powders, was dropped to keep the process as simple as possible.

Table 2 shows the sintered density of blanks pressed at 85,000 psi. Sintering at 1900C for 1 hour in vacuum yielded densities in the desired range for the jet vane. Blend A was dropped at this point because no particular advantage in pressing or sintering seemed to be gained from this blend compared to the unblended powders.

Infiltration. Sintered blanks made from the 3.5 - 4.0 micron powder and 5 -6 micron powder were then investigated for copper infiltration response.

FIGURE 1. COPPER INFILTRATED TUNGSTEN JET VANES
The vanes are 4 inches in length.

TABLE 1. COLD ISOSTATIC PRESSING STUDIES ON TUNGSTEN POWDER

Particle Size, microns	Vibratory Density, percent	Green Pressed Density, percent		
		60,000 psi	85,000 psi	150,000 psi
0.5 - 1.5	18.8		63.8	
3.5 - 4.0	29.2	68.6	75.7	81.9
5.0 - 6.0	30.5	69.6	77.2	83.4
Blend A	28.8		76.6	82.4
Blend B			75.6	82.2

Blend A = 50w/o 3.5 - 4.0 micron powder, 50w/o 5.0 - 6.0 micron powder

Blend B = 12w/o 0.5 - 1.5 micron powder, 44w/o 3.5 - 4.0 micron powder 44w/o 5.0 - 6.0 micron powder

TABLE 2. DENSITY OF TUNGSTEN AFTER SINTERING AT 1900°C IN VACUUM FOR 1 HOUR[a]

Particle Size, microns	Density, percent
	Average of three specimens
3.5 - 4.0	80.4
5.0 - 6.0	80.4
Blend A	79.7

(a) Specimens had been cold isostatically pressed at 85,000 psi prior to sintering.

The initial trials were carried out in a hydrogen atmosphere at 1200C. Two problems were initially encountered with infiltrating the sintered skeleton with copper. The first problem was the rapid coating of only the surface of the blanks with the copper, thereby trapping some of the hydrogen within the part and preventing full infiltration. To overcome this problem the exterior of the part, with the exception of the bottom, was coated with a brazing stop-off. This caused the copper to enter from the bottom of the part only. Although this modification showed some improvement, full infiltration was still not obtained. Further improvement was gained by increasing the temperature from 1200C to 1260C. These two modifications resulted in filling over 90 percent of the porosity with copper.

Tensile strengths and percent infiltration of well infiltrated parts are shown in Table 3. The tensile strength of poorly infiltrated material is about 65,000 psi. The microstructures of these two materials are shown in Figures 2 and 3. The 3.5 - 4.0 micron material has a noticably finer pore size than the 5 to 6 micron material. Good infiltration was observed in each of the materials.

Because of the slightly better tensile values of the parts made from the 3.5 - 4.0 micron material and the fact that this powder is more commercially available, it was selected for the vane fabrication studies.

Vane Fabrication and Machining Studies. To meet the objectives of this program to reduce the amount of input material and machining required, the part was cold isostatic pressed in a bag to near net shape of the vane. The material used for the pressing bag was a plastisol. A heated from was used to make the bag. The heated form was dipped into the plastisol for about 15 seconds and then cured on the form. The form is sized to allow for compaction of the powder during pressing and for shrinkage during sintering, with some excess on each surface for machining stock. The bag is supported during loading in a perforated metal cage and is sealed at the top with two metal bars. A loaded part ready for processing is shown in Figure 4. The bag is left in the cage for support during pressing. After pressing, the bag is removed from the cage and the part separated from the bag.

After pressing, the part was machined in the green condition using specially designed fixtures. These fixtures have carbide tool inserts brazed to the surfaces to establish the standoff of the fixture from the abrasive belt during machining. It was first placed in a fixture (Figure 5) to machine the narrow top and bottom surfaces and establish the overall height. After machining these surfaces, the part was placed in another fixture to maching the flat faces on the sides of the part (Figure 6). The top and bottom surfaces of the vane were used as reference surfaces, and the part rested on a cradle to position it for the proper stock removal. After machining of the two surfaces on one side, the cradle was raised about 0.50 inch and the other side was machined. The final part thickness allowed for the shrinkage that occurs during sintering which is about two percent. A small amount of excess material was also left on each end and on the bottom for machining after sintering and infiltration. This allowed for the small amount of distortion that occurs during sintering and enables the bottom to be machined flat. The total time for green machining is about ten minutes.

After machining, the part was sintered at 1900C for 1 hour in vacuum. Following sintering, the part was infiltrated with copper. The surfaces of the vane with the exception of the bottom, were coated with brazing stop-off

TABLE 3. TENSILE STRENGTH OF COPPER-INFILTRATED TUNGSTEN

Powder Size, microns	Sintered Density, g/cc	Sintered Density, percent	Percent Infiltration	Ultimate Tensile Strength,[a] psi
3.5 - 4.0	15.54	80.5	96.0	111,600
				109,800
5.0 - 6.0	15.34	79.5	92.5	108,700
				99,600

(a) Two test specimens for each powder type.

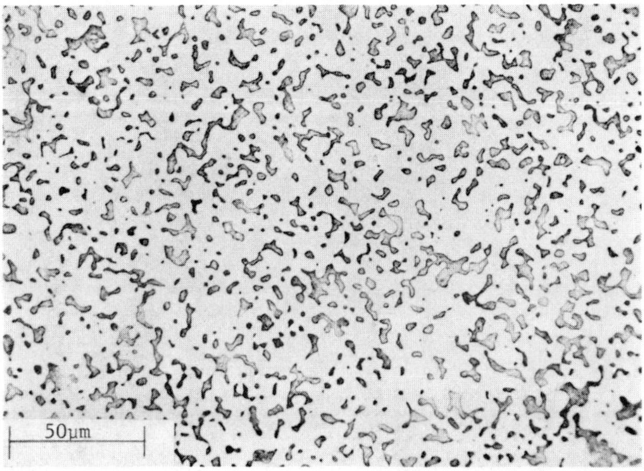

FIGURE 2. 3.5-4.0 MICRON TUNGSTEN AFTER SINTERING IN VACUUM AT 1900C FOR 1 HOUR AND INFILTRATION IN HYDROGEN AT 1040°C FOR 1.5 HOUR, THEN 1260°C FOR 0.5 HOUR.

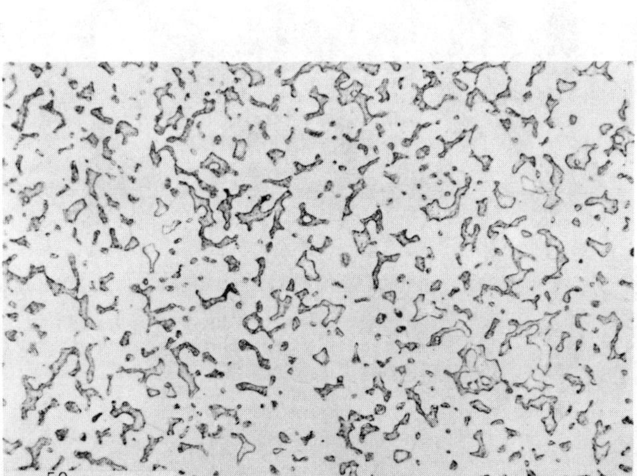

FIGURE 3. 5.0-6.0 MICRON TUNGSTEN AFTER SINTERING IN VACUUM AT 1900C FOR 1 HOUR AND INFILTRATION IN HYDROGEN AT 1040°C FOR 1.5 HOURS, THEN 1260°C FOR 0.5 HOUR.

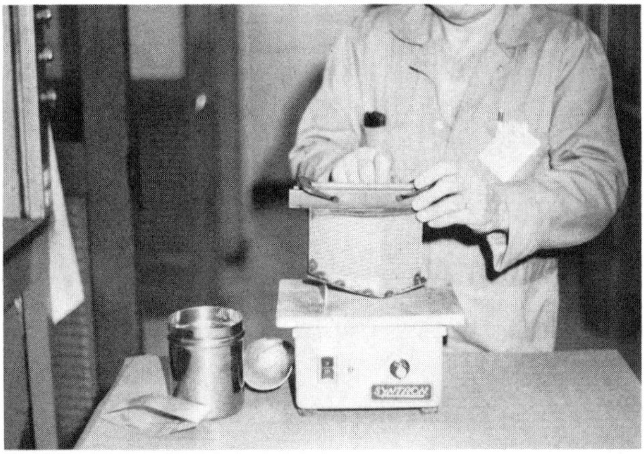

FIGURE 4. LOADED BAG ASSEMBLY READY FOR PRESSING

The bag is sealed by two metal bars clamped to the top of the bag by "C" clamps.

FIGURE 5. FIXTURE FOR MACHINING OF THE TOP AND BOTTOM SURFACES OF THE VANE.

FIGURE 6. FIXTURE FOR MACHINING THE SIDES OF THE VANE

prior to infiltration. The density and percent infiltration were determined after infiltration and the vanes were radiographed to inspect for any internal defects.

Final machining was then performed on the sintered and infiltrated vane. First a small amount of stock was removed from the bottom of the vane. Then a 0.5 inch diameter hole for the actuator shaft was drilled using carbide tooling, and two small attachment holes were drilled and tapped using conventional high speed steel tools. The vane was then placed in a fixture that used the 0.5 inch diameter hole as a reference, and the ends were machined to length on a belt sander. Again carbide pads were used to limit the movement of the fixture into the belt. The ends were radiused by hand on a belt sander. The vane was then dye penetrant inspected and the dimensions recorded.

A summary of the processing steps for the jet vanes is listed below:

1. Press the tungsten powder (3.5 - 4.0 micron particle size) to near-net shape at 80,000 psi

2. Machine the pressed part

3. Sinter in vacuum at 1900°C for 1 hour

4. Infiltrate with copper in a hydrogen atmosphere at 1260-1280°C for 0.5 hours after a hold at 1040°C for 1.5 hours

5. Inspect for density, percent infiltration, and internal quality

6. Machine base to obtain final height, drill and tap holes in the base, and radius the ends of the vane

7. Inspect again for the surface quality using dye penetrant inspection

8. Record part dimensions.

Evaluation of the Jet Vanes. Four vanes are located at the exit of the nozzle in a missile using this type of guidance control. Firing tests to evaluate the performance of the vanes simulate very closely actual firing conditions. The vane is mounted on a zirconia-coated-stainless-steel platform which protects the servo beneath the vane. A 17-4PH shaft extends into the vane and connects the vane to the servo. The temperature of the propulsive gas is about 5000F and the burn time is in the range of 4 to 5 seconds. A vane after a firing test with these conditions is shown in Figure 7.

Test have also been made at longer burn times of up to about nine seconds. At these longer burn times, there has been some melting of 17-4PH shaft toward the leading edge and more loss of the infiltrated tungsten on the leading edge. A photomicrograph of the infiltrated tungsten adjacent to the leading edge after a nine second burn time is shown in Figure 8. There has been some loss of copper from the pores, and the pore size has decreased considerably due to the exposure to the very high temperatures. The structure just behind this area has the normal pore size for this material. The copper was molten during the firing test.

The vanes produced on this program performed better than the vanes which had been machined from the commercially available material. This difference

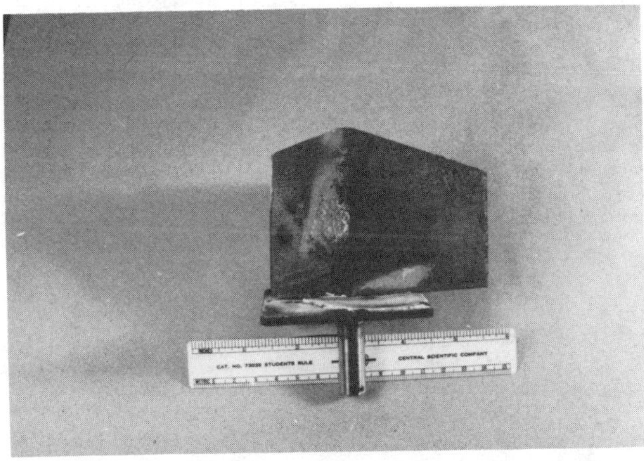

FIGURE 7. JET VANE AFTER TEST FIRING FOR
4 to 5 SECONDS

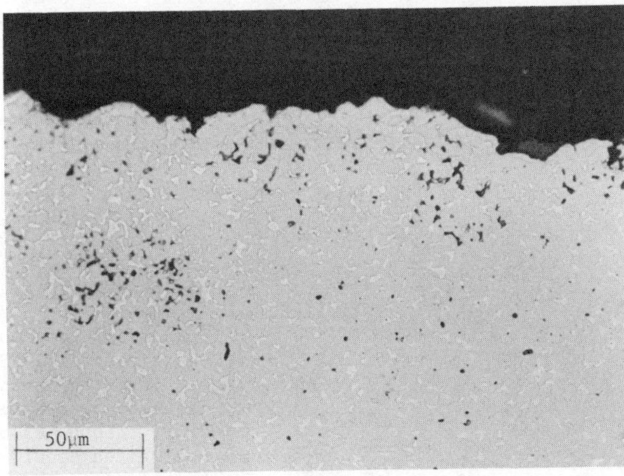

FIGURE 8. MICROSTRUCTURE OF LEADING EDGE OF
JET VANE AFTER 9 SECOND TEST FIRING

is noticable only in firing tests of longer duration, such as the nine second test. The primary difference between these materials is the particle size of the tungsten powder used and the subsequent pore size after sintering. The commercially available infiltrated tungsten appears to be fabricated from tungsten powder that has a particle size of about six microns since the microstructure is very similar to the 5 - 6 micron material evaluated by Battelle. The smaller pore size which results from the use of the 3.5 - 4.0 micron powder appears to have an advantage in this application.

Summary

A lower cost process for the fabrication of the jet vanes has been developed as a result of the experimental work during this study. The cost reduction has been accomplished by reducing the amount of tungsten used by pressing to near net shape and by developing a simple, rapid machining process performed in the green pressed condition for most of the vane surfaces. In addition to the successful achievement of the primary program goal, a vane with improved performance characteristics has resulted from this work.

Acknowledgement

The assistance and support of the Naval Weapons Center, China Lake, California, in the performance of this work and for the firing test evaluations is greatly appreciated.

Subject Index

Author Index